Murray C. Kemp
Yoshio Kimura

Introduction to Mathematical Economics

Springer-Verlag
New York Heidelberg Berlin

Murray C. Kemp
School of Economics
University of New South Wales
Kensington, NSW 2033
Australia

Yoshio Kimura
Faculty of Economics
Nagoya City University
Mizuhocho Mizuhoku
Nagoya
Japan

AMS Subject Classifications
15A48, 15A57, 90Axx, 90C05, 90C30, 90C50, 93D05, 93D99

Library of Congress Cataloging in Publication Data

Kemp, Murray C.
Introduction to mathematical economics.

Bibliography: p.
Includes index.
1. Economics, Mathematical. I. Kimura, Yoshio, 1932– joint author. II. Title.
HB135.K46 330'.01'51 77-26117

Printed in the United States of America

9 8 7 6 5 4 3 2 1

ISBN 0-387-90304-6 Springer-Verlag New York
ISBN 3-540-90304-6 Springer-Verlag Berlin Heidelberg

Preface

Our objectives may be briefly stated. They are two. First, we have sought to provide a compact and digestible exposition of some sub-branches of mathematics which are of interest to economists but which are underplayed in mathematical texts and dispersed in the journal literature. Second, we have sought to demonstrate the usefulness of the mathematics by providing a systematic account of modern neoclassical economics, that is, of those parts of economics from which jointness in production has been excluded. The book is introductory not in the sense that it can be read by any high-school graduate but in the sense that it provides some of the mathematics needed to appreciate modern general equilibrium economic theory. It is aimed primarily at first-year graduate students and final-year honors students in economics who have studied mathematics at the university level for two years and who, in particular, have mastered a full-year course in analysis and calculus.

The book is the outcome of a long correspondence punctuated by periodic visits by Kimura to the University of New South Wales. Without those visits we would never have finished. They were made possible by generous grants from the Leverhulme Foundation, Nagoya City University, and the University of New South Wales. Equally indispensible were the expert advice and generous encouragement of our friends Martin Beckmann, Takashi Negishi, Ryuzo Sato, and Yasuo Uekawa.

Contents

Linear Inequalities 1

The essentially nonnegative character of many economic variables combines with the traditional assumption of constant returns to give many of the models in economics the structure of a system of linear inequalities. The present chapter is devoted to some of the properties of such systems. For a sampling of the models the reader may turn to Chipman [1965–1966], Dorfman, Samuelson and Solow [1958], Gale [1960], Koopmans [1957], McKenzie [1960], Morishima [1964], and Nikaido [1968].

1 Some Fundamental Theorems

This introductory section is devoted to a small number of fundamental propositions. We begin with a lemma of Tucker's which has been a fruitful generator of propositions of interest to economists.

Lemma 1 (Tucker [1956]). *Let $\boldsymbol{A}$ be an $m \times n$ real matrix. The system of linear inequalities**

$$\boldsymbol{A}\boldsymbol{x} = \boldsymbol{0}, \qquad \boldsymbol{x} \geqq \boldsymbol{0}, \qquad \boldsymbol{A}'\boldsymbol{u} \geqq \boldsymbol{0} \tag{1}$$

* The vector orderings which will be used are defined as follows. Let $\boldsymbol{x}$ and $\boldsymbol{y}$ be any two n-dimensional vectors. Then,

$$\begin{aligned} \boldsymbol{x} > \boldsymbol{y} \quad &\Leftrightarrow \quad x_i > y_i \quad \text{for all } i \\ \boldsymbol{x} \geqq \boldsymbol{y} \quad &\Leftrightarrow \quad x_i \geqq y_i \quad \text{for all } i \\ \boldsymbol{x} = \boldsymbol{y} \quad &\Leftrightarrow \quad x_i = y_i \quad \text{for all } i \end{aligned}$$

and

$$\boldsymbol{x} \geq \boldsymbol{y} \quad \Leftrightarrow \quad \boldsymbol{x} \geqq \boldsymbol{y} \quad \text{but } \boldsymbol{x} \neq \boldsymbol{y}$$

where x_i and y_i denote the ith elements of $\boldsymbol{x}$ and $\boldsymbol{y}$ respectively.

has solutions $\boldsymbol{x}$ *and* $\boldsymbol{u}$ *such that*

$$x_1 + \boldsymbol{a}'_1 \boldsymbol{u} > 0. \tag{2}$$

PROOF. The proof is by induction on n, the number of columns. When $n = 1$ the proposition is trivial: if $\boldsymbol{A}\ (=\boldsymbol{a}_1) = \boldsymbol{0}$ then any $\boldsymbol{x}\ (=x_1) > \boldsymbol{0}$ and $\boldsymbol{u}$ satisfy (1) and (2), and if $\boldsymbol{A} \neq \boldsymbol{0}$ then $\boldsymbol{x} = \boldsymbol{0}$ and $\boldsymbol{u} = \boldsymbol{A}$ suffice.

Suppose that the proposition is valid for a matrix $\boldsymbol{A}$ with n columns and consider the augmented matrix $\tilde{\boldsymbol{A}} = (\boldsymbol{a}_1, \ldots, \boldsymbol{a}_n, \boldsymbol{a}_{n+1}) = (\boldsymbol{A}, \boldsymbol{a}_{n+1})$. From the induction assumption, there exist $\boldsymbol{x}$ and $\boldsymbol{u}$ such that $\boldsymbol{A}\boldsymbol{x} = \boldsymbol{0}$, $\boldsymbol{x} \geqq \boldsymbol{0}$, $\boldsymbol{A}'\boldsymbol{u} \geqq \boldsymbol{0}$, and $x_1 + \boldsymbol{a}'_1\boldsymbol{u} > 0$. If $\boldsymbol{a}'_{n+1}\boldsymbol{u} \geqq 0$ then $\tilde{\boldsymbol{x}}' = (\boldsymbol{x}', 0)$ and $\boldsymbol{u}$ are the desired solution. Suppose alternatively that $\boldsymbol{a}'_{n+1}\boldsymbol{u} < 0$. In this case, we define

$$\lambda_j = \frac{-\boldsymbol{a}'_j\boldsymbol{u}}{\boldsymbol{a}'_{n+1}\boldsymbol{u}} \geqq 0 \qquad (j = 1, \ldots, n) \tag{3}$$

and

$$\boldsymbol{B} = (\boldsymbol{b}_1, \ldots, \boldsymbol{b}_n) = (\boldsymbol{a}_1 + \lambda_1 \boldsymbol{a}_{n+1}, \ldots, \boldsymbol{a}_n + \lambda_n \boldsymbol{a}_{n+1}) \tag{4}$$

so that $\boldsymbol{B}'\boldsymbol{u} = \boldsymbol{0}$. Applying the induction assumption again, this time to $\boldsymbol{B}$, there exist $\boldsymbol{y}$ and $\boldsymbol{z}$ such that $\boldsymbol{B}\boldsymbol{y} = \boldsymbol{0}$, $\boldsymbol{y} \geqq \boldsymbol{0}$, $\boldsymbol{B}'\boldsymbol{z} \geqq \boldsymbol{0}$, and $y_1 + \boldsymbol{b}'_1\boldsymbol{z} > 0$. Now let us define $\boldsymbol{\eta}' = (\boldsymbol{y}', \sum_{j=1}^n \lambda_j y_j)$ and $\boldsymbol{w} = \boldsymbol{z} + \mu\boldsymbol{u}$, where $\mu = -\boldsymbol{a}'_{n+1}\boldsymbol{z}/\boldsymbol{a}'_{n+1}\boldsymbol{u}$. Then

$$\begin{aligned}
\tilde{\boldsymbol{A}}\boldsymbol{\eta} &= \sum_{j=1}^n \boldsymbol{a}_j y_j + \boldsymbol{a}_{n+1}\left(\sum_{j=1}^n \lambda_j y_j\right) \\
&= \sum_{j=1}^n (\boldsymbol{a}_j + \lambda_j \boldsymbol{a}_{n+1}) y_j \\
&= \boldsymbol{B}\boldsymbol{y} = \boldsymbol{0}, \\
\boldsymbol{\eta} &\geqq \boldsymbol{0}, \\
\tilde{\boldsymbol{A}}'\boldsymbol{w} &= \begin{pmatrix} \boldsymbol{b}'_1 - \lambda_1 \boldsymbol{a}'_{n+1} \\ \vdots \\ \boldsymbol{b}'_n - \lambda_n \boldsymbol{a}'_{n+1} \\ \boldsymbol{a}'_{n+1} \end{pmatrix} (\boldsymbol{z} + \mu\boldsymbol{u}) \\
&= \begin{pmatrix} \boldsymbol{B}'\boldsymbol{z} \\ 0 \end{pmatrix} \qquad [\text{since } -\lambda_j \boldsymbol{a}'_{n+1}(\boldsymbol{z} + \mu\boldsymbol{u}) = 0, j = 1, \ldots, n] \\
&\geqq \boldsymbol{0},
\end{aligned}$$

and

$$\eta_1 + \boldsymbol{a}'_1\boldsymbol{w} = y_1 + \boldsymbol{b}'_1\boldsymbol{z} > 0.$$

Thus the proposition is valid for $n + 1$. □

The choice of subscript in (2) was quite arbitrary; the lemma is valid for any subscript $j, j = 1, \ldots, n$. Let $x_j^{(j)}$ denote the jth element of the vector $\boldsymbol{x}^{(j)}$. Then we have the following formal extension of Lemma 1.

Lemma 1′. *System* (1) *has solutions* $\boldsymbol{x}^{(j)}$ *and* $\boldsymbol{u}^{(j)}$ *such that*

$$x_j^{(j)} + \boldsymbol{a}'_j \boldsymbol{u}^{(j)} > 0 \qquad (j = 1, \ldots, n).$$

PROOF. Let $\boldsymbol{P} = (\boldsymbol{e}_j, \boldsymbol{e}_2, \ldots, \boldsymbol{e}_{j-1}, \boldsymbol{e}_1, \boldsymbol{e}_{j+1}, \ldots, \boldsymbol{e}_n)$. Applying Lemma 1 to the system $(\boldsymbol{AP})\boldsymbol{x} = \boldsymbol{0}$, $\boldsymbol{x} \geqq \boldsymbol{0}$, $\boldsymbol{P}'\boldsymbol{A}'\boldsymbol{u} \geqq \boldsymbol{0}$, there exist solutions $\boldsymbol{x}$ and $\boldsymbol{u}$ such that $x_1 + \boldsymbol{a}'_j\boldsymbol{u} > 0$. Writing $\boldsymbol{Px} = \boldsymbol{x}^{(j)}$ and $\boldsymbol{u} = \boldsymbol{u}^{(j)}$, and noting that $x_j^{(j)} = x_1$, $\boldsymbol{P} \geqq \boldsymbol{0}$, and $\boldsymbol{PP}' = \boldsymbol{P}'\boldsymbol{P} = \boldsymbol{I}$, we obtain $(\boldsymbol{AP})\boldsymbol{x} = \boldsymbol{Ax}^{(j)} = \boldsymbol{0}$, $\boldsymbol{x}^{(j)} \geqq \boldsymbol{0}$, $\boldsymbol{A}'\boldsymbol{u}^{(j)} \geqq \boldsymbol{0}$, and $x_j^{(j)} + \boldsymbol{a}'_j\boldsymbol{u}^{(j)} > 0$. □

It is now a simple matter to prove

Theorem 1. *System* (1) *has solutions* $\boldsymbol{x}^*$ *and* $\boldsymbol{u}^*$ *such that* $\boldsymbol{x}^* + \boldsymbol{A}'\boldsymbol{u}^* > \boldsymbol{0}$.

PROOF. Let us define $\boldsymbol{x}^* = \sum_{j=1}^n \boldsymbol{x}^{(j)}$ and $\boldsymbol{u}^* = \sum_{j=1}^n \boldsymbol{u}^{(j)}$, where $\boldsymbol{x}^{(j)}$ and $\boldsymbol{u}^{(j)}$ are solutions of (1), their existence following from Lemma 1. Then

$$x_i^* + (\boldsymbol{A}'\boldsymbol{u}^*)_i \geqq x_i^{(i)} + \boldsymbol{a}'_i\boldsymbol{u}^{(i)} > 0 \qquad (i = 1, \ldots, n).$$

We now turn to a series of propositions which flow from Lemma 1 and Theorem 1.

Theorem 2 (Stiemke [1915]). *There exists a vector* $\boldsymbol{x} > \boldsymbol{0}$ *such that* $\boldsymbol{Ax} = \boldsymbol{0}$ *if and only if* $\boldsymbol{A}'\boldsymbol{u} \geq \boldsymbol{0}$ *has no solution.*

PROOF. (*Necessity*). Suppose the contrary. Then there exists $\boldsymbol{u}$ such that $\boldsymbol{A}'\boldsymbol{u} \geq \boldsymbol{0}$. Since $\boldsymbol{Ax} = \boldsymbol{0}$ has a solution $\boldsymbol{x} > \boldsymbol{0}$, we obtain

$$\begin{aligned} (\boldsymbol{u}'\boldsymbol{A})\boldsymbol{x} &> 0 \qquad [\text{since } \boldsymbol{u}'\boldsymbol{A} \geq \boldsymbol{0} \text{ and } \boldsymbol{x} > \boldsymbol{0}] \\ = (\boldsymbol{u}')\boldsymbol{Ax} &= 0 \qquad [\text{because } \boldsymbol{Ax} = \boldsymbol{0}], \end{aligned}$$

which is a contradiction.

(*Sufficiency*). From the assumption and Theorem 1, any solution to $\boldsymbol{A}'\boldsymbol{u} \geqq \boldsymbol{0}$ must be such that $\boldsymbol{A}'\boldsymbol{u} = \boldsymbol{0}$. It follows from Theorem 1 that $\boldsymbol{x}^* > \boldsymbol{0}$ and $\boldsymbol{Ax} = \boldsymbol{0}$ has a positive solution. □

Theorem 3 (Farkas [1902], Minkowski [1910]). *The system* $\boldsymbol{Ax} = \boldsymbol{b} \neq \boldsymbol{0}$ *has a nonnegative solution* $\boldsymbol{x} \geqq \boldsymbol{0}$ *if and only if* $\boldsymbol{y}'\boldsymbol{b} \geqq \boldsymbol{0}$ *for any solution of* $\boldsymbol{A}'\boldsymbol{y} \geqq \boldsymbol{0}$.

PROOF (*Necessity*). Let $\boldsymbol{y}$ be any solution of $\boldsymbol{A}'\boldsymbol{y} \geqq \boldsymbol{0}$, so that $\boldsymbol{y}'\boldsymbol{A} \geqq \boldsymbol{0}'$. Then $0 \leqq \boldsymbol{y}'\boldsymbol{Ax} = \boldsymbol{y}'\boldsymbol{b}$.

(*Sufficiency*). From Theorem 1, the system

$$(-\boldsymbol{b}, \boldsymbol{A})'\boldsymbol{u} \geqq \boldsymbol{0}, \qquad (-\boldsymbol{b}, \boldsymbol{A})\begin{pmatrix} x_0 \\ \boldsymbol{x} \end{pmatrix} = \boldsymbol{0}, \qquad \begin{pmatrix} x_0 \\ \boldsymbol{x} \end{pmatrix} \geqq \boldsymbol{0},$$

has solutions

$$\begin{pmatrix} x_0^* \\ \boldsymbol{x}^* \end{pmatrix} \quad \text{and} \quad \boldsymbol{u}^*$$

such that $x_0^* - \boldsymbol{b}'\boldsymbol{u}^* > 0$. By assumption, $\boldsymbol{b}'\boldsymbol{u}^* \geq 0$, since $\boldsymbol{A}'\boldsymbol{u}^* \geqq \boldsymbol{0}$. Hence $x_0^* > \boldsymbol{b}'\boldsymbol{u}^* \geqq \boldsymbol{0}$. Hence, finally, $\bar{\boldsymbol{x}} = \boldsymbol{x}^*/x_0^*$ is a nonnegative solution of $\boldsymbol{Ax} = \boldsymbol{b}$. □

Theorem 3 is usually called the Minkowski–Farkas Lemma.

As a straightforward corollary to Theorem 3 we obtain Gale's Duality Theorem.

Theorem 4 (Gale [1960]). *Exactly one of the following alternatives holds: either the inequality* $\boldsymbol{Ax} > \mathbf{0}$ *has a nonnegative solution; or the inequality* $\boldsymbol{u}'\boldsymbol{A} \leqq \mathbf{0}$ *has a semipositive solution.*

PROOF. Suppose that both statements are true. Then there exist $\boldsymbol{x} \geqq \mathbf{0}$ and $\boldsymbol{u} \geq \mathbf{0}$ such that $\boldsymbol{b} = \boldsymbol{Ax} > \mathbf{0}$ and $\boldsymbol{A}'\boldsymbol{u} \leqq \mathbf{0}$. Hence $0 < \boldsymbol{b}'\boldsymbol{u} \leqq 0$, a self-contradiction.

The proof is completed by showing that if the second alternative is false then the first is true. Noting that $\boldsymbol{u} \geq \mathbf{0}$ is equivalent to $\boldsymbol{u} \geqq \mathbf{0}$ and $\sum_i u_i > 0$, the second alternative is false if and only if the system

$$(\boldsymbol{u}', \boldsymbol{w}')\begin{pmatrix} \boldsymbol{A} & -\boldsymbol{l} \\ \boldsymbol{I} & \mathbf{0} \end{pmatrix} = (\mathbf{0}, -t),$$

where t is an arbitrary positive number and $\boldsymbol{l}$ is a vector of ones, has no nonnegative solution. From the Minkowski–Farkas Lemma (Theorem 3) this system has no nonnegative solution if there exists $(\boldsymbol{x}', z)$ such that

$$\begin{pmatrix} \boldsymbol{A} & -\boldsymbol{l} \\ \boldsymbol{I} & \mathbf{0} \end{pmatrix}\begin{pmatrix} \boldsymbol{x} \\ z \end{pmatrix} \geqq \mathbf{0} \quad \text{and} \quad (\mathbf{0}, -t)\begin{pmatrix} \boldsymbol{x} \\ z \end{pmatrix} < 0,$$

from which it follows that $\boldsymbol{Ax} > \mathbf{0}$ and $\boldsymbol{x} > \mathbf{0}$, since $-tz < 0$ and $\boldsymbol{Ax} \geqq \boldsymbol{l}z > \mathbf{0}$. □

By considering in turn the special cases $\boldsymbol{A} = (-\boldsymbol{B})'$, $\boldsymbol{A} = \boldsymbol{B}'$, and $\boldsymbol{A} = -\boldsymbol{B}$, we obtain the following.

Corollary 1. (i) *Either* $\boldsymbol{Bu} \geqq \mathbf{0}$ *has a semipositive solution or* $\boldsymbol{x}'\boldsymbol{B} < \mathbf{0}'$ *has a nonnegative solution.* (ii) *Either* $\boldsymbol{Bu} \leqq \mathbf{0}$ *has a semipositive solution or* $\boldsymbol{x}'\boldsymbol{B} > \mathbf{0}'$ *has a nonnegative solution.* (iii) *Either* $\boldsymbol{u}'\boldsymbol{B} \geqq \mathbf{0}'$ *has a semipositive solution or* $\boldsymbol{Bx} < \mathbf{0}$ *has a nonnegative solution.*

Thus, given the task of finding a semipositive (nonnegative) solution of a system of inequalities, we may construct the *dual* problem of finding a nonnegative (semipositive) solution of a new system obtained from the old or primal problem by transposing the vector of variables, converting weak (strict) inequalities into strict (weak) and reversing the direction of the inequalities. Theorem 4 tells us that, given any pair of problems, the primal and its dual, one and only one possesses a solution.

Remark. Theorem 3 is a spring from which many rivers flow. Not only Theorem 4 and Corollary 1 but several other useful propositions may be derived from it. Here we state and prove three such propositions without dignifying them as numbered theorems. In each case we omit the introductory "Exactly one of the following alternatives holds."

i. Either the inequality

$$\boldsymbol{Ax} \geqq \boldsymbol{c}$$

has a solution or the system

$$\boldsymbol{A}'\boldsymbol{u} = \boldsymbol{0}, \qquad \boldsymbol{c}'\boldsymbol{u} > 0,$$

has a nonnegative solution.

ii. Either the inequality

$$\boldsymbol{Ax} \leqq \boldsymbol{b}$$

has a nonnegative solution or the inequalities

$$\boldsymbol{A}'\boldsymbol{u} \geqq \boldsymbol{0}, \qquad \boldsymbol{b}'\boldsymbol{u} < 0,$$

have a semipositive solution.

iii. Either the equation

$$\boldsymbol{Ax} = \boldsymbol{0}$$

has a semipositive solution or the inequality

$$\boldsymbol{A}'\boldsymbol{u} > \boldsymbol{0}$$

has a solution.

PROOF. It is easy to see that, in each of the three cases (i)–(iii), it is not possible for both statements to hold simultaneously. It therefore suffices to show that if one of the alternatives does not hold then the other must hold.

i. Let t be any positive number. Then the system

$$\boldsymbol{A}'\boldsymbol{u} = \boldsymbol{0}, \qquad \boldsymbol{c}'\boldsymbol{u} > 0,$$

has no nonnegative solution if and only if the equation

$$\begin{pmatrix} \boldsymbol{c}' \\ \boldsymbol{A}' \end{pmatrix} \boldsymbol{u} = \begin{pmatrix} t \\ \boldsymbol{0} \end{pmatrix}$$

has no nonnegative solution. This by virtue of Theorem 3, implies that there exists a vector $(v_0, \boldsymbol{v}')$ such that

$$(\boldsymbol{c}, \boldsymbol{A})\begin{pmatrix} v_0 \\ \boldsymbol{v} \end{pmatrix} \geqq \boldsymbol{0}, \qquad (t, \boldsymbol{0})\begin{pmatrix} v_0 \\ \boldsymbol{v} \end{pmatrix} < 0.$$

From the second of this pair of inequalities, $v_0 < 0$. Let us now define $\boldsymbol{x}_0 = \boldsymbol{v}/(-v_0)$. Then, from the first of the above pair of inequalities, the inequality $\boldsymbol{Ax} \geqq \boldsymbol{c}$ has a solution $\boldsymbol{x}_0$.

ii. The inequality $\boldsymbol{Ax} \leqq \boldsymbol{b}$ has a nonnegative solution if and only if the equation $(\boldsymbol{A}, \boldsymbol{I})\binom{\boldsymbol{x}}{\boldsymbol{y}} = \boldsymbol{b}$ has a nonnegative solution, where $\boldsymbol{I}$ is the $m \times m$

identity matrix and y is an $m \times 1$ vector. It follows that if $Ax \leqq b$ has no nonnegative solution then the inequalities

$$\begin{pmatrix} A' \\ I \end{pmatrix} u \geqq 0, \qquad b'u < 0,$$

have a solution. Hence the inequalities

$$A'u \geqq 0, \qquad b'u < 0,$$

have a nonnegative solution.

iii. The equation $Ax = 0$ has no semipositive solution if and only if the equation

$$\begin{pmatrix} l' \\ A \end{pmatrix} x = \begin{pmatrix} t \\ 0 \end{pmatrix}$$

has no nonnegative solution for any positive t. It follows that if $Ax = 0$ has no semipositive solution then

$$(l, A')\begin{pmatrix} v_0 \\ v \end{pmatrix} \geqq 0, \qquad (t, 0)\begin{pmatrix} v_0 \\ v \end{pmatrix} < 0,$$

has a solution. By reasoning similar to that used in proving (i) it can be shown that $A'(-v/v_0) \geqq l > 0$. □

We have been concerned with the solubility of systems of linear equations and inequalities with nonnegativity restrictions on some variables. We now provide an existence condition for linear equations without such restrictions.

Theorem 5. *The equation*

$$Ax = b$$

has a solution if and only if $b'u \leqq 0$ *for any* u *such that* $A'u = 0$.

PROOF (*Necessity*). Let x be a solution of $Ax = b$ and suppose that there exists u such that $b'u > 0$, $A'u = 0$. Then

$$0 < b'u = (x'A')u = x'(A'u) = 0,$$

which is a contradiction.

(*Sufficiency*). Let us denote the rank of A by r. Then, without loss of generality, we can assume that the first r columns $a_1, \ldots, a_r$ of A are linearly independent. Suppose that the equation $Ax = b$ has no solution. Then the vectors $a_1, \ldots, a_r$ and b are also linearly independent. Let $\hat{A} \equiv (a_1, \ldots, a_r, b)$. Without loss of generality we may take the first $r + 1$ rows of $\hat{A}$ to be linearly independent. Let us therefore partition $\boldsymbol{a}_i$, $i = 1, \ldots, r$, and $\boldsymbol{b}$ after the $(r + 1)$th element, so that

$$\boldsymbol{a}_i = \begin{pmatrix} \boldsymbol{a}_{i1} \\ \cdots \\ \boldsymbol{a}_{i2} \end{pmatrix} \qquad (i = 1, \ldots, r),$$

$$\boldsymbol{b} = \begin{pmatrix} \boldsymbol{b}_1 \\ \cdots \\ \boldsymbol{b}_2 \end{pmatrix}.$$

Then $\tilde{\boldsymbol{A}} \equiv (\boldsymbol{a}_{11}, \ldots, \boldsymbol{a}_{r1}, \boldsymbol{b}_1)$ is a nonsingular matrix of order $r + 1$ so that, for any positive t, the equation

$$\tilde{\boldsymbol{A}}'\boldsymbol{v} = \begin{pmatrix} \boldsymbol{0} \\ t \end{pmatrix}$$

has a unique solution $\hat{\boldsymbol{v}}$. Thus the vector

$$\boldsymbol{y} = \begin{pmatrix} \hat{\boldsymbol{v}} \\ \boldsymbol{0} \end{pmatrix}$$

satisfies $\boldsymbol{A}'\boldsymbol{y} = \boldsymbol{0}$ and $\boldsymbol{b}'\boldsymbol{y} = t > 0$, which is the contraposition of the sufficiency condition. □

We now turn our attention to an important result known as Hadamard's Theorem.

Definition 1. An $n \times n$ complex matrix $\boldsymbol{A} = (a_{ij})$ is said to possess a *dominant diagonal* (dd) if there exist $d_i > 0$, $i = 1, \ldots, n$, such that

$$d_i |a_{ii}| > \sum_{j \neq i} d_j |a_{ij}| \qquad (i = 1, \ldots, n). \tag{5}$$

If, in addition, $a_{ii} > 0$, $i = 1, \ldots, n$, then $\boldsymbol{A}$ is said to possess a *positive dominant diagonal* (pdd). Similarly, if $\boldsymbol{A}$ has a dd which is negative then it is said to possess a *negative dominant diagonal* (ndd).

Theorem 6. *Let $\boldsymbol{A}$ be an $n \times n$ matrix defined over the complex field.*

i. *if $\boldsymbol{A}$ has a* dd *then it is nonsingular (Hadamard* [1903]).
ii. *if $\boldsymbol{A}$ is real and has a* pdd *then all principal minors of $\boldsymbol{A}$ are positive.*

In Chapter 3, we shall refer to matrices with property (ii) as P-matrices.

PROOF. (i) Suppose that the implication is false, so that there exists a nonzero vector $\boldsymbol{q}' = (q_1, \ldots, q_n)$ such that

$$\boldsymbol{A}\boldsymbol{q} = \boldsymbol{0}.$$

Then

$$|a_{ii}q_i| = \left| - \sum_{j \neq i} a_{ij}q_j \right| \leqq \sum_{j \neq i} |a_{ij}| \cdot |q_j| \qquad (i = 1, \ldots, n).$$

Defining $p_j = q_j/d_j$, $j = 1, \ldots, n$, the above inequalities reduce to

$$d_i|a_{ii}| \cdot |p_i| \leqq \sum_{j \neq i} d_j|a_{ij}| \cdot |p_j|.$$

Assume that $|p_k| = \max_{1 \leqq j \leqq n}|p_j|$. Then, dividing both sides of the kth member of the above inequalities by $|p_k|$, we obtain

$$d_k|a_{kk}| \leqq \sum_{j \neq k} d_j|a_{kj}|,$$

in contradiction of our hypothesis.

(ii) Since by hypothesis every principal submatrix of $\boldsymbol{A}$ also possesses a pdd, it suffices to show that det $\boldsymbol{A} > 0$. Suppose the contrary. Then, in view of assertion (i), det $\boldsymbol{A} < 0$. There therefore exists a $\lambda_0 > 0$ such that $\det(\lambda_0 \boldsymbol{I} + \boldsymbol{A}) = 0$. Since $a_{jj} > 0$, $j = 1, \ldots, n$; and since $\lambda_0 > 0$,

$$d_i|\lambda_0 + a_{ii}| = d_i(\lambda_0 + a_{ii}) > \sum_{j \neq i} d_j|a_{ij}| \qquad (i = 1, \ldots, n).$$

This, together with assertion (i), implies that $\lambda_0 \boldsymbol{I} + \boldsymbol{A}$ is nonsingular, contradicting the definition of λ_0. □

Suppose that some of the strict inequalities of Definition 1 are replaced by equalities. What then happens to assertion (i) of Theorem 6? To answer this question we introduce

Definition 2. An $n \times n$ matrix $\boldsymbol{A}$ defined over the complex field is said to possess a *semidominant diagonal* (sdd) if there exist $d_i > 0$, $i = 1, \ldots, n$, such that

$$d_i|a_{ii}| \geqq \sum_{j \neq i} d_j|a_{ij}| \qquad (i = 1, \ldots, n)$$

with at least one strict inequality.

Corollary 2 (Taussky [1949]). *If the matrix $\boldsymbol{A}$ in Definition 2 possesses a* sdd *and is indecomposable, that is, for any nonempty subset J of $N = \{1, \ldots, n\}$ there exist indices $i \in J$ and $j \notin J$ such that $a_{ij} \neq 0$, then it is nonsingular.*

Proof. We note first that all diagonal elements of $\boldsymbol{A}$ are nonzero. Suppose the contrary. Then there exists an index i_0 for which

$$0 = d_{i_0}|a_{i_0 i_0}| \geqq \sum_{j \neq i_0} d_j|a_{i_0 j}| \geqq 0,$$

which further implies that

$$a_{i_0 j} = 0 \qquad (j \neq i_0),$$

so that the subset $J = \{i_0\}$ of N fails to satisfy the condition of indecomposability. Hence $a_{ii} \neq 0$ for all i.

Let $J = \{i \in N: d_i|a_{ii}| > \sum_{j \neq i} d_j|a_{ij}|\}$. By hypothesis, J is nonvacuous. Since the proposition is true for $J = N$ it suffices to verify it for the case in which J is properly contained in N. Let $\bar{J}$ denote the set complementary to J relative to N. Then $\bar{J}$ is also a nonempty proper subset of N. In view of the indecomposability of $\boldsymbol{A}$, there exist indices $i_1 \in \bar{J}$ and $j_1 \notin \bar{J}$ such that $a_{i_1 j_1} \neq 0$. It is therefore possible to convert the i_1th equality into a strict inequality without disturbing any of the pre-existing strict inequalities. By repeating this procedure, $\boldsymbol{A}$ can be shown to possess a dd and hence to be nonsingular. □

We turn next to the well-known "Hawkins–Simon Condition." The satisfaction of this condition is both necessary and sufficient for an open statical Leontief system to possess a unique nonegative solution, given any nonegative final demand.

Theorem 7 (Hawkins and Simon [1949]). *Let $\boldsymbol{A}$ be an $n \times n$ matrix with nonpositive off-diagonal elements. Then the following four statements are mutually equivalent.*

i. *For any given $\boldsymbol{b} > \boldsymbol{0}$ there exists an $\boldsymbol{x} > \boldsymbol{0}$ such that $\boldsymbol{A}\boldsymbol{x} = \boldsymbol{b}$. Equivalently, the inequality $\boldsymbol{A}\boldsymbol{x} > \boldsymbol{0}$ has a positive solution.*

ii. $$\det\begin{pmatrix} a_{11} & \cdots & a_{1k} \\ \vdots & & \vdots \\ a_{k1} & \cdots & a_{kk} \end{pmatrix} > 0 \qquad (k = 1, \ldots, n).$$

iii. *For any $\boldsymbol{b} \geqq \boldsymbol{0}$ the equation $\boldsymbol{A}\boldsymbol{x} = \boldsymbol{b}$ possesses a nonnegative solution.*
iv. $\boldsymbol{A}^{-1} \geq \boldsymbol{0}$.

If $\boldsymbol{A}$ is indecomposable, statements (iii) and (iv) may be strengthened.

iii′. For any $\boldsymbol{b} \geq \boldsymbol{0}$ the equation $\boldsymbol{A}\boldsymbol{x} = \boldsymbol{b}$ has a positive solution.
iv′. $\boldsymbol{A}^{-1} > \boldsymbol{0}$.

Remark. The "Hawkins–Simon Condition" is the double implication (i)⇔(ii).

Proof. The equivalence of statements (i)–(iv) is established by showing that (i) ⇒ (ii) ⇒ (iii) ⇒ (iv) ⇒ (i).

(i) ⇒ (ii). Since $a_{ij} \leqq 0\ (i \neq j)$, (i) implies that $\boldsymbol{A}$ has a dd. Moreover, it is easy to verify that $a_{jj} > 0$, $j = 1, \ldots, n$. Suppose the contrary. Then $a_{ii} \leqq 0$ for some i. Hence

$$0 \geqq \sum_j a_{ij}x_j > 0,$$

a contradiction. Thus $\boldsymbol{A}$ possesses a pdd and assertion (ii) follows immediately from (ii) of the previous theorem.

(ii) ⇒ (iii). We proceed by induction on n. When $n = 1$, the system $\boldsymbol{Ax} = \boldsymbol{b}$ reduces to $ax = b$, $a > 0$ and $b \geqq 0$, with the nonnegative solution $x = b/a$.

Since $a_{11} > 0$ by assumption, we can specify a_{11} as a pivotal element in the familiar method of elimination. Thus the original equation $\boldsymbol{Ax} = \boldsymbol{b}$ reduces to

$$\begin{pmatrix} 1 & a_{12}^* & \cdots & a_{1n}^* \\ 0 & a_{22}^* & \cdots & a_{2n}^* \\ \vdots & \vdots & & \vdots \\ 0 & a_{n2}^* & \cdots & a_{nn}^* \end{pmatrix} \boldsymbol{x} = \begin{pmatrix} b_1^* \\ b_2^* \\ \vdots \\ b_n^* \end{pmatrix} \tag{6}$$

where

$$a_{1j}^* = \frac{a_{1j}}{a_{11}} \qquad (j = 2, \ldots, n),$$

$$b_1^* = \frac{b_1}{a_{11}},$$

$$a_{ij}^* = a_{ij} - a_{1j}^* a_{i1} \qquad (i = 2, \ldots, n; j = 1, \ldots, n),$$

and

$$b_i^* = b_i - b_1^* a_{i1} \qquad (i = 2, \ldots, n).$$

By the nature of the elimination method utilized, we have

$$\det\begin{pmatrix} a_{11} & \cdots & a_{1k} \\ \vdots & & \vdots \\ a_{k1} & \cdots & a_{kk} \end{pmatrix} = a_{11} \det\begin{pmatrix} a_{22}^* & \cdots & a_{2k}^* \\ \vdots & & \vdots \\ a_{k2}^* & \cdots & a_{kk}^* \end{pmatrix} > 0 \qquad (k = 2, \ldots, n),$$

from which it follows that

$$\det\begin{pmatrix} a_{22}^* & \cdots & a_{2k}^* \\ \vdots & & \vdots \\ a_{k2}^* & \cdots & a_{kk}^* \end{pmatrix} > 0 \qquad (k = 2, \ldots, n).$$

Moreover, $a_{ij}^* \leqq 0\,(i \neq j; i, j = 2, \ldots, n)$ and $b_i^* \geqq 0\,(i = 1, \ldots, n)$ for $a_{ij} \leqq$ $(i \neq j)$, $b_i \geqq 0$ $(i = 1, \ldots, n)$ and $a_{11} > 0$. Hence the induction assumption asserts that there exists a $\boldsymbol{y} = (y_2, \ldots, y_n)' \geqq \boldsymbol{0}$ such that

$$\boldsymbol{A}^* \boldsymbol{y} = \boldsymbol{b}^*$$

where

$$\boldsymbol{A}^* = \begin{pmatrix} a_{22}^* & \cdots & a_{2k}^* \\ \vdots & & \vdots \\ a_{n2}^* & \cdots & a_{nn}^* \end{pmatrix} \quad \text{and} \quad \boldsymbol{b}^* = (b_2^*, \ldots, b_n^*)'.$$

Let us define $y_1 = (1/a_{11})(b_1 - \sum_{j=2}^{n} a_{1j}y_j)$ and $\tilde{\boldsymbol{y}} = (y_1, y_2, \ldots, y_n)'$. Then $\tilde{\boldsymbol{y}} \geqq \boldsymbol{0}$ and a direct calculation shows that

$$\boldsymbol{A}\tilde{\boldsymbol{y}} = \boldsymbol{b}.$$

(iii) ⇒ (iv). $\boldsymbol{Ax} = \boldsymbol{b}$ possesses a nonnegative solution for any nonnegative $\boldsymbol{b}$, in particular for the ith unit vector $\boldsymbol{e}_i$, $i = 1, \ldots, n$. Let us denote by $\boldsymbol{x}_i$ the nonnegative solution of $\boldsymbol{Ax} = \boldsymbol{e}_i$. Then every $\boldsymbol{x}_i$ must be semipositive, for otherwise there would exist an index i_0 such that $\boldsymbol{x}_{i_0} = \boldsymbol{0}$. However, this implies that

$$\boldsymbol{0} = \boldsymbol{A}\boldsymbol{x}_{i_0} = \boldsymbol{e}_{i_0} \geq \boldsymbol{0},$$

a contradiction. Hence

$$\boldsymbol{A}^{-1} = (\boldsymbol{x}_1, \ldots, \boldsymbol{x}_n) \geq \boldsymbol{0}.$$

The proof that (iv) ⇒ (i) is obvious and hence is omitted.

It remains to verify that if $\boldsymbol{A}$ is indecomposable then (iii) and (iv) can be replaced by (iii′) and (iv′). The implications (iii′) ⇒ (iv′) ⇒ (i) are quite straightforward; moreover, (i) implies (ii) whether or not $\boldsymbol{A}$ is indecomposable. It therefore suffices to demonstrate that (ii) ⇒ (iii′).

Since (ii) implies (iii), for any $\boldsymbol{b} \geq \boldsymbol{0}$ there exists $\boldsymbol{x} \geqq \boldsymbol{0}$ such that $\boldsymbol{Ax} = \boldsymbol{b}$. We first suppose that $\boldsymbol{x} = \boldsymbol{0}$. Then

$$\boldsymbol{0} \leq \boldsymbol{b} = \boldsymbol{Ax} = \boldsymbol{0},$$

a self-contradiction. Hence $\boldsymbol{x} \geq \boldsymbol{0}$. Define the set $I_{\boldsymbol{x}} = \{i \in N : x_i = 0\}$. If $\boldsymbol{x} > \boldsymbol{0}$ then there remains nothing to prove. Suppose therefore that $I_{\boldsymbol{x}} \neq \varnothing$. Then, for any $i \in I_{\boldsymbol{x}}$,

$$0 \geqq \sum_{j \notin I_{\boldsymbol{x}}} a_{ij}x_j - b_i \geqq 0.$$

Hence $a_{ij} = 0$ for all $i \in I_{\boldsymbol{x}}$ and all $j \notin I_{\boldsymbol{x}}$, contradicting the assumption that $\boldsymbol{A}$ is indecomposable. □

Remark Concerning Theorem 7. It has been shown (in the proof that (i) ⇒ (ii)) that (i) implies that all principal minors of $\boldsymbol{A}$ (not just the leading principal minors) are positive. Thus, if we are given a square matrix with nonpositive off-diagonal elements, the Hawkins–Simon Condition is equivalent to the statement that the matrix under consideration is a P-matrix.

Theorem 6 and Corollary 2 run in terms of what might be called row dominant diagonals (dd) and row semidominant diagonals (sdd). With the aid of Theorem 7 we now can verify analogous propositions for column dd and column sdd.

Theorem 6′. *Let* $\boldsymbol{A}$ *be as defined in Theorem* 6. *If* $\boldsymbol{A}$ *has a column* dd, *that is, there exists* $d_j > 0, j = 1, \ldots, n$, *such that*

$$d_j|a_{jj}| > \sum_{i \neq j} d_i|a_{ij}| \qquad (j = 1, \ldots, n), \tag{7}$$

then $\boldsymbol{A}$ *has a row* dd *and hence is nonsingular.*

PROOF. Corresponding to $\boldsymbol{A}$ define a matrix $\boldsymbol{B} = (b_{ij})$ such that

$$b_{ij} = -|a_{ij}| \qquad (i \neq j),$$
$$b_{ii} = |a_{ii}|.$$

Then $\boldsymbol{A}$ possesses a column dd if and only if $\boldsymbol{B}'$ possesses a row dd with the same d_j's. Since $\boldsymbol{B}'$ and $\boldsymbol{B}$ contain nonpositive off-diagonal elements, Theorem 7 is equally applicable to both. Since by hypothesis $\boldsymbol{B}'$ satisfies condition (i) of Theorem 7, every leading principal minor of $\boldsymbol{B}'$ is positive. Hence the same holds for every leading principal minor of $\boldsymbol{B}$. Thus there exists an $\boldsymbol{x} > \boldsymbol{0}$ such that $\boldsymbol{Bx} > \boldsymbol{0}$; equivalently, $\boldsymbol{A}$ possesses a row dd. By virtue of Theorem 6, $\boldsymbol{A}$ is nonsingular. □

Similarly, we have

Corollary 2′. *If the matrix* $\boldsymbol{A}$ *defined in Corollary* 2 *is indecomposable and if there exist* $d_j > 0, j = 1, \ldots, n$, *such that*

$$d_j|a_{jj}| \geqq \sum_{i \neq j} d_i|a_{ij}| \qquad (j = 1, \ldots, n) \tag{8}$$

with at least one strict inequality then $\boldsymbol{A}$ *is nonsingular.*

It is now clear that the distinction between matrices with row dd (sdd) and column dd (sdd) is superfluous and can be abandoned.

The essence of Corollaries 2 and 2′ lies in their assertion that an indecomposable square matrix with a sdd possesses a dd. With this result at our beck, the Brauer–Solow Theorem on Leontief matrices can be easily obtained.

Theorem 8 (Brauer [1946], Solow [1952]). *If* $\boldsymbol{A}$ *is an indecomposable* $n \times n$ *matrix with nonnegative off-diagonal elements and if* $\boldsymbol{l}'\boldsymbol{A} \leq \boldsymbol{l}'$ *then* $(\boldsymbol{I} - \boldsymbol{A})^{-1} > \boldsymbol{0}$ *where* $\boldsymbol{l}' = (1, \ldots, 1)$ *is the sum vector consisting of n ones.*

PROOF. Since $\boldsymbol{A}$ is indecomposable, so is $(\boldsymbol{I} - \boldsymbol{A})$. Moreover, from the hypothesis that $\boldsymbol{l}'\boldsymbol{A} \leq \boldsymbol{l}'$, $(\boldsymbol{I} - \boldsymbol{A})$ has a sdd. Hence $(\boldsymbol{I} - \boldsymbol{A})$ has a dd; or, equivalently, there exists an $\boldsymbol{x} > \boldsymbol{0}$ such that $(\boldsymbol{I} - \boldsymbol{A})\boldsymbol{x} > \boldsymbol{0}$. Since all off-diagonal elements of $(\boldsymbol{I} - \boldsymbol{A})$ are nonpositive, the equivalence of (i) and (iv′) in Theorem 7 ensures that $(\boldsymbol{I} - \boldsymbol{A})^{-1} > \boldsymbol{0}$. □

Remark Concerning Theorem 8. Without the indecomposability of $\boldsymbol{A}$, Theorem 8 could be restated as:

Theorem 8′. *If the matrix* $\boldsymbol{A}$ *in Theorem* 8 *satisfies* $\boldsymbol{l}'\boldsymbol{A} < \boldsymbol{l}'$ *then* $(\boldsymbol{I} - \boldsymbol{A})^{-1} > \boldsymbol{0}$.

The proof is immediate and so it is left to the reader.

McKenzie has proposed yet another concept of diagonal dominance, *viz.* quasi-diagonal dominance.

Definition 3 (McKenzie [1960; revised 1971]). An $n \times n$ matrix defined over the complex field is said to possess a *quasi-dominant diagonal* (qdd) if all of its principal submatrices possess sdd. (McKenzie's revision of his 1960 definition may be found in Uekawa [1971; Footnote 5].)

Our next proposition elucidates the relationships between the three concepts.

Theorem 9. *An $n \times n$ matrix $\boldsymbol{A}$ defined over the complex field has a quasi-dominant diagonal if and only if it has a dominant diagonal.*

PROOF. The proof of sufficiency is trivial. We therefore confine ourselves to the verification of necessity. If $\boldsymbol{A}$ is indecomposable then it satisfies the condition of Corollary 2 and is therefore diagonally dominant. If $\boldsymbol{A}$ is decomposable, there exists a permutation matrix $\boldsymbol{P}$ such that

$$\boldsymbol{PAP}' = \begin{pmatrix} \boldsymbol{A}_{11} & \boldsymbol{A}_{12} & \cdots & \boldsymbol{A}_{1m} \\ & \boldsymbol{A}_{22} & \cdots & \boldsymbol{A}_{2m} \\ & & \ddots & \vdots \\ \boldsymbol{0} & & & \boldsymbol{A}_{mm} \end{pmatrix} \tag{9}$$

where $\boldsymbol{A}_{vv}$, $v = 1, \ldots, m$, is square and indecomposable as well as semi-diagonally dominant. Again from Corollary 2, $\boldsymbol{A}_{vv}$, $v = 1, \ldots, m$, are diagonally dominant. To proceed further, let us redenote the matrix $\boldsymbol{B}$ defined in the proof of Theorem 6′ by $\boldsymbol{B}(\boldsymbol{A})$. Then it is clear that $\boldsymbol{B}(\boldsymbol{PAP}') = \boldsymbol{PB}(\boldsymbol{A})\boldsymbol{P}'$ and that $\boldsymbol{B}(\boldsymbol{A}_{vv})$, $v = 1, \ldots, m$, satisfy condition (ii) of Theorem 7. Therefore, in view of the decomposition (9), $\boldsymbol{B}(\boldsymbol{P}'\boldsymbol{AP}') = \boldsymbol{PB}(\boldsymbol{A})\boldsymbol{P}'$ satisfies the same condition. From Theorem 7 again, there therefore exists a $\boldsymbol{y} > \boldsymbol{0}$ such that $\boldsymbol{PB}(\boldsymbol{A})\boldsymbol{P}'\boldsymbol{y} > \boldsymbol{0}$, from which it follows that $\boldsymbol{B}(\boldsymbol{A})\boldsymbol{P}'\boldsymbol{y} > \boldsymbol{0}$. The proof is completed by noticing that $\boldsymbol{P}'\boldsymbol{y} > \boldsymbol{0}$. □

A generalization of the assertion (i) of Theorem 6 can be achieved by introducing the notion of dominant diagonal *blocks*. Before attempting a formal definition, we provide some preliminary concepts. Let p be an arbitrary integer not larger than n; and let I_i, $i = 1, \ldots, p$, be a partition of the set $N = \{1, \ldots, n\}$, so that

$$I_i \cap I_j = \varnothing \qquad (i \neq j)$$

and

$$\bigcup_{j=1}^{p} I_j = N.$$

Let A be an $m \times n$ complex matrix and let

$$\|A\| = \sup_{x \neq 0} \frac{h(Ax)}{h(x)}$$

where $h(\cdot)$ denotes any vector norm defined in Appendix A. We follow tradition in referring to $\|A\|$ as the Minkowski norm of A induced by the vector norm $h(\cdot)$. Among the properties of the Minkowski norm the following are of immediate interest.

M.P.1. $\|A\| = \max_{h(x)=1} h(Ax)$.
M.P.2. For any n-vector x, $\|x\| = h(x)$.
M.P.3. For any n-vector x, $\|Ax\| \leqq \|A\| \cdot \|x\|$.

By repeated application of (M.P.3) we then find that $\|A \cdot B\| \leqq \|A\| \cdot \|B\|$, provided that the matrices A and B are conformable in that order. We proceed to verify these three properties.

(M.P.1) For any nonzero n-vector y, define $x = y/h(y)$. Then, by virtue of property (3) of vector norms, $h(x) = 1$ and

$$\frac{h(Ay)}{h(y)} = \frac{h(Ax)}{h(x)} = h(Ax).$$

Since $h(Ax)$ is continuous[1] on a compact set $X = \{x : h(x) = 1\}$, $h(Ax)$ attains its maximum on that set (see Theorem 9 of Appendix A). This implies that

$$\|A\| = \sup_{h(x)=1} h(Ax) = \max_{h(x)=1} h(Ax).$$

Thus $\|A\|$ does exist.

(M.P.2) This follows from (M.P.1).

(M.P.3) From the definition of $\|A\|$ and from (M.P.2),

$$\|A\| \cdot \|x\| = \|A\| \cdot h(x) \geqq h(Ax) = \|Ax\|$$

for any $x \neq 0$.

Definition 1′. An $n \times n$ complex matrix A is said to possess *dominant diagonal blocks* if there exists a partition I_i, $i = 1, \ldots, p$, of N and $d_{I_i} > 0$, $i = 1, \ldots, p$, such that

$$d_{I_i} > \sum_{I_j \neq I_i} d_{I_j} \|A_{I_i I_i}^{-1} \cdot A_{I_i I_j}\| \qquad (I_i = I_1, \ldots, I_p) \tag{10}$$

where $A_{I_i I_j}$ is a submatrix of A with typical element a_{rs} such that $r \in I_i$ and $s \in I_j$.

Theorem 6″ (Pearce [1974] and Okuguchi [1976]). *If A is an $n \times n$ complex matrix with dominant diagonal blocks then A is nonsingular.*

PROOF. Suppose the contrary. Then there exists a vector $x = (x'_{I_1}, \ldots, x'_{I_p})' \neq 0$ such that $Ax = 0$. Hence, drawing on (M.P.2), (M.P.3) and property (2) of vector norms,

$$h(x_{I_i}) \leq \sum_{I_j \neq I_i} \|A_{I_i I_i}^{-1} \cdot A_{I_i I_j}\| h(x_{I_j}).$$

Let us define $d_{I_i} y_{I_i} = x_{I_i}$, $I_i = I_1, \ldots, I_p$. Then, for I_k such that

$$\max_{I_i} h(y_{I_i}) = h(y_{I_k}),$$

we have

$$d_{I_k} \cdot h(y_{I_k}) \leqq \sum_{I_j \neq I_k} d_{I_j} \cdot \|A_{I_k I_k}^{-1} \cdot A_{I_k I_j}\| \cdot h(y_{I_j})$$

Since by definition $y_{I_k} \neq 0$, the above inequality implies that

$$d_{I_k} \leqq \sum_{I_j \neq I_k} d_{I_j} \cdot \|A_{I_k I_k}^{-1} \cdot A_{I_k I_j}\|$$

This, however, contradicts the hypothesis. □

Remark concerning Theorem 6″. If each I_i consists of a single index, the condition (10) clearly reduces to the condition (5), since then $\|A_{I_i I_i}^{-1} \cdot A_{I_i I_j}\| = |a_{ij}|/|a_{ii}|$.

2 Linear Programming and Matrix Games

In spite of considerable recent progress in the development of more general techniques of mathematical programming, linear programming remains of special interest to economists. Not only is it widely used in the solution of practical problems (within the firm, for example), but it also plays a central role in the theoretical analysis of linear economic models. The present section is devoted to the properties of linear programs and their relation to matrix games. Throughout the section, matrices and vectors are defined on the field of real numbers.

Definition 4. Let A be an $m \times n$ matrix, x and c $n \times 1$ vectors, and u and b $m \times 1$ vectors. Then problems (I) and (II) below comprise a pair of dual linear programming problems.

(I) $\max_{x} c'x$ subject to $Ax \leqq b$, $x \geqq 0$

(II) $\min_{u} u'b$ subject to $u'A \geqq c'$, $u \geqq 0$.

The set $F_{\mathrm{I}} = \{\boldsymbol{x}: \boldsymbol{A}\boldsymbol{x} \leqq \boldsymbol{b}, \boldsymbol{x} \geqq \boldsymbol{0}\}$ is called the *feasible set of problem* (I), and $F_{\mathrm{II}} = \{\boldsymbol{u}: \boldsymbol{u}'\boldsymbol{A} \geqq \boldsymbol{c}', \boldsymbol{u} \geqq \boldsymbol{0}\}$ the *feasible set of problem* (II). If F_{I} is nonempty, problem (I) is said to be *feasible* and the elements of F_{I} are called *feasible vectors*. Similarly for F_{II}.

Problem (I) is usually referred to as the primal problem, problem (II) as the dual problem. However a problem of type (I) can be converted into a problem of type (II) simply by substituting $-\boldsymbol{d} \equiv \boldsymbol{c}$, $-\boldsymbol{e} \equiv \boldsymbol{b}$ and $-\boldsymbol{B} \equiv \boldsymbol{A}$, and vice versa. Thus it is only a matter of convenience which problem is called primal and which dual.

In the first part of this section we provide a series of propositions concerning the existence and general properties of solutions of problems (I) and (II). These propositions culminate in Theorems 10–12. We then move on to consider further properties of the solutions to linear problems and to establish the relation between linear programming problems and matrix games.

We begin our analysis with a fundamental proposition based on Tucker's Lemma and Theorem 1 of Section 1.

Theorem 10. *Let* $\boldsymbol{K}$ *be an* $n \times n$ *skew-symmetric matrix, so that* $\boldsymbol{K}' = -\boldsymbol{K}$. *Then the system of linear inequalities*

$$\boldsymbol{K}\boldsymbol{w} \geqq \boldsymbol{0},$$

$$\boldsymbol{w} \geqq \boldsymbol{0}$$

possesses a solution $\boldsymbol{w}^*$ *such that*

$$\boldsymbol{K}\boldsymbol{w}^* + \boldsymbol{w}^* > \boldsymbol{0}.$$

PROOF. Consider the system of inequalities

$$(\boldsymbol{I}, \boldsymbol{K}')\begin{pmatrix}\boldsymbol{x}\\ \boldsymbol{u}\end{pmatrix} = \boldsymbol{0}, \qquad \begin{pmatrix}\boldsymbol{x}\\ \boldsymbol{u}\end{pmatrix} \geqq \boldsymbol{0},$$

$$(\boldsymbol{I}, \boldsymbol{K}')'\boldsymbol{v} \geqq \boldsymbol{0},$$

where $\boldsymbol{I}$ is the identity matrix of order n, and $\boldsymbol{x}$, $\boldsymbol{u}$ and $\boldsymbol{v}$ are $n \times 1$ vectors. From Theorem 1, this system has a solution $\boldsymbol{x}^*$, $\boldsymbol{u}^*$, $\boldsymbol{v}^*$ such that

$$\begin{pmatrix}\boldsymbol{I}\\ \boldsymbol{K}\end{pmatrix}\boldsymbol{v}^* + \begin{pmatrix}\boldsymbol{x}^*\\ \boldsymbol{u}^*\end{pmatrix} > \boldsymbol{0}, \tag{11}$$

whence

$$\boldsymbol{x}^* + \boldsymbol{K}'\boldsymbol{u}^* = \boldsymbol{x}^* - \boldsymbol{K}\boldsymbol{u}^* = \boldsymbol{0}, \tag{12}$$

$$\boldsymbol{x}^* \geqq \boldsymbol{0}, \qquad \boldsymbol{u}^* \geqq \boldsymbol{0}, \tag{13}$$

$$\boldsymbol{v}^* \geqq \boldsymbol{0}, \qquad \boldsymbol{K}\boldsymbol{v}^* \geqq \boldsymbol{0}. \tag{14}$$

Let us now define $w^* = u^* + v^*$. Then

$$w^* = u^* + v^* \geqq 0 \qquad \text{[from (13) and (14)]},$$

$$\begin{aligned} Kw^* + w^* &= Ku^* + Kv^* + u^* + v^* \\ &= x^* + v^* + Kv^* + u^* \qquad \text{[from (12)]} \\ &> 0 \qquad \text{[from (11)]}, \end{aligned}$$

$$\begin{aligned} Kw^* &= Ku^* + Kv^* \\ &= x^* + Kv^* \qquad \text{[from (12)]} \\ &\geqq 0 \qquad \text{[from (13) and (14)]}. \end{aligned}$$

□

Definition 5. A feasible vector x_0 for problem (I) is said to be *optimal* if, for any $x \in F_{\text{I}}$, $c'x_0 \geqq c'x$. Similarly, a feasible vector u_0 for problem (II) is said to be *optimal* if, for any $u \in F_{\text{II}}$, $u_0'b \leqq u'b$. The pair (x_0, u_0) of optimal vectors is called an *optimal program.*

Lemma 2. *Let* x *and* u *be any feasible vectors for problems* (I) *and* (II), *respectively. Then* (i)

$$c'x \leqq u'b.$$

(ii) *If there exist* $x_0 \in F_{\text{I}}$ *and* $u_0 \in F_{\text{II}}$ *such that*

$$c'x_0 \geqq u_0'b$$

then the pair (x_0, u_0) *is an optimal program.*

PROOF. (i) Since $x \in F_{\text{I}}$, $Ax \leqq b$. Premultiplying both sides by $u' \geqq 0$, we obtain $u'Ax \leqq u'b$. On the other hand, $u'A \geqq c'$ and $x \geqq 0$, so that $u'Ax \geqq c'x$. Hence $u'b \geqq c'x$.

(ii) From (i), just verified, $c'x_0 \geqq u_0'b \geqq cx'$ for any $x \in F_{\text{I}}$. Thus x_0 is an optimal vector. In similar fashion, u_0 can be shown to be optimal. □

To enable us to deal with dual linear problems systematically and symmetrically, we consider the system

$$Kw \geqq 0,$$
$$w \geqq 0,$$

where

$$K = \begin{pmatrix} 0 & -A & b \\ A' & 0 & -c \\ -b' & c' & 0 \end{pmatrix}$$

is a skew-symmetric matrix of order $m + n + 1$ and

$$w = \begin{pmatrix} u \\ x \\ t \end{pmatrix}.$$

Applying Theorem 10 to this system, there exists $\boldsymbol{w}^*$ such that $\boldsymbol{K}\boldsymbol{w}^* \geqq \boldsymbol{0}$, $\boldsymbol{w}^* \geqq \boldsymbol{0}$, and $\boldsymbol{K}\boldsymbol{w}^* + \boldsymbol{w}^* > \boldsymbol{0}$. More specifically,

$$\boldsymbol{A}\boldsymbol{x}^* \leqq \boldsymbol{b}t^*, \tag{15}$$

$$\boldsymbol{A}'\boldsymbol{u}^* \geqq \boldsymbol{c}t^*, \tag{16}$$

$$\boldsymbol{c}'\boldsymbol{x}^* \geqq \boldsymbol{b}'\boldsymbol{u}^*, \tag{17}$$

$$\boldsymbol{u}^* + \boldsymbol{b}t^* > \boldsymbol{A}\boldsymbol{x}^*, \tag{18}$$

$$\boldsymbol{A}'\boldsymbol{u}^* + \boldsymbol{x}^* > \boldsymbol{c}t^*, \tag{19}$$

$$\boldsymbol{c}'\boldsymbol{x}^* > \boldsymbol{b}'\boldsymbol{u}^* - t^*, \tag{20}$$

$$\boldsymbol{x}^* \geqq \boldsymbol{0}, \tag{21}$$

$$\boldsymbol{u}^* \geqq \boldsymbol{0}, \tag{22}$$

$$t^* \geqq 0. \tag{23}$$

Bearing these inequalities in mind, we can proceed to Lemma 3.

Lemma 3. *Consider the system* (15)–(23). *Then*

i. *both F_{I} and F_{II} are nonempty if and only if $t^* > 0$, and*
ii. *neither* (I) *nor* (II) *has an optimal program if $t^* = 0$.*

PROOF. (i) (*Sufficiency*). Since t^* is positive, we can define $\boldsymbol{x}_0$ and $\boldsymbol{u}_0$ by $\boldsymbol{x}_0 = \boldsymbol{x}^*/t^*$ and $\boldsymbol{u}_0 = \boldsymbol{u}^*/t^*$. From (15), (16), (21), and (22), $\boldsymbol{x}_0$ and $\boldsymbol{u}_0$ are feasible.

(i) (*Necessity*). Let $\boldsymbol{x}$ and $\boldsymbol{u}$ be feasible vectors for (I) and (II) respectively, and suppose that $t^* = 0$. Then, from (16) with $t^* = 0$,

$$\boldsymbol{A}'\boldsymbol{u}^* \geqq \boldsymbol{0}. \tag{24}$$

On the other hand, the assumed feasibility of $\boldsymbol{u}$ implies that

$$0 \geqq \boldsymbol{u}'\boldsymbol{A}\boldsymbol{x}^* \geqq \boldsymbol{c}'\boldsymbol{x}^*$$

which, together with (20), yields

$$0 \geqq \boldsymbol{c}'\boldsymbol{x}^* > \boldsymbol{b}'\boldsymbol{u}^*. \tag{25}$$

In view of the remark concerning Theorem 3, especially (ii), the inequalities (24) and (25) imply that F_{I} is empty, contradicting the hypothesis.

(ii) In view of (i), either both F_{I} and F_{II} are empty or just one of them, say F_{I}, is nonempty. From the very definition of an optimal program, a problem with an empty feasible set has no optimal program; hence it suffices to consider the second possibility. Let us define $\boldsymbol{x}_\lambda = \boldsymbol{x} + \lambda\boldsymbol{x}^*$, where λ is nonnegative and $\boldsymbol{x} \in F_{\mathrm{I}}$. Then, clearly, $\boldsymbol{x}_\lambda \geqq \boldsymbol{0}$; moreover, from (15) with $t^* = 0$, $\boldsymbol{A}\boldsymbol{x}_\lambda = \boldsymbol{A}\boldsymbol{x} + \lambda\boldsymbol{A}\boldsymbol{x}^* \leqq \boldsymbol{A}\boldsymbol{x} \leqq \boldsymbol{b}$. Hence $\boldsymbol{x}_\lambda \in F_{\mathrm{I}}$ for any nonnegative λ.

Premultiplying both sides of $Ax \leqq b$ by $(u^*)'$ and taking into account (24), we obtain

$$0 \leqq b'u^*,$$

which, by virtue of (20), asserts that $c'x^* > 0$. Hence, $c'x_\lambda = c'x + \lambda \cdot c'x^*$ becomes indefinitely large as λ goes to infinity and no optimal program for (I) exists. Needless to say, a similar proof is available for the case $F_{\mathrm{I}} = \varnothing$ but $F_{\mathrm{II}} \neq \varnothing$. □

Theorem 11 (Existence). *For dual problems* (I) and (II) *there exists an optimal program if and only if both F_{I} and F_{II} are nonempty.*

PROOF. Since necessity is obvious, we concentrate on the proof of sufficiency. From (17) and Lemma 3(i) and from Lemma 2(ii), x_0 and u_0 are optimal. □

Remark. As an immediate corollary to Theorem 11, Lemma 3(ii) can be strengthened as follows:

Neither the primary nor dual problem has an optimal programme, if and only if $t^* = 0$.

PROOF. It sufficies to prove the converse of Lemma 3(ii). Suppose the contrary, then, by Lemma 3(i) and Theorem 11, x_0 and u_0 are optimal, which violates the assumption. □

Remark. In view of (ii) in the remark following Corollary 1, both F_{I} and F_{II} are nonempty if and only if the system

$$A'u \geqq 0, \qquad b'u < 0$$

has no nonnegative (in fact semipositive) solution and the system

$$Aw \leqq 0, \qquad c'w > 0$$

has no nonnegative solution.

Theorem 12 (Duality). *Let F_{I} and F_{II} be nonempty. Then:* (i) $x^0 \in F_{\mathrm{I}}$ *is optimal if and only if there exists* $u^0 \in F_{\mathrm{II}}$ *such that* $c'x^0 = (u^0)'b$; *and* (ii) $u^* \in F_{\mathrm{II}}$ *is optimal if and only if there exists* $x^* \in F_{\mathrm{I}}$ *such that* $(u^*)'b = c'x^*$.

PROOF. Since both statements can be verified in essentially the same way, it suffices to prove (i). Sufficiency having been established in Lemma 2(ii), it remains to prove necessity. Since x^0 is optimal by assumption, $t^* > 0$ from Lemma 3. Thus the pair (x_0, u_0), defined in Lemma 3, is an optimal program. Hence, from (17), $c'x^0 = c'x_0 \geqq b'u_0$. From the feasibility of x_0 and u_0, on the other hand, $c'x_0 \leqq b'u_0$. Hence, identifying u_0 with u^0, there exists a vector $u_0 \in F_{\mathrm{II}}$ that satisfies $c'x^0 = c'x_0 = b'u_0$. □

Remark. The strict inequalities (18) and (19) have played no essential role in the proofs of Theorems 11 and 12. However, they do help to establish the properties of an optimal program $(\boldsymbol{x}_0, \boldsymbol{u}_0)$. Rewriting (18) and (19) we obtain

$$\boldsymbol{u}_0 + \boldsymbol{b} > A\boldsymbol{x}_0, \tag{18'}$$

$$\boldsymbol{u}_0' A + \boldsymbol{x}_0' > \boldsymbol{c}'. \tag{19'}$$

Element by element comparison reveals:

$$(\boldsymbol{u}_0)_i = 0 \text{ implies } (A\boldsymbol{x}_0)_i < b_i; \qquad (A\boldsymbol{x}_0)_i = b_i \text{ implies } (\boldsymbol{u}_0)_i > 0; \tag{18''}$$

$$(\boldsymbol{x}_0)_j = 0 \text{ implies } (\boldsymbol{u}_0' A)_j > c_j; \qquad (\boldsymbol{u}_0' A)_j = c_j \text{ implies } (\boldsymbol{x}_0)_j > 0. \tag{19''}$$

In general, the $(m + n)$ pairs of constraints

$$(A\boldsymbol{x})_i \leqq b_i, \qquad u_i \geqq 0 \qquad (i = 1, \ldots, m),$$
$$(\boldsymbol{u}'A)_j \geqq c_j, \qquad x_j \geqq 0 \qquad (j = 1, \ldots, n),$$

are called dual constraints. Alternatively, we may refer to each of

$$A\boldsymbol{x} \leqq \boldsymbol{b}, \qquad \boldsymbol{u} \geqq \boldsymbol{0},$$

and

$$\boldsymbol{u}'A \geqq \boldsymbol{c}, \qquad \boldsymbol{x} \geqq \boldsymbol{0},$$

as a pair of dual (matrix) constraints.

Corollary 3. *Suppose that both problems,* (I) *and* (II), *are feasible. Then*:

i. *for every optimal* $\boldsymbol{x}$ *and* $\boldsymbol{u}$, $\boldsymbol{c}'\boldsymbol{x} = \boldsymbol{b}'\boldsymbol{u}$;

iia. *if there exists an optimal* $\boldsymbol{x}$ *such that* $(A\boldsymbol{x})_i < b_i$, $i \in \{1, \ldots, m\}$, *then* $u_i = 0$ *for every optimal* $\boldsymbol{u}$;

iib. *if, for every optimal* $\boldsymbol{x}$, $(A\boldsymbol{x})_i = b_i$ *then there exists an optimal* $\boldsymbol{u}$ *such that* $u_i > 0$.

The possibilities described in (iia) and (iib) are, of course, mutually exclusive.

PROOF. (i) From Theorem 12, there exists a feasible $\boldsymbol{u}$, say $\boldsymbol{u}_x$, such that $\boldsymbol{c}'\boldsymbol{x} = \boldsymbol{b}'\boldsymbol{u}_x$. Suppose now that there exists an optimal $\boldsymbol{u}$ such that $\boldsymbol{b}'\boldsymbol{u}_x \neq \boldsymbol{b}'\boldsymbol{u}$. Then $\boldsymbol{c}'\boldsymbol{x} = \boldsymbol{b}'\boldsymbol{u}_x > \boldsymbol{b}'\boldsymbol{u}$, which contradicts Lemma 2(i).

(iia) Let us assume the contrary, that there exists an optimal $\boldsymbol{u}$ for which $u_i > 0$. Then it is easy to see that

$$\sum_k u_k b_k > \sum_k (A\boldsymbol{x})_k u_k = \sum_k (\boldsymbol{u}'A)_k x_k \geqq \sum_k c_k x_k,$$

which, in view of (i), contradicts the optimality of $\boldsymbol{x}$ and $\boldsymbol{u}$.

(iib) Since $(A\boldsymbol{x})_i = b_i$ for every optimal $\boldsymbol{x}$, it must be true that $(A\boldsymbol{x}_0)_i = b_i$. From the remarks following the proof of Theorem 12, $\boldsymbol{u}_0$ possesses the desired property. □

Let us consider now the function

$$L(x, u) = c'x + u'(b - Ax) \tag{26}$$
$$= u'b + (c' - u'A)x. \tag{26'}$$

This is the Lagrangian function familiar from the classical problem of optimizing a linear function of variables that are constrained to satisfy a system of linear equalities. In its form (26) the function is associated with problem (I), with constraint equalities instead of inequalities; and in its form (26′) the function is associated with problem (II), again with constraint equalities. In the former case, u is a vector of Lagrangian multipliers; in the later case x contains the multipliers.

Definition 6. Let X and U be an n-dimensional vector space and an m-dimensional vector space respectively, and let $F(x, u)$ be a real-valued function defined on the Cartesian product space $X \times U$. Then the point $(x^*, u^*) \in X \times U$ is said to be a *saddle point* if for given $x^* \geqq 0$, $u^* \geqq 0$ and any $x \geqq 0$, $u \geqq 0$,

$$F(x, u^*) \leqq F(x^*, u^*) \leqq F(x^*, u).$$

Theorem 13. *Nonnegative x^* and u^* are optimal vectors for the dual problems* (I) *and* (II) *if and only if (x^*, u^*) is a saddle point of $L(x, u)$. Furthermore, if x^* and u^* are optimal vectors for* (I) *and* (II) *then $L(x^*, u^*)$ is the common optimal value for the two problems.*

PROOF. Suppose that x^* and u^* are optimal vectors. Then, for any $x \geqq 0$ and $u \geqq 0$,

$$\begin{aligned} L(x^*, u^*) - L(x, u^*) &= c'x^* - c'x - (u^*)'Ax^* + (u^*)'Ax \\ &\geqq c'x^* - (u^*)'b \qquad [\text{because } (u^*)'A \geqq c',\ Ax^* \leqq b] \\ &= 0 \qquad [\text{from Corollary 3(i)}]. \end{aligned}$$

Similarly, $L(x^*, u) - L(x^*, u^*) \geqq c'x^* - (u^*)'b = 0$.

Conversely, if (x^*, u^*) is a saddle point of $L(x, u)$ then every nonnegative x and u must satisfy

$$0 \leqq L(x^*, u^*) - L(x, u^*) = ((u^*)'A - c')(x - x^*) \tag{27}$$

and

$$0 \leqq L(x^*, u) - L(x^*, u^*) = (b - Ax^*)(u - u^*). \tag{28}$$

Since (27) and (28) must be satisfied by any nonnegative x and u, they must be satisfied by $x = x^* + e_j$, $j = 1, \ldots, n$, and $u = u^* + e_i$, $i = 1, \ldots, m$. Substituting in (27) and (28), we find that $(u^*)'A \geqq c'$ and $Ax^* \leqq b$. (x^*, u^*) being a saddle point,

$$L(0, u^*) = (u^*)' \cdot b \leqq L(x^*, u^*) \leqq L(x^*, 0) = c' \cdot x^*,$$

which, by Lemma 2(ii), implies the optimality of $\boldsymbol{x}^*$ and $\boldsymbol{u}^*$. Moreover, from Corollary 3(i) and the above inequality, the saddle value of the Lagrangian function is the common optimal value for two problems. □

So far, our attention has been focused on the most general properties of the optimal vectors for dual linear programming problems. We now enter into more detail. The following simple example suggests the kind of question we shall be interested in.

EXAMPLE.

$$\max_{x_1, x_2} c_1 x_1 + c_2 x_2$$

$$\text{subject to} \quad x_1 + 2x_2 \leqq 10, \tag{29}$$

$$2x_1 + x_2 \leqq 8, \tag{30}$$

$$x_1, x_2 \geqq 0, \tag{31}$$

where $c_1, c_2 > 0$. As Figure 1.1 makes clear, the optimal vectors depend on the value of c_1/c_2. The details of this dependence are contained in Table 1. It will be noticed that in cases I, III and V, there is a unique optimal vector while in cases II and IV there is an infinity of optimal vectors. In case II, for example, any vector which can be written in the form $\sigma A + (1 - \sigma)B$, with $0 \leqq \sigma \leqq 1$, is an optimal vector. Evidently the "extreme points" A, B, and C of the feasible set $0ABC$ are of special interest. These points correspond to solutions of subsets of constraints (29), (30) that hold with equality. For example, point A is the solution of the pair of equations $x_1 + 2x_2 = 10$, $x_2 = 0$.

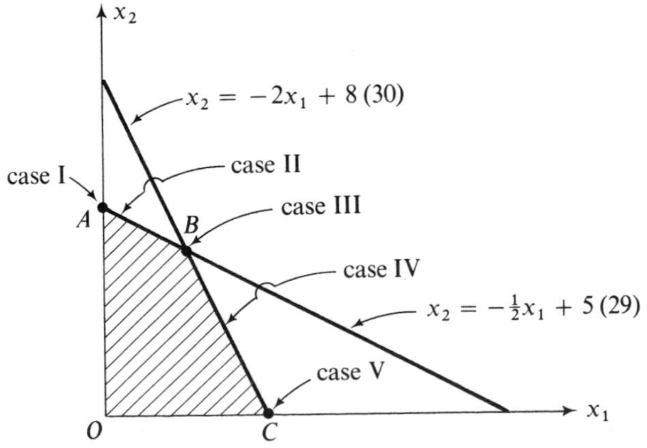

Figure 1.1

TABLE 1

Case	Optimal Vector	Constraint Equality	Zero Variable in Optimal Vector
I: $\frac{c_1}{c_2} < \frac{1}{2}$	A	(29)	x_1
II: $\frac{c_1}{c_2} = \frac{1}{2}$	$\overline{AB}$	(29)[(30) only at B]	none
III: $\frac{1}{2} < \frac{c_1}{c_2} < 2$	B	(30)	nonc
IV: $\frac{c_1}{c_2} = 2$	$\overline{BC}$	(30)[(29) only at B]	none
V: $\frac{c_1}{c_2} > 2$	C	(30)	x_2

Definition 7. Let M_i, $i = 1, 2$, and $N_j, j = 1, 2$, be arbitrary subsets of $M = \{1, \ldots, m\}$ and $N = \{1, \ldots, n\}$, respectively. Of course, $N = N_1 \cup N_2$ and $M = M_1 \cup M_2$. Then each of

$$(\mathrm{I}^\circ) \qquad \begin{cases} \sum_{j \in N_1} a_{ij}x_j = b_i & (i \in M_1), \\ x_j = 0 & (j \in N_2), \end{cases}$$

and

$$(\mathrm{II}^\circ) \qquad \begin{cases} \sum_{i \in M_1} u_i a_{ij} = c_j & (j \in N_1), \\ u_i = 0 & (i \in M_2), \end{cases}$$

is called a *system of "equated" constraints.* Systems (I°) and (II°) are said to be dual if $M_2 = M - M_1$ and $N_2 = N - N_1$. Let A_1 be the matrix of coefficients a_{ij} which appear in (I°). If A_1 is nonsingular then (I°) and (II°) are called *nonsingular systems.*

Points A, B, and C of Figure 1.1 have been described as extreme points. We now offer a general definition of the concept.

Definition 8. A vector $\boldsymbol{x}$ is said to be an *extreme feasible vector* if it is feasible and is not the mean of other feasible vectors. That is, if $\boldsymbol{x}$ is the mean of feasible vectors, say $\boldsymbol{x}_1$ and $\boldsymbol{x}_2$, then $\boldsymbol{x} = \boldsymbol{x}_1 = \boldsymbol{x}_2$. An *extreme optimal vector* is defined in a similar way. Needless to say, the set of extreme optimal vectors is included in the set of extreme feasible vectors. Everywhere in this definition $\boldsymbol{x}$ may be replaced by $\boldsymbol{u}$.

Theorem 14. *A nonzero feasible vector* $\mathbf{x}$ *(or* $\mathbf{u}$*) is an extreme feasible vector if and only if it satisfies a nonsingular system of equated constraints.*

Proof (*Necessity*). Let $\boldsymbol{x}$ be an extreme feasible vector. Since $\boldsymbol{x} \neq \mathbf{0}$, the set $N_1 = \{j : x_j > 0\}$ is nonempty. Let us define vectors $\boldsymbol{x}_1$ and $\boldsymbol{x}_2$ by

$$x_{j1} = x_{j2} = x_j = 0 \quad \text{for all } j \notin N_1$$

$$\left.\begin{aligned} x_{j1} &= x_j + \varepsilon \\ x_{j2} &= x_j - \varepsilon \end{aligned}\right\} \quad \text{for all } j \in N_1$$

where ε is an arbitrary positive number. Suppose now that $\sum_j a_{ij} x_j < b_i$, $i = 1, \ldots, m$. Then, by choosing ε sufficiently small we can ensure that $\boldsymbol{x}_1$ and $\boldsymbol{x}_2$ are feasible vectors. Moreover, $\boldsymbol{x} = \frac{1}{2}(\boldsymbol{x}_1 + \boldsymbol{x}_2)$, which contradicts the assumption that $\boldsymbol{x}$ is an extreme feasible vector. Thus the set $I_x = \{i : \sum_j a_{ij} x_j = b_i\}$ is nonempty. Let $N_2 = N - N_1$, so that $N_2 = \{j : x_j = 0\}$. Delete the rows of $\boldsymbol{A}$ that are not in I_x and the columns that are in N_2. All columns of the resulting matrix $\bar{\boldsymbol{A}}$ are linearly independent. To see this, suppose the contrary. Then there exists a nonzero vector $\boldsymbol{v}$ such that $\bar{\boldsymbol{A}}\boldsymbol{v} = \mathbf{0}$. Let us now define $\bar{\boldsymbol{x}}_1 = \bar{\boldsymbol{x}} + \varepsilon \boldsymbol{v}$ and $\bar{\boldsymbol{x}}_2 = \bar{\boldsymbol{x}} - \varepsilon \boldsymbol{v}$, where $\bar{\boldsymbol{x}}$ is a vector consisting of the positive elements of $\boldsymbol{x}$ and ε is an arbitrary positive number, and let $\bar{\boldsymbol{b}}$ be a vector consisting of those elements of $\boldsymbol{b}$ with suffix in I_x. Then clearly

$$\bar{\boldsymbol{A}}\bar{\boldsymbol{x}}_j = \bar{\boldsymbol{A}}\bar{\boldsymbol{x}} \pm \varepsilon \bar{\boldsymbol{A}}\boldsymbol{v} = \bar{\boldsymbol{b}} \quad \text{for any } \varepsilon > 0.$$

Let $\varepsilon = (\frac{1}{2}) \cdot (\min_{i \in v^*} x_i / |v_i|)$, where $v^* = \{i \in N_1 : v_i \neq 0\}$. Then $\bar{\boldsymbol{x}}_j \geqq 0$, $j = 1, 2$, $\bar{\boldsymbol{x}}_1 \neq \bar{\boldsymbol{x}}_2$, and $\bar{\boldsymbol{x}} = \frac{1}{2}(\bar{\boldsymbol{x}}_1 + \bar{\boldsymbol{x}}_2)$. This contradicts the assumption that $\boldsymbol{x}$ is an extreme feasible vector; for elements of $\boldsymbol{x}$ not in $\bar{\boldsymbol{x}}$ are all zero, whence

$$\begin{pmatrix} \bar{\boldsymbol{x}} \\ \mathbf{0} \end{pmatrix} = \boldsymbol{x} = \frac{1}{2}\begin{pmatrix} \bar{\boldsymbol{x}}_1 \\ \mathbf{0} \end{pmatrix} + \frac{1}{2}\begin{pmatrix} \bar{\boldsymbol{x}}_2 \\ \mathbf{0} \end{pmatrix}.$$

Thus all columns of $\bar{\boldsymbol{A}}$ are linearly independent, implying that $\bar{\boldsymbol{A}}$ can be reduced to a nonsingular matrix $\boldsymbol{A}_1$ by deleting redundant rows (if any). It is evident, therefore, that $\boldsymbol{A}_1 \bar{\boldsymbol{x}} = \boldsymbol{b}_1$, where $\boldsymbol{b}_1$ is an appropriate subvector of $\boldsymbol{b}$.

(*Sufficiency*). Without loss of generality, we can assume that $M_1 = N_1 = \{1, \ldots, p\}$. Suppose that there exist two feasible vectors $\boldsymbol{x}_i$, $i = 1, 2$, such that $\boldsymbol{x} = \frac{1}{2}(\boldsymbol{x}_1 + \boldsymbol{x}_2)$. We have $x_{ji} \geqq 0$ $(i = 1, 2; j = 1, \ldots, n)$ and, by definition, $x_j = 0$ for $j > p$; hence, for $j > p$, $x_{j1} = x_{j2} = 0$. Let us denote $(x_{1i}, \ldots, x_{pi})'$,

$i = 1, 2$, $(x_1, \ldots, x_p)'$, and $(b_1, \ldots, b_p)'$ by $\bar{x}_i$, $\bar{x}$, and $\bar{b}$, respectively. Then $A_1\bar{x}_j \leqq \bar{b}, j = 1, 2$, and $\frac{1}{2}A_1(\bar{x}_1 + \bar{x}_2) = \bar{b}$. Hence, $A_1\bar{x}_j = \bar{b}_j, j = 1, 2$. Thus the nonsingularity of A means that $\bar{x}_1 = \bar{x}_2 = \bar{x}$; this, in turn, shows that x is an extreme feasible vector. □

Theorem 15. *Feasible nonzero vectors* x *and* u *are extreme optimal vectors if and only if they satisfy a nonsingular dual system of equated constraints.*

PROOF (*Sufficiency*). Suppose that a feasible nonzero x satisfies a nonsingular system of equated constraints with associated matrix A_1. Since the system is dual, we have

$$\begin{aligned}\sum_{j=1}^{n} c_j x_j &= \sum_{j \in N_1} c_j x_j \\ &= \sum_{j \in N_1}\left(\sum_{i \in M_1} u_i a_{ij}\right) x_j \\ &= \sum_{i \in M_1} u_i\left(\sum_{j \in N_1} a_{ij} x_j\right) = \sum_{i \in M_1} u_i b_i = \sum_{i=1}^{n} u_i b_i,\end{aligned}$$

which, in view of Lemma 2(ii), ensures the optimality of x and u. And, from Theorem 14, x and u are extreme feasible vectors. Thus x and u are extreme optimal vectors.

(*Necessity*). Suppose that $x \neq 0$ and $u \neq 0$ are extreme optimal vectors. Since they are extreme feasible vectors, x and u must satisfy some constraints as equalities (Theorem 14). Let us suppose that

$$\sum_{j=1}^{n} a_{ij} x_j = b_i \qquad (i = 1, \ldots, p)$$

and that

$$\sum_{i=1}^{m} u_i a_{ij} = c_j \qquad (j = 1, \ldots, q).$$

Then, Corollary 3(iia) shows that $u_i = 0$ for $i = p + 1, \ldots, m$, since $\sum_j a_{ij} x_j < b_i$ for $i > p$. Similarly $x_j = 0$ for $j > q$. Furthermore, some x_j, $j = 1, \ldots, q$, and $u_i, i = 1, \ldots, p$, may vanish. We therefore can renumber the variables so that $x_j > 0$ for $i = 1, \ldots, \bar{q} \leqq q$ and $u_i > 0$ for $i = 1, \ldots, \bar{p} \leqq p$. Since by assumption x and u are nonzero, $\bar{p}$ and $\bar{q}$ must be positive. Let us define

$$\hat{A} = (a_{ij}) \qquad (i = 1, \ldots, p; j = 1, \ldots, q).$$

Then, by reasoning similar to that used in the necessity proof of Theorem 14, the first $\bar{q}$ columns and $\bar{p}$ rows of $\hat{A}$ are linearly independent. Let r denote the rank of $\hat{A}$. Then clearly

$$0 < \max\{\bar{p}, \bar{q}\} \leqq r \leqq \min\{p, q\}.$$

Suppose without loss of generality that the first r rows and columns of $\hat{A}$ are linearly independent. Then $\boldsymbol{x}$ and $\boldsymbol{u}$ must satisfy the following non-singular dual system of equated constraints:

$$\begin{cases} \sum_{j=1}^{r} a_{ij}x_j = b_i & (i = 1, \ldots, r) \\ x_j = 0 & (j = r+1, \ldots, n), \end{cases}$$

$$\begin{cases} \sum_{i=1}^{r} u_i a_{ij} = c_j & (j = 1, \ldots, r) \\ u_i = 0 & (i = r+1, \ldots, m). \end{cases}$$

□

From the point of view of computation, Theorems 14 and 15 are especially important. As an immediate consequence of them, we can assert that the sets of feasible and optimal vectors have finitely many extreme points; and this in turn ensures that the familiar simplex method, which searches for optimal vectors among extreme feasible vectors, will never be bogged down in closed loops.

Having essentially finished our theoretical examination of linear programming problems, we turn now to the relation between linear programming and matrix games. We consider first the notion of a matrix game.

In general, a game is a situation in which each of several persons called players must select one of several alternative courses of action without knowing the choice of every other player and in which for each player the outcome of at least one course of action depends on the choice of at least one other player. The rules of a game specify the gain or loss which will accrue to each player for each combination of choices by the several players. Evidently most individual and collective decision-making occurs in game theoretic situations defined broadly in this way.

Here, however, we shall confine ourselves to zero-sum, two-person games, that is, to matrix games. Such games are played by two players, I and II, with m and n possible courses of action, respectively. They are zero-sum games in the sense that one player's gain (respectively, loss) is exactly equal to the other player's loss (respectively, gain). Let p_{ij} be the amount which II must pay I when I adopts his ith course of action and II chooses his jth action. The matrix $\boldsymbol{P} = (p_{ij})$, where $i = 1, \ldots, m$ and $j = 1, \ldots, n$, is called the *payoff matrix* of the game and contains all the information relevant to a rational analysis of the game. We shall sometimes refer to a game by its payoff matrix.

We now define several terms which will facilitate our later analysis.

Definition 9. A nonnegative vector with elements which add to one is called a *probability vector*. Let x_i be the relative frequency with which player I selects his ith alternative. Then the probability row vector $\boldsymbol{x}' = (x_1, \ldots, x_m)$

is said to be a *mixed strategy* for player I. Similarly, a probability column vector $\mathbf{y} = (y_1, \ldots, y_n)'$ represents a mixed strategy for player II. When all but one of the elements of $\mathbf{x}$ (or $\mathbf{y}$) are zero, the remaining element being one, we have a *pure strategy*, a special case of a mixed strategy. For example, the ith unit row vector $\mathbf{e}^i = (0, \ldots, 0, 1, 0, \ldots, 0)$ in $\mathbb{R}^m$ represents the ith pure strategy of player I. Let $\mathbf{x}$ and $\mathbf{y}$ be any probability vectors in $\mathbb{R}^m$ and $\mathbb{R}^n$, respectively. Then $E(\mathbf{x}, \mathbf{y}) = \mathbf{x}'\mathbf{P}\mathbf{y}$ is the *expected payoff* to I when mixed strategies $\mathbf{x}$ and $\mathbf{y}$ are adopted by I and II, respectively. Of course, $-E(\mathbf{x}, \mathbf{y})$ is the expected payoff to II. If there exists a number v and mixed strategies $\mathbf{x}^*$ and $\mathbf{y}^*$ such that

$$E(\mathbf{x}, \mathbf{y}^*) \leqq v \leqq E(\mathbf{x}^*, \mathbf{y}) \quad \text{for any probability vectors } \mathbf{x} \text{ and } \mathbf{y} \tag{32}$$

then v is called the *value of the game* and $\mathbf{x}^*$ and $\mathbf{y}^*$ are called the *optimal strategies* for I and II, respectively. For, by employing $\mathbf{x}^*$ ($\mathbf{y}^*$), I (II) can ensure that his expected pay-off is at least equal to v ($-v$), whatever the strategy employed by II (I). Evidently $E(\mathbf{x}^*, \mathbf{y}^*) = v$ if v, $\mathbf{x}^*$ and $\mathbf{y}^*$ exist.

For (32) to hold it is necessary that

$$E(\mathbf{e}^i, \mathbf{y}^*) \leqq v \leqq E(\mathbf{x}^*, \mathbf{e}^j) \qquad (i = 1, \ldots, m; j = 1, \ldots, n). \tag{33}$$

On the other hand, if (33) holds then, for any probability vectors $\mathbf{x}$ and $\mathbf{y}$,

$$E(\mathbf{x}, \mathbf{y}^*) = \left(\sum_i x_i \mathbf{e}^i\right)\mathbf{P}\mathbf{y}^* = \sum_i x_i E(\mathbf{e}^i, \mathbf{y}^*) \leqq v \sum_i x_i = v$$

and

$$E(\mathbf{x}^*, \mathbf{y}) = (\mathbf{x}^*)'\mathbf{P}\left(\sum_{j=1}^n y_j \mathbf{e}^j\right) = \sum_{j=1}^n y_j E(\mathbf{x}^*, \mathbf{e}^j) \geqq v \sum_{j=1}^n y_j = v.$$

Thus (32) and (33) are equivalent. Let us define

$$S_k = \left\{\mathbf{x} \in \mathbb{R}^k : \mathbf{x} \geqq \mathbf{0}, \sum_{i=1}^k x_i = 1\right\},$$

the set of all probability vectors of order k. S_k is often called the k-dimensional simplex. Let us also denote by $\mathbf{l}_m$ and $\mathbf{l}_n$ the m- and n-dimensional vectors consisting of ones. Then (33) and therefore (32) are equivalent to

$$\mathbf{P}\mathbf{y}^* \leqq v\mathbf{l}_m \quad \text{and} \quad (\mathbf{x}^*)'\mathbf{P} \geqq v\mathbf{l}_n'. \tag{34}$$

As we shall see later, the inequalities (34) enable us to convert a matrix game into a linear programming problem. Henceforth the payoff matrix $\mathbf{P}$ is assumed to be $m \times n$.

Lemma 4. *If a game $\mathbf{P}$ has a value v and optimal strategies $\mathbf{x}^*$ and $\mathbf{y}^*$ then a game $\mathbf{A} = \mathbf{P} + \alpha\mathbf{L}$, where α is an arbitrary real number and $\mathbf{L}$ is an $m \times n$ matrix of ones, has a value $v + \alpha$ and optimal strategies $\mathbf{x}^*$ and $\mathbf{y}^*$.*

Proof. Let us denote the expected payoff of $\boldsymbol{P}$ by $E_{\boldsymbol{P}}$ and the expected payoff of $\boldsymbol{A}$ by $E_{\boldsymbol{A}}$. Then

$$\begin{aligned} E_{\boldsymbol{A}}(\boldsymbol{x}, \boldsymbol{y}) &= \boldsymbol{x}'(\boldsymbol{P} + \alpha \boldsymbol{L})\boldsymbol{y} \\ &= \boldsymbol{x}'\boldsymbol{P}\boldsymbol{y} + \alpha \boldsymbol{x}'\boldsymbol{L}\boldsymbol{y} \\ &= E_{\boldsymbol{P}}(\boldsymbol{x}, \boldsymbol{y}) + \alpha. \end{aligned}$$

Hence

$$E_{\boldsymbol{A}}(\boldsymbol{x}, \boldsymbol{y}^*) \leqq v + \alpha \leqq E_{\boldsymbol{A}}(\boldsymbol{x}^*, \boldsymbol{y}).$$ □

Lemma 5. *Every matrix game* $\boldsymbol{P}$ *has at most one value.*

Proof. Let v_i, $\boldsymbol{x}_i$ and $\boldsymbol{y}_i$, $i = 1, 2$, be the values of $\boldsymbol{P}$ and the optimal strategies associated with v_i, respectively. Since $v_1 \leqq E(\boldsymbol{x}_1, \boldsymbol{y})$ for any $\boldsymbol{y} \in S_n$ and $\boldsymbol{y}_2 \in S_n$, and since $E(\boldsymbol{x}, \boldsymbol{y}_2) \leqq v_2$ for any $\boldsymbol{x} \in S_m$ and $\boldsymbol{x}_1 \in S_m$, we have

$$v_1 \leqq E(\boldsymbol{x}_1, \boldsymbol{y}_2) \leqq v_2.$$

Similarly,

$$v_2 \leqq E(\boldsymbol{x}_2, \boldsymbol{y}_1) \leqq v_1.$$

Hence $v_1 = v_2$. □

Lemma 6. *Let* $\boldsymbol{P}$ *be a matrix game. If every column sum of* $\boldsymbol{P}$ *is positive then* $\boldsymbol{P}$ *has a value and there exist optimal strategies.*

Proof. Consider the dual linear programs

$$\max_{\boldsymbol{\eta}} \boldsymbol{l}_n'\boldsymbol{\eta} \qquad \min_{\boldsymbol{\xi}} \boldsymbol{\xi}'\boldsymbol{l}_m$$

$$\text{subject to } \boldsymbol{P}\boldsymbol{\eta} \leqq \boldsymbol{l}_m \qquad \text{subject to } \boldsymbol{\xi}'\boldsymbol{P} \geqq \boldsymbol{l}_n'$$

$$\boldsymbol{\eta} \geqq \boldsymbol{0} \qquad \boldsymbol{\xi} \geqq \boldsymbol{0}.$$

Let $\sigma = \min_{1 \le j \le n} \sum_{i=1}^{n} p_{ij}$, so that $\sigma > 0$; and let $\boldsymbol{\xi} = \sigma^{-1}\boldsymbol{l}_m$. Then

$$(\boldsymbol{\xi}'\boldsymbol{P})_j = \sigma^{-1} \cdot \sum_{i=1}^{n} p_{ij} \geqq 1 \qquad (j = 1, \ldots, n).$$

Since $\sigma^{-1}\boldsymbol{l}_m$ is positive, the minimizing problem has a feasible vector $\sigma^{-1}\boldsymbol{l}_m$. Moreover, $\boldsymbol{\eta} = \boldsymbol{0}$ is clearly a feasible vector for the maximizing problem; hence, recalling Theorem 11, for the above dual linear problems there exist optimal vectors $\boldsymbol{\eta}^*$ and $\boldsymbol{\xi}^*$. Clearly, $\boldsymbol{\xi}^* \geq \boldsymbol{0}$ for $\boldsymbol{\xi}^* \neq \boldsymbol{0}$ (since $\boldsymbol{\xi}^* \geqq \boldsymbol{0}$ is obvious); otherwise, $\boldsymbol{0}' = (\boldsymbol{\xi}^*)'\boldsymbol{P} < \boldsymbol{l}_n'$, which is a contradiction. Thus $\boldsymbol{\eta}^* \geq \boldsymbol{0}$; for, from Corollary 3(i), $s = \boldsymbol{l}_n'\boldsymbol{\eta}^* = \boldsymbol{l}_m'\boldsymbol{\xi}^* > 0$. Let us now define $\boldsymbol{x}^* = \boldsymbol{\xi}^*/s$ and $\boldsymbol{y}^* = \boldsymbol{\eta}^*/s$. Obviously $\boldsymbol{x}^* \in S_m$, $\boldsymbol{y}^* \in S_n$, and we can easily obtain $\boldsymbol{P}\boldsymbol{y}^* \leq (1/s)\boldsymbol{l}_m$ and $(\boldsymbol{x}^*)'\boldsymbol{P} \geqq (1/s)\boldsymbol{l}_n'$. Thus, from (34), $\boldsymbol{P}$ has at least a value s^{-1} and optimal strategies $\boldsymbol{x}^*$ and $\boldsymbol{y}^*$. □

Theorem 16. *For every matrix game* $\boldsymbol{P}$ *there exists a unique value and optimal strategies.*

PROOF. Clearly there exists $\alpha > 0$ such that

$$\boldsymbol{Q} = \boldsymbol{P} + \alpha\boldsymbol{L} > \boldsymbol{0}.$$

Hence, from Lemma 6, for $\boldsymbol{Q}$ there exists a value w and optimal strategies $\boldsymbol{x}^*$ and $\boldsymbol{y}^*$. Applying Lemma 4, $\boldsymbol{P}$ has a value $(w - \alpha)$ and optimal strategies $\boldsymbol{x}^*$ and $\boldsymbol{y}^*$. From Lemma 5, $(w - \alpha)$ is unique. □

Let us now pause to summarize the argument to this point. Given a matrix game $\boldsymbol{P}$ we can define

$$\boldsymbol{Q} = \boldsymbol{P} + \alpha\boldsymbol{L}, \qquad \alpha > 0,$$

and construct the dual linear problems

$$\text{(G)} \qquad \begin{array}{ll} \max\limits_{\boldsymbol{v}} \boldsymbol{l}_n'\boldsymbol{v} & \min\limits_{\boldsymbol{u}} \boldsymbol{u}'\boldsymbol{l}_m \\ \text{subject to } \boldsymbol{Q}\boldsymbol{v} \leqq \boldsymbol{l}_m & \text{subject to } \boldsymbol{u}'\boldsymbol{Q} \geqq \boldsymbol{l}_n' \\ \qquad\qquad \boldsymbol{v} \geqq \boldsymbol{0} & \qquad\qquad \boldsymbol{u} \geqq \boldsymbol{0}. \end{array}$$

The proof of Lemma 6 implies that for the dual problems (G) there exist optimal vectors $\boldsymbol{v}^*$ and $\boldsymbol{u}^*$. Thus the value of the matrix game $\boldsymbol{P}$ is given by the optimal value $\boldsymbol{l}_n'\boldsymbol{v}^* = \boldsymbol{l}_m'\boldsymbol{u}^*$ of the dual linear problems (G), and $\boldsymbol{x}^* = (\boldsymbol{l}_m'\boldsymbol{u}^*)^{-1} \cdot \boldsymbol{u}^*$ and $\boldsymbol{y}^* = (\boldsymbol{l}_n'\boldsymbol{v}^*)^{-1} \cdot \boldsymbol{v}^*$ are optimal strategies for $\boldsymbol{P}$. Therefore, with a given matrix game there is associated a pair of dual linear problems which generates the value and optimal strategies of the game. That is, a matrix game can be converted into a linear programming problem. Is it always true that, conversely, to a given pair of dual linear problems there corresponds a matrix game which provides us with optimal vectors of the dual problems? The answer is, not necessarily. The converse proposition is valid only if certain conditions are satisfied. Those conditions are stated in

Theorem 17. *For the dual linear problems*

$$\begin{array}{ll} \max\limits_{\boldsymbol{x}} \boldsymbol{c}'\boldsymbol{x} & \min\limits_{\boldsymbol{u}} \boldsymbol{u}'\boldsymbol{b} \\ \textit{subject to } \boldsymbol{A}\boldsymbol{x} \leqq \boldsymbol{b} & \textit{subject to } \boldsymbol{u}'\boldsymbol{A} \geqq \boldsymbol{c}' \\ \qquad\qquad \boldsymbol{x} \geqq \boldsymbol{0} & \qquad\qquad \boldsymbol{u} \geqq \boldsymbol{0} \end{array}$$

there exist optimal vectors $\boldsymbol{x}^*$ *and* $\boldsymbol{u}^*$ *if and only if the matrix game*

$$\boldsymbol{P} = \begin{pmatrix} \boldsymbol{A} & -\boldsymbol{b} \\ -\boldsymbol{c}' & (\boldsymbol{u}^*)'\boldsymbol{A}\boldsymbol{x}^* \end{pmatrix}$$

has a value of zero and optimal strategies

$$\lambda\begin{pmatrix} \boldsymbol{u}^* \\ 1 \end{pmatrix} \quad \textit{and} \quad \mu\begin{pmatrix} \boldsymbol{x}^* \\ 1 \end{pmatrix},$$

where $\lambda, \mu > 0$.

PROOF (*Sufficiency*). From the definition of the value of a game,

$$E\left(\lambda\begin{pmatrix}\boldsymbol{u}^*\\1\end{pmatrix}, \boldsymbol{e}^j\right) \geqq 0 \qquad (j = 1, \ldots, n)$$

and

$$E\left(\boldsymbol{e}^i, \mu\begin{pmatrix}\boldsymbol{x}^*\\1\end{pmatrix}\right) \leqq 0 \qquad (i = 1, \ldots, m).$$

It follows that

$$(\lambda(\boldsymbol{u}^*)', \lambda)\begin{pmatrix}\boldsymbol{A} & -\boldsymbol{b}\\ -\boldsymbol{c}' & (\boldsymbol{u}^*)'\boldsymbol{A}\boldsymbol{x}^*\end{pmatrix} \geqq \boldsymbol{0}$$

and

$$\begin{pmatrix}\boldsymbol{A} & -\boldsymbol{b}\\ -\boldsymbol{c}' & (\boldsymbol{u}^*)'\boldsymbol{A}\boldsymbol{x}^*\end{pmatrix}\begin{pmatrix}\mu\boldsymbol{x}^*\\ \mu\end{pmatrix} \leqq \boldsymbol{0}$$

so that

$$\begin{aligned}(\boldsymbol{u}^*)'\boldsymbol{A} &\geqq \boldsymbol{c}',\\ \boldsymbol{A}\boldsymbol{x}^* &\leq \boldsymbol{b},\\ \boldsymbol{c}'\boldsymbol{x}^* &\geqq (\boldsymbol{u}^*)'\boldsymbol{A}\boldsymbol{x}^* \geqq (\boldsymbol{u}^*)'\boldsymbol{b}.\end{aligned} \tag{35}$$

From (35) and the nonnegativity of $\boldsymbol{x}^*$ and $\boldsymbol{u}^*$ we see that $\boldsymbol{x}^*$ and $\boldsymbol{u}^*$ are optimal.

(*Necessity*). Let $\boldsymbol{x}^*$ and $\boldsymbol{u}^*$ be optimal vectors for the dual linear problems described in the statement of the theorem. From Corollary 3(i), $\boldsymbol{c}'\boldsymbol{x}^* = (\boldsymbol{u}^*)'\boldsymbol{b}$. Hence

$$(\boldsymbol{u}^*)'\boldsymbol{A}\boldsymbol{x}^* \geqq \boldsymbol{c}'\boldsymbol{x}^* = (\boldsymbol{u}^*)'\boldsymbol{b} \geqq (\boldsymbol{u}^*)'\boldsymbol{A}\boldsymbol{x}^*$$

so that

$$(\boldsymbol{u}^*)'\boldsymbol{b} = \boldsymbol{c}'\boldsymbol{x}^* = (\boldsymbol{u}^*)'\boldsymbol{A}\boldsymbol{x}^*.$$

Therefore

$$\begin{aligned}E\left(\lambda\begin{pmatrix}\boldsymbol{u}^*\\1\end{pmatrix}, \mu\begin{pmatrix}\boldsymbol{x}^*\\1\end{pmatrix}\right) &= (\lambda(\boldsymbol{u}^*)', \lambda)\boldsymbol{P}\begin{pmatrix}\mu\boldsymbol{x}^*\\ \mu\end{pmatrix}\\ &= (\lambda(\boldsymbol{u}^*)'\boldsymbol{A} - \lambda\boldsymbol{c}', 0)\begin{pmatrix}\mu\boldsymbol{x}^*\\ \mu\end{pmatrix}\\ &= \lambda\mu((\boldsymbol{u}^*)'\boldsymbol{A}\boldsymbol{x}^* - \boldsymbol{c}'\boldsymbol{x}^*)\\ &= 0,\\ E\left(\lambda\begin{pmatrix}\boldsymbol{u}^*\\1\end{pmatrix}, \begin{pmatrix}\boldsymbol{x}\\ x_{n+1}\end{pmatrix}\right) &= (\lambda((\boldsymbol{u}^*)'\boldsymbol{A} - \boldsymbol{c}'), 0)\begin{pmatrix}\boldsymbol{x}\\ x_{n+1}\end{pmatrix}\\ &= \lambda((\boldsymbol{u}^*)'\boldsymbol{A} - \boldsymbol{c}')\boldsymbol{x}\\ &\geqq 0 \quad \text{for all } \begin{pmatrix}\boldsymbol{x}\\ x_{n+1}\end{pmatrix} \in S_{n+1},\end{aligned}$$

and, finally,

$$E\left(\begin{pmatrix} \boldsymbol{u} \\ u_{m+1} \end{pmatrix}, \mu\begin{pmatrix} \boldsymbol{x}^* \\ 1 \end{pmatrix}\right) = (\boldsymbol{u}', u_{m+1})\begin{pmatrix} \mu(\boldsymbol{A}\boldsymbol{x} - \boldsymbol{b}) \\ 0 \end{pmatrix}$$
$$= \mu \cdot \boldsymbol{u}'(\boldsymbol{A}\boldsymbol{x}^* - \boldsymbol{b})$$
$$\leqq 0 \quad \text{for all} \begin{pmatrix} \boldsymbol{u} \\ u_{m+1} \end{pmatrix} \in S_{m+1}. \qquad \square$$

3 Convex Sets and Convex Cones

We have discussed linear inequalities from an algebraic point of view. In the present section, we shall be mainly concerned with the geometric properties of the set of solutions of systems of linear inequalities. For this purpose, the concepts of convex set and convex cone are inevitable. Throughout the section, our attention will be restricted to finite-dimensional real vector spaces.

Definition 10. A subset S of $\mathbb{R}^n$ is said to be a *convex set* if

$$\lambda \boldsymbol{x} + (1 - \lambda)\boldsymbol{y} \in S \quad \text{for all } \lambda \in [0, 1] \quad \text{and} \quad \boldsymbol{x}, \boldsymbol{y} \in S.$$

That is, a convex set is a set such that the line segment joining any two points in the set is also in the set. As a generalization of the line segment joining two points, we have the notion of a convex linear combination of finitely many points. Let $\boldsymbol{x}_i$, $i = 1, \ldots, K$, be points in $\mathbb{R}^n$. Then $\sum_{i=1}^{K} \lambda_i \boldsymbol{x}_i$ is called a *convex linear combination* of $\boldsymbol{x}_1, \ldots, x_K$ if $\sum_{i=1}^{K} \lambda_i = 1$ and $\lambda_i \geqq 0$ for $i = 1, \ldots, K$.

As examples of convex sets we have the feasible sets of dual linear problems. Let $F_{\mathrm{I}} = \{\boldsymbol{x} : \boldsymbol{A}\boldsymbol{x} \leqq \boldsymbol{b}, \boldsymbol{x} \geqq \boldsymbol{0}\}$ and $F_{\mathrm{II}} = \{\boldsymbol{u} : \boldsymbol{u}'\boldsymbol{A} \geqq \boldsymbol{c}', \boldsymbol{u} \geqq \boldsymbol{0}\}$. Then, for any $\lambda \in [0, 1]$ and $\boldsymbol{x}_i \in F_{\mathrm{I}}$, $i = 1, 2$,

$$\lambda \boldsymbol{x}_1 + (1 - \lambda)\boldsymbol{x}_2 \geqq \boldsymbol{0},$$
$$\boldsymbol{A}(\lambda \boldsymbol{u}_1 + (1 - \lambda)\boldsymbol{x}_2) = \lambda \boldsymbol{A}\boldsymbol{x}_1 + (1 - \lambda)\boldsymbol{A}\boldsymbol{x}_2$$
$$\leqq \lambda \boldsymbol{b} + (1 - \lambda)\boldsymbol{b} = \boldsymbol{b}.$$

The convexity of F_{II} may be verified in a similar manner.

Definition 11. A subset C of $\mathbb{R}^n$ is called a *convex cone* if any nonnegative linear combination of any two points $\boldsymbol{x}$ and $\boldsymbol{y}$ of C also belongs to C, that is, if

$$(\alpha \boldsymbol{x} + \beta \boldsymbol{y}) \in C \quad \text{for all } \alpha \geqq 0, \beta \geqq 0 \quad \text{and all} \quad \boldsymbol{x}, \boldsymbol{y} \in C.$$

A *polar cone* of a convex cone C is a set C^* such that

$$C^* = \{\boldsymbol{y} : \boldsymbol{y}'\boldsymbol{x} \leqq 0 \quad \text{for all} \quad \boldsymbol{x} \in C\}.$$

Geometrically, C^* is the set of vectors each of which forms with every vector of C an angle not less than $\pi/2$. The set C of all nonnegative linear combinations of finitely many vectors $\boldsymbol{a}_1, \ldots, \boldsymbol{a}_K$ is called a *finite cone* or *convex polyhedral cone*. Symbolically,

$$C = \{\boldsymbol{y}:\boldsymbol{y} = \boldsymbol{A}\boldsymbol{x}, \boldsymbol{x} \geqq \boldsymbol{0}\} \quad \text{or} \quad \{\boldsymbol{A}\boldsymbol{x}:\boldsymbol{x} \geqq \boldsymbol{0}\},$$

where $\boldsymbol{A} = \{\boldsymbol{a}_1, \ldots, \boldsymbol{a}_K\}$. The set $(\boldsymbol{a})$ of all nonnegative scalar multiples of a given point $\boldsymbol{a}$ is called the *half-line* generated by $\boldsymbol{a}$; that is,

$$(\boldsymbol{a}) = \{\lambda\boldsymbol{a}:\lambda \geqq 0\}.$$

In this terminology a finite cone C can be represented by

$$C = (\boldsymbol{a}_1) + \cdots + (\boldsymbol{a}_K).$$

Lemma 7. *Let L be a linear subspace. Then L^* is the orthogonal complement of L, that is,*

$$L^* = \{\boldsymbol{y}:\boldsymbol{y}'\boldsymbol{x} = 0 \textit{ for all } \boldsymbol{x} \in L\}.$$

PROOF. From the definition of L^*,

$$L^* = \{\boldsymbol{y}:\boldsymbol{y}'\boldsymbol{x} \leqq 0 \text{ for all } \boldsymbol{x} \in L\}.$$

Since L is a linear subspace, $-\boldsymbol{x} \in L$ for any $\boldsymbol{x} \in L$, so that

$$\boldsymbol{y}'\boldsymbol{x} \leqq 0 \quad \text{and} \quad \boldsymbol{y}'(-\boldsymbol{x}) \leqq 0 \quad \text{for all } \boldsymbol{y} \in L^*, \boldsymbol{x} \in L.$$

Hence $\boldsymbol{y}'\boldsymbol{x} = 0$ for all $\boldsymbol{y} \in L^*$ and

$$\{\boldsymbol{y}:\boldsymbol{y}'\boldsymbol{x} \leqq 0 \text{ for all } \boldsymbol{x} \in L\} \subseteq \{\boldsymbol{y}:\boldsymbol{y}'\boldsymbol{x} = 0 \text{ for all } \boldsymbol{x} \in L\}.$$

The converse inclusion is obvious. □

Lemma 8. *If C is a finite cone then $(C^*)^* \equiv C^{**} = C$. This property is called the duality of finite cones.*

PROOF. Let $\boldsymbol{y}$ be any element of C. Then, from the definition of C^* and the fact that $\boldsymbol{y}$ is in C,

$$0 \geqq \boldsymbol{z}'\boldsymbol{y} = \boldsymbol{y}'\boldsymbol{z} \quad \text{for any } \boldsymbol{z} \in C^*.$$

From this inequality, it follows that $C \subseteq C^{**}$. (Notice that this is true for every convex cone, not just for finite cones.)

To establish the converse set inclusion, suppose the contrary. Then there exists $\boldsymbol{b} \in C^{**}$ such that $\boldsymbol{b} \notin C$. This implies that $\boldsymbol{A}\boldsymbol{x} = \boldsymbol{b}$ has no nonnegative solution, for $C = \{\boldsymbol{y}:\boldsymbol{A}\boldsymbol{x} = \boldsymbol{y} \text{ for any } \boldsymbol{x} \geqq \boldsymbol{0}\}$. According to Theorem 3, there exists $\boldsymbol{y}$ such that $\boldsymbol{A}'\boldsymbol{y} \geqq \boldsymbol{0}$ and $\boldsymbol{b}'\boldsymbol{y} < 0$. Since $\boldsymbol{A}'\boldsymbol{y} \geqq \boldsymbol{0}$,

$$(-\boldsymbol{y}')\boldsymbol{A}\boldsymbol{x} \leqq 0 \quad \text{for any } \boldsymbol{x} \geqq \boldsymbol{0},$$

whence $-\boldsymbol{y} \in C^*$. Recalling that $\boldsymbol{b} \in C^{**}$, we see that

$$\boldsymbol{b}'(-\boldsymbol{y}) = -\boldsymbol{b}'\boldsymbol{y} \leqq 0,$$

which contradicts the fact that $\boldsymbol{b}'\boldsymbol{y} < 0$. □

Lemma 9. (i) *A linear subspace L is a finite cone and* (ii) *the set of all nonnegative solutions of $\boldsymbol{A}\boldsymbol{x} = \boldsymbol{0}$ is a finite cone, where $\boldsymbol{A}$ is assumed to be $m \times n$.*

PROOF. (i) Let $\boldsymbol{a}_j, j = 1, \ldots, r$, be a basis of L. Then, for any $\boldsymbol{x} \in L$, there exist $\alpha_j \in \mathbb{R}$ such that $\boldsymbol{x} = \sum_{j=1}^{r} \alpha_j \boldsymbol{a}_j$, Since the α_j are real they can be written

$$\alpha_j = \hat{\alpha}_j - \check{\alpha}_j \quad (j = 1, \ldots, r)$$

where $\hat{\alpha}_j, \check{\alpha}_j \geqq 0, j = 1, \ldots, r$. Hence

$$\boldsymbol{x} = \sum_{j=1}^{r} \hat{\alpha}_j \boldsymbol{a}_j + \sum \check{\alpha}_j(-\boldsymbol{a}_j) = (\boldsymbol{A}, -\boldsymbol{A})\begin{pmatrix} \hat{\boldsymbol{\alpha}} \\ \check{\boldsymbol{\alpha}} \end{pmatrix}.$$

(ii) If $\boldsymbol{A}\boldsymbol{x} = \boldsymbol{0}$ has only the null solution, the assertion is immediate, for $\boldsymbol{0}$ can be represented as $(\boldsymbol{0})$.

Before taking the next step, we introduce some additional terminology. Consider a consistent system of linear equations $\boldsymbol{A}\boldsymbol{x} = \boldsymbol{b}$. (By "consistent" we mean only that the system has a solution.) Let $\boldsymbol{x}$ be a solution and let us define $\boldsymbol{J}_{\boldsymbol{x}} = \{j : x_j \neq 0\} = \{j_1, \ldots, j_r\}$. A solution $\boldsymbol{x}$ is said to be a *basic solution* if $\boldsymbol{a}_{j_1}, \ldots, \boldsymbol{a}_{j_r}$ are linearly independent. Any matrix has finitely many linearly independent columns; hence any consistent system of nonhomogeneous linear equations has finitely many basic solutions.

Let $\boldsymbol{x}$ be any semipositive solution of $\boldsymbol{A}\boldsymbol{x} = \boldsymbol{0}$. Then $\boldsymbol{\eta} = (\sum_{j=1}^{n} x_j)^{-1} \cdot \boldsymbol{x}$ clearly satisfies the equation

$$\boldsymbol{y}' \cdot (\boldsymbol{A}', \boldsymbol{l}_n) = (\boldsymbol{0}', 1) \tag{36}$$

and vice versa.

Thus, it suffices to show that every nonnegative solution of (36) can be represented as a nonnegative linear combination of all nonnegative basic solutions of (36), for then $\boldsymbol{x} = (\sum_i x_i) \cdot \boldsymbol{\eta}$ can be also represented as a nonnegative linear combination of them. The proof of the assertion is by induction on n. Let $\boldsymbol{y}$ be any nonnegative solution of (36). Then, without loss of generality, $\boldsymbol{y}$ can be assumed to be positive, for otherwise those columns of the coefficient matrix which correspond to zero-components of $\boldsymbol{y}$ can be deleted. If $\boldsymbol{y}$ is a basic solution we can proceed to the next step; hence we can assume that $\boldsymbol{y}$ is nonbasic, which, in turn, implies that the coefficient matrix has a column which can be represented as a linear combination of the remaining columns. Therefore, there exists a solution $\boldsymbol{\xi}$ of

$$\begin{pmatrix} \boldsymbol{A} \\ \boldsymbol{l}'_n \end{pmatrix} \boldsymbol{z} = \begin{pmatrix} \boldsymbol{0} \\ 0 \end{pmatrix}$$

which contains both negative and positive elements. Define $J = \{j \in N : \xi_j > 0\}$ and $\theta = \min_{j \in J} y_j/\xi_j$, where y_j denotes the jth component of $\boldsymbol{y}$. Then, clearly, $\boldsymbol{y}_1 = \boldsymbol{y} - \theta \cdot \boldsymbol{\xi}$ is a semipositive solution of (36) and, from the construction, $\boldsymbol{y}_1$ has at least one zero-component. If $\boldsymbol{y}_1$ is not yet a basic solution the same procedure as above can be repeated, until a nonnegative (in fact, semipositive) basic solution will be obtained; hence we can safely regard $\boldsymbol{y}_1$ as a nonnegative basic solution of (36).

Since $\boldsymbol{y}_1$ has a zero-component and solves (36), there exists an index $k \in N$ such that

$$y_k < y_{k_1}, \tag{37}$$

where $\boldsymbol{y}_1' = (y_{11}, \ldots, y_{n1})$. From (37), $y_{k_1}/y_k > 1$; hence $\beta = \max_j y_{j_1}/y_j = y_{h_1}/y_h \geqq y_{k_1}/y_k > 1$, and

$$\boldsymbol{y}_2 = \frac{1}{\beta - 1}(\beta \cdot \boldsymbol{y} - \boldsymbol{y}_1) \tag{38}$$

is a nonnegative solution of (36) with at most $n - 1$ positive elements. (Recall that the hth element of $\boldsymbol{y}_2$ is zero.) Thus, from the induction assumption, $\boldsymbol{y}_2$ can be represented as a nonnegative linear combination of nonnegative basic solutions. Furthermore, from (38),

$$\boldsymbol{y} = \frac{1}{\beta} \cdot \boldsymbol{y}_1 + \left(1 - \frac{1}{\beta}\right) \cdot \boldsymbol{y}_2. \tag{39}$$

Since $\boldsymbol{y}_1$ is a nonnegative basic solution and $\boldsymbol{y}_2$ can be represented as a nonnegative linear combination of nonnegative basic solutions, (39) ensures that $\boldsymbol{y}$ is a nonnegative linear combination of nonnegative basic solutions. (Note that, by definition, $1 > 1/\beta > 0$.)

Finally, when $n = 2$ the assertion is trivial. □

Lemma 10. *The intersection of a linear subspace L with the nonnegative orthant $\mathbb{R}^n_+$ (the set of all nonnegative vectors) is a finite cone.*

Proof. From Lemma 7, L^* is a linear subspace; hence it has a basis $\boldsymbol{b}_i$, $i = 1, \ldots, r$. Let $\boldsymbol{B}$ be the matrix with $\boldsymbol{b}_i$, $i = 1, \ldots, r$, as the ith column, and let us define $C = \{\boldsymbol{x} \geqq \boldsymbol{0} : \boldsymbol{B}'\boldsymbol{x} = \boldsymbol{0}\}$. Then, from Lemma 9(ii), C is a finite cone. Thus it suffices to show that $L \cap \mathbb{R}^n_+ = C$.

Let $\boldsymbol{x}$ be any vector in $L \cap \mathbb{R}^n_+$ so that $\boldsymbol{x} \geqq \boldsymbol{0}$ and $\boldsymbol{x} \in L$. Since $\boldsymbol{b}_i \in L^*$, $i = 1, \ldots, r$, we have $\boldsymbol{B}'\boldsymbol{x} = \boldsymbol{0}$ and, therefore, $\boldsymbol{x} \in C$.

Conversely, let $\boldsymbol{x}$ be any vector in C. Then, for any $\boldsymbol{z} \in L^*$, $\boldsymbol{z}'\boldsymbol{x} = 0$; for $\boldsymbol{z} = \boldsymbol{B}\boldsymbol{y}$, so that $\boldsymbol{z}'\boldsymbol{x} = \boldsymbol{y}'(\boldsymbol{B}'\boldsymbol{x}) = 0$. Hence $\boldsymbol{x} \in C$ is in $L^{**} = L$; and evidently $\boldsymbol{x} \geqq \boldsymbol{0}$. Thus $\boldsymbol{x} \in L \cap \mathbb{R}^n_+$. □

Theorem 18. *The intersection of finitely many, say m, half-spaces $\{\boldsymbol{x} : \boldsymbol{A}\boldsymbol{x} \leqq \boldsymbol{0}\}$ is a finite cone. Conversely, with a given finite cone $K(\boldsymbol{B}) = \{\boldsymbol{B}\boldsymbol{z} : \boldsymbol{z} \geqq \boldsymbol{0}\}$ there is associated a matrix $\boldsymbol{A}$ with the property that $K(\boldsymbol{B}) = \{\boldsymbol{x} : \boldsymbol{A}\boldsymbol{x} \leqq \boldsymbol{0}\}$.*

Proof. We first show that the set of solutions of a system of linear homogeneous inequalities is a finite cone. Let L be the subspace defined by $L = \{\boldsymbol{A}\boldsymbol{x} : \boldsymbol{x} \in \mathbb{R}^n\}$. From Lemma 10, $L \cap \mathbb{R}^m_+$ is a finite cone; hence $L \cap \mathbb{R}^m_+ = \{\boldsymbol{Y}\boldsymbol{x} : \boldsymbol{x} \geqq \boldsymbol{0}\}$ where $\boldsymbol{Y} = (\boldsymbol{y}_1, \ldots, \boldsymbol{y}_r)$. Let us choose $\boldsymbol{x}_i$, $i = 1, \ldots, r$, such that $-\boldsymbol{y}_i = \boldsymbol{A}\boldsymbol{x}_i$ and define $\hat{X} = \{\boldsymbol{x} : \boldsymbol{A}\boldsymbol{x} = \boldsymbol{0}\}$. $\hat{X}$ is a linear subspace; from

Lemma 9(i), therefore, there exist $\hat{x}_1, \ldots, \hat{x}_s$ such that $\hat{X} = (\hat{x}_1) + \cdots + (\hat{x}_s)$. Let us define $B = (x_1, \ldots, x_r, \hat{x}_1, \ldots, \hat{x}_s)$. Then for all $y \in K(B)$,

$$\begin{aligned} Ay &= A(x_1, \ldots, x_r)z_1 + A(\hat{x}_1, \ldots, \hat{x}_s)z_2 \\ &= -(y_1, \ldots, y_r)z_1 \\ &\leqq 0. \qquad [\text{because } (y_1, \ldots, y_r)z_1 \in L \cap \mathbb{R}^m_+]. \end{aligned}$$

Conversely, if $Ax \leqq 0$ then $-Ax \in L \cap \mathbb{R}^m_+$. Hence there exists $\xi \geqq 0$ such that $-Ax = Y\xi$. Therefore $Ax = -\sum_{i=1}^r \xi_i y_i = \sum_{i=1}^r \xi_i(Ax_i)$ so that $A(x - \sum_{i=1}^r \xi_i x_i) = 0$. It follows that $(x - \sum_{i=1}^r \xi_i x_i) \in \hat{X}$ and

$$x - \sum_{i=1}^{r} \xi_i x_i = \sum_{i=1}^{s} \mu_i \hat{x}_i \quad \text{for some } \mu_i \geqq 0.$$

Thus

$$\begin{aligned} x &= \sum_{i=1}^{r} \xi_i x_i + \sum_{i=1}^{s} \mu_i \hat{x}_i \\ &= B\begin{pmatrix} \xi \\ \mu \end{pmatrix} \in K(B). \end{aligned}$$

Turning to the second half of the theorem, we notice that $(K(B))^* = \{z : z'B \leqq 0\}$. Let z be any vector in $(K(B))^*$. Then $z'(Bx) \leqq 0$ for all $x \geqq 0$. Hence $z'(Be_j) \leqq 0, j = 1, \ldots, n$, which implies that $z'B \leqq 0$. If z is any vector such that $z'B \leqq 0$ then, clearly,

$$z'(Bx) = (z'B)x \leqq 0 \quad \text{for all } x \geqq 0.$$

From Lemma 8,

$$K(B) = ((K(B))^*)^* = \{z : z'B \leqq 0\}^* = \{z : B'z \leqq 0\}^*.$$

From the first part of the theorem, there exists a matrix C such that $\{z : B'z \leqq 0\} = K(C)$, whence

$$K(B) = (K(C))^* = \{x : C'x \leqq 0\} \qquad \square$$

Theorem 18 establishes a one-to-one correspondence between the set of systems of homogeneous linear inequalities and the set of finite cones. Our next theorem, which is an application of Theorem 18, reveals the geometric structure of the set of solutions of a system of nonhomogeneous linear inequalities.

Theorem 19. *If a system of nonhomogeneous linear inequalities $Ax \leqq b$ has a solution then there exist solutions x_i of $Ax \leqq 0, i = 1, \ldots, h$, and solutions v_j of $Ax \leqq b, j = h + 1, \ldots, s$, such that*

$$\{x : Ax \leqq b\} = \left\{x = \sum_{i=1}^{h} \lambda_i x_i + \sum_{j=h+1}^{s} \beta_j v_j : \lambda_i \geqq 0, \beta_j \geqq 0, \sum_{j=h+1}^{s} \beta_j = 1\right\}$$

where s is a finite integer.

PROOF. Consider the system

$$\begin{pmatrix} A & -b \\ 0' & -1 \end{pmatrix} \begin{pmatrix} x \\ t \end{pmatrix} \leqq \begin{pmatrix} 0 \\ 0 \end{pmatrix}. \tag{40}$$

Let x_0 be any solution of $Ax \leqq b$. Then, for any positive t, $t \cdot (x_0', 1)'$ is a solution of (40). In view of Theorem 18, there exist finitely many solutions $(x_i', t_i)'$, $i = 1, \ldots, s$, such that any solution $(x', t)'$ of (40) can be represented as a nonnegative linear combination of them:

$$(x', t)' = \sum_{i=1}^{s} \lambda_i \cdot (x_i', t_i)'.$$

Since the system (40) has a solution in which $t > 0$, there exists an index i_0 of $\{1, \ldots, s\}$ such that $t_{i_0} > 0$. Let us renumber the

$$\begin{pmatrix} x_i \\ t_i \end{pmatrix}$$

so that $t_i = 0$ for $i = 1, \ldots, h$, and $t_i > 0$ for $i = h + 1, \ldots, s$. Then

$$\begin{aligned}
\begin{pmatrix} x \\ t \end{pmatrix} &= \sum_{i=1}^{h} \lambda_i \begin{pmatrix} x_i \\ 0 \end{pmatrix} + \sum_{i=h+1}^{s} \lambda_i \begin{pmatrix} x_i \\ t_i \end{pmatrix} \\
&= \sum_{i=1}^{h} \lambda_i \begin{pmatrix} x_i \\ 0 \end{pmatrix} + \sum_{i=h+1}^{s} \frac{\lambda_i}{t_i} \begin{pmatrix} x_i/t_i \\ 1 \end{pmatrix} \\
&= \sum_{i=1}^{h} \lambda_i \begin{pmatrix} x_i \\ 0 \end{pmatrix} + \sum_{i=h+1}^{s} \beta_i \begin{pmatrix} v_i \\ 1 \end{pmatrix}.
\end{aligned} \tag{41}$$

The solution of $Ax \leqq b$ can be obtained by setting $t = 1$ in (41); hence $\sum_{i=h+1}^{s} \beta_i = 1$. Furthermore, direct substitution reveals that $Ax_i \leqq 0$, $i = 1, \ldots, h$, and that $Av_i \leqq b$, $i = h + 1, \ldots, s$. □

For completeness, we refer to the geometry of solutions of linear equations. Let $Ax = 0$ be a system of linear homogeneous equations and let x and y be any two solutions of $Ax = 0$. Then every linear combination $\alpha x + \beta y$ of x and y is also a solution of the above system. Hence the set of solutions of $Ax = 0$ forms a linear subspace. On the other hand, if a system of linear nonhomogeneous equations $Ax = b$ has a solution $\bar{x}$ then the set of solutions of $Ax = b$ forms a displaced linear subspace. Define $z = x - \bar{x}$, where x is any solution of $Ax = b$. Then $Az = 0$. Hence, by the foregoing argument, the set $\{z : Az = 0\}$ forms a linear subspace, whence the set under consideration is a displaced (displaced by $\bar{x}$) linear subspace which is often called a *linear variety* (Berge [1963; p. 138]).

Finally, we state and prove a proposition which will be put to work in Chapter 6.

Lemma 11 (Mukherji [1973]). *Let A be an $n \times n$ real matrix with the property that there exists $x \neq 0$ such that*

$$x'A = 0'$$

and

$$Ax = 0.$$

Then

$$\frac{A_{ij}}{x_i x_j} = \frac{A_{rs}}{x_r x_s} \quad \textit{for all } i, j, r, s \in J_x \equiv \{k : x_k \neq 0\}$$

and

$$A_{ij} = 0 \quad \textit{if } i \notin J_x \textit{ or if } j \notin J_x$$

*where A_{ij} is the cofactor of a_{ji}, the (ji)*th *element of A.*

PROOF. If $\rho(A)$, the rank of A, is not greater than $n - 2$, the assertion is immediate, for then $A_{ij} = 0$ for all i and j. Thus it suffices to verify the assertion on the supposition that $\rho(A) = n - 1$. Let adj $A = (A_{ij})$. Then, as is well known,

$$(\text{adj } A) \cdot A = A \cdot (\text{adj } A) = 0.$$

Since $\rho(A) = n - 1$, there exist t_j and τ_i such that

$$(\text{adj } A)^i = \tau_i x' \tag{42-a}$$

and

$$(\text{adj } A)_j = t_j x \tag{42-b}$$

where $(\text{adj } A)_j$ and $(\text{adj } A)^i$ denote respectively the jth column and the ith row of adj A. Thus the second half of the assertion follows directly from (42). Moreover, Equation (42-a) implies that

$$\frac{A_{ij}}{x_i x_j} = \frac{\tau_i}{x_i} \quad \text{for any } i, j \in J_x. \tag{42'-a}$$

Similarly,

$$\frac{A_{rs}}{x_r x_s} = \frac{t_s}{x_s} \quad \text{for any } r, s \in J_x. \tag{42'-b}$$

Since the indices i and s are in J_x,

$$A_{is} = t_s x_i = \tau_i x_s,$$

which in turn implies that

$$\frac{\tau_i}{x_i} = \frac{t_s}{x_s}.$$

The assertion then follows from (42'). □

4 Nonlinear Programming

In the present section we shall be concerned with the typical nonlinear programming problem, a generalized version of the classical constrained optimization problem. This kind of problem is central to much of modern economic theory and plays an important role in the theory of optimal control. (Among many available references on the optimal control problem, Arrow and Kurz [1970; especially Chapter 2], and Hadley and Kemp [1971; especially Chapters 2 through 4] are especially useful for economists.)

We begin by proving a separation theorem for convex sets. This theorem will facilitate the later proof of the important Kuhn–Tucker Theorem of nonlinear programming.

Definition 12. For any given vector $\boldsymbol{p}$ and scalar γ, the set of vectors $\boldsymbol{x}$ such that $\boldsymbol{p}'\boldsymbol{x} = \gamma$ is called a *hyperplane*. Let S and T be convex sets. If there is a hyperplane such that

$$\boldsymbol{p}'\boldsymbol{x} \geqq \gamma \quad \text{for any } \boldsymbol{x} \in S$$

and

$$\boldsymbol{p}'\boldsymbol{y} \leqq \gamma \quad \text{for any } \boldsymbol{y} \in T$$

then the hyperplane is said to be a *separating hyperplane* of S and T. Furthermore, a hyperplane is said to be a *supporting hyperplane* of a set S at point $\boldsymbol{p}$ if $\boldsymbol{p}$ is contained in the hyperplane and the set S, and S is contained in one of the closed half-spaces determined by the hyperplane. Formally, if $\boldsymbol{x}$ is the coordinate vector of $\boldsymbol{p}$ and if

$$\boldsymbol{p}'\boldsymbol{x} = \gamma \qquad (\boldsymbol{x} \in S),$$

$$\boldsymbol{p}'\boldsymbol{y} \leqq \gamma \quad (\text{or } \boldsymbol{p}'\boldsymbol{y} \geqq \gamma) \quad \text{for any } \boldsymbol{y} \in S,$$

then the hyperplane determined by $\boldsymbol{p}$ and γ is the supporting hyperplane of S at $\boldsymbol{p}$.

Lemma 12. *Let S be a nonempty convex set in $\mathbb{R}^n$. If S does not contain the origin $\mathbf{0}$, then there exists a nonzero vector $\boldsymbol{c} \in \mathbb{R}^n$ such that*

$$\boldsymbol{c}' \cdot \boldsymbol{x} \geqq 0 \quad \textit{for all } \boldsymbol{x} \in S.$$

PROOF. We notice first that every convex linear combination of finitely many points $\boldsymbol{x}_1, \ldots, \boldsymbol{x}_m$ of S is itself contained in S. This can be proved by induction on m. When $m = 2$, the assertion trivially follows from the convexity of S. Let $\sum_{i=1}^m \lambda_i \cdot \boldsymbol{x}_i$ be any convex linear combination of $\boldsymbol{x}_i$ ($i = 1, \ldots, m$). Since $\sum_{i=1}^m \lambda_i = 1$ and $\lambda_i \geqq 0$ ($i = 1, \ldots, m$), there is an index i_0 such that $1 \geqq \lambda_{i_0} > 0$ and $m \geqq i_0 \geqq 1$. If $\lambda_{i_0} = 1$, the convex linear combination clearly lies in S, for it is nothing else than $\boldsymbol{x}_{i_0}$. Otherwise,

$$\sum_{i \neq i_0} \lambda_i = 1 - \lambda_{i_0} > 0,$$

whence

$$\sum_{i=1}^{m} \lambda_i \cdot \boldsymbol{x}_i = \lambda_{i_0}\boldsymbol{x}_{i_0} + (1 - \lambda_{i_0}) \cdot \left(\sum_{j \neq i_0} \frac{\lambda_i}{\sum_{i \neq i_0} \lambda_i} \cdot \boldsymbol{x}_j \right).$$

Since

$$\sum_{j \neq i_0} \frac{\lambda_i}{\sum_{i \neq i_0} \lambda_i} \cdot \boldsymbol{x}_j$$

is a convex linear combination of $m - 1$ points of S, the induction assumption ensures that it is a point of S. Thus $\sum_{i=1}^{m} \lambda_i \cdot \boldsymbol{x}_i$ can be expressed as a convex linear combination of two points in S. From the assumed convexity of S, therefore, it belongs to S.

Let $\boldsymbol{x}$ be an arbitrary point of S. Corresponding to the point $\boldsymbol{x}$, define a nonempty set $A_{\boldsymbol{x}} = \{\boldsymbol{y} \in \mathbb{R}^n : \boldsymbol{y}' \cdot \boldsymbol{y} = 1 \text{ and } \boldsymbol{x}' \cdot \boldsymbol{y} \geqq 0\}$. From Theorem 9 in Appendix A, $A_{\boldsymbol{x}}$ is a closed subset of $\mathbb{R}^n$. Suppose that $\boldsymbol{X}$ is a matrix the columns of which consist of any finite number of points $\boldsymbol{x}_1, \ldots, \boldsymbol{x}_m$ of S. Then the equation

$$\begin{pmatrix} \boldsymbol{X} \\ \boldsymbol{l}' \end{pmatrix} \cdot \boldsymbol{u} = \begin{pmatrix} \boldsymbol{0} \\ 1 \end{pmatrix}$$

has no nonnegative solution. For otherwise there would exist a nonnegative $\boldsymbol{u}$ such that $\boldsymbol{0} = \boldsymbol{X} \cdot \boldsymbol{u} \in S$ and $\boldsymbol{l}' \cdot \boldsymbol{u} = 1$, a contradiction. Thus, by Theorem 3, there exist an n-dimensional vector $\boldsymbol{z}$ and a real number z_{n+1} such that

$$z_{n+1} < 0 \quad \text{and} \quad \boldsymbol{X}' \cdot \boldsymbol{z} \geq -\boldsymbol{l} \cdot z_{n+1},$$

which, together, imply that

$$\boldsymbol{x}_i' \cdot \boldsymbol{z} > 0, i = 1, \ldots, m.$$

Obviously, $\boldsymbol{z}$ being nonzero, $\boldsymbol{y} = \boldsymbol{z}/(\boldsymbol{z}' \cdot \boldsymbol{z})^{1/2}$ belongs to the intersection of $A_{\boldsymbol{x}_i} (i = 1, \ldots, m)$. Thus the intersection of every finite family of closed subsets of a compact set $A = \{\boldsymbol{y} \in \mathbb{R}^n : \boldsymbol{y}' \cdot \boldsymbol{y} = 1\}$ is nonempty. Hence, from Theorem 12 and Corollary 4.1 of Appendix A, $\bigcap_{x \in S} A_{\boldsymbol{x}}$ is also nonempty. Let $\boldsymbol{c}$ be an element of the intersection of all $A_{\boldsymbol{x}}$. Then $\boldsymbol{c}$ is clearly nonzero and $\boldsymbol{c}' \cdot \boldsymbol{x} \geqq 0$ for all $\boldsymbol{x}$ of S. □

Theorem 20. *Let S and T be two nonempty disjoint convex sets in* $\mathbb{R}^n$. *Then there exists a separating hyperplane of S and T.*

Proof. Let Γ be the set $S - T$, that is, $\Gamma = \{\boldsymbol{x} - \boldsymbol{y} : \boldsymbol{x} \in S, \boldsymbol{y} \in T\}$. Then Γ is a nonempty convex set which does not contain the origin. Let $\boldsymbol{x}_i - \boldsymbol{y}_i (i = 1, 2)$ be any two points of Γ. Then, for every $\lambda \in [0, 1]$, we have

$$\begin{aligned} &\lambda \cdot (\boldsymbol{x}_1 - \boldsymbol{y}_1) + (1 - \lambda) \cdot (\boldsymbol{x}_2 - \boldsymbol{y}_2) \\ &\qquad = (\lambda \cdot \boldsymbol{x}_1 + (1 - \lambda) \cdot \boldsymbol{x}_2) - (\lambda \cdot \boldsymbol{y}_1 + (1 - \lambda) \cdot \boldsymbol{y}_2). \end{aligned}$$

The assumed convexity of S and T ensures that the points on the right-hand side of the above expression, that is, $(\lambda \cdot \boldsymbol{x}_1 + (1 - \lambda) \cdot \boldsymbol{x}_2)$ and $(\lambda \cdot \boldsymbol{y}_1 + (1 - \lambda) \cdot \boldsymbol{y}_2)$, belong to S and T respectively. Hence their difference must be in Γ. Thus Γ is convex. Suppose that Γ contains the origin. Then there exist $\boldsymbol{x}$ of S and $\boldsymbol{y}$ of T such that $\boldsymbol{x} - \boldsymbol{y} = \boldsymbol{0}$, from which it follows that $\boldsymbol{x} = \boldsymbol{y}$. However, this is absurd, since S and T are disjoint.

Thus, applying Lemma 12 to the set Γ, there exists a vector $\boldsymbol{c} \in \mathbb{R}^n$ such that $\boldsymbol{c}' \cdot \boldsymbol{z} \geqq 0$ for all $\boldsymbol{z} \in \Gamma$. Noticing that with every $\boldsymbol{z}$ of Γ there are associated $\boldsymbol{x}$ of S and $\boldsymbol{y}$ of T such that $\boldsymbol{z} = \boldsymbol{x} - \boldsymbol{y}$, we obtain

$$\boldsymbol{c}' \cdot \boldsymbol{x} \geqq \boldsymbol{c}' \cdot \boldsymbol{y} \quad \text{for all } \boldsymbol{x} \in S \quad \text{and} \quad \boldsymbol{y} \in T. \tag{43}$$

Fix $\boldsymbol{x}$ at an arbitrary point in S. Then (43) implies that the set

$$R_T = \{\boldsymbol{c}' \cdot \boldsymbol{y} : \boldsymbol{y} \in T\}$$

is bounded above. In view of Theorem 1 in Appendix A, $\beta = \sup R_T$ surely exists. Similarly, the set $R_S = \{\boldsymbol{c}' \cdot \boldsymbol{x} : \boldsymbol{x} \in S\}$ is bounded below; hence $\gamma = \inf R_s$ exists.

To obtain a contradiction, suppose that there exists an $\boldsymbol{x}_0 \in S$ such that $\boldsymbol{c}' \cdot \boldsymbol{x}_0 < \beta$. Then to $\varepsilon = \beta - \boldsymbol{c}' \cdot \boldsymbol{x}_0 > 0$ there corresponds an element $\boldsymbol{y}_\varepsilon$ of T such that $\boldsymbol{c}' \cdot \boldsymbol{y}_\varepsilon > \beta - \varepsilon = \boldsymbol{c}' \cdot \boldsymbol{x}_0$, contradicting (43). Thus we have

$$\boldsymbol{c}' \cdot \boldsymbol{x} \geqq \beta \geqq \boldsymbol{c}' \cdot \boldsymbol{y} \quad \text{for any } \boldsymbol{x} \in S \quad \text{and} \quad \boldsymbol{y} \in T.$$

Similarly,

$$\boldsymbol{c}' \cdot \boldsymbol{x} \geqq \gamma \geqq \boldsymbol{c}' \cdot \boldsymbol{y} \quad \text{for any } \boldsymbol{x} \in S \quad \text{and} \quad \boldsymbol{y} \in T.$$

Therefore the hyperplane $\boldsymbol{c}' \cdot \boldsymbol{z} = \gamma(\beta)$ is a separating hyperplane of S and T. □

Geometrically speaking, Theorem 20 asserts that one of the convex sets S and T is contained in one of the half-spaces determined by the separating hyperplane, while the other set is contained in the half-space on the opposite side. A glance at Figure 1.2b enables us to see that even when S and

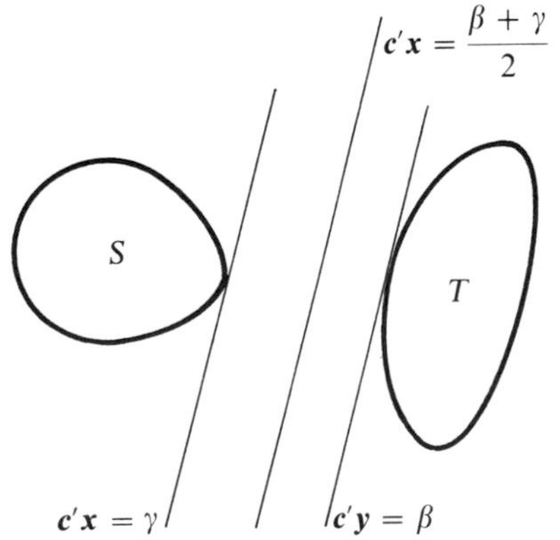

Figure 1.2a

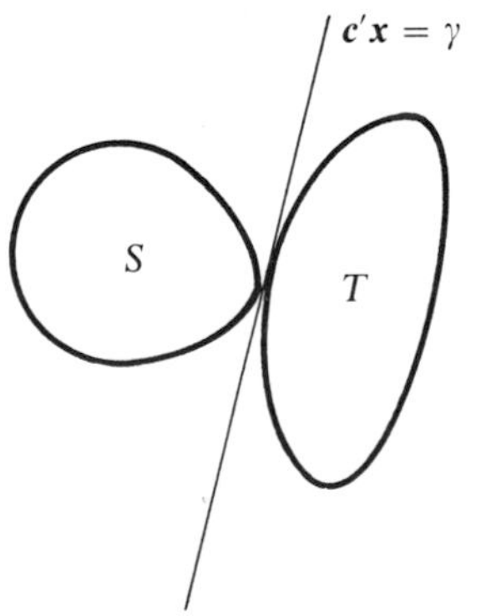

Figure 1.2b

T have at least one point in common on their boundaries Theorem 20 still holds *mutatis mutandis*. In fact Theorem 20 can be generalized as follows.

Theorem 20′. *Let S and T be two convex sets in a finite-dimensional normed vector space. If S has interior points and T contains no interior points of S, then there exists a separating hyperplane of S and T.*

Theorem 20′, which is the most general separation theorem, stems from the so-called Hahn–Banach Theorem, the proof of which relies on Zorn's Lemma. Expositions of the Hahn–Banach Theorem and Zorn's Lemma are far beyond the scope of this chapter. However, interested readers may consult Halmos [1960; Zorn's Lemma], Simmons [1963; Hahn–Banach Theorem] and Luenberger [1969; Separation Theorem].

The essence of the classical optimizing procedure of setting the derivatives of the objective function equal to zero lies in finding a supporting (or tangential) hyperplane; the crucial point is not the application of the calculus but the existence of such a hyperplane. On the other hand, a glance at Figure 1.2 enables us to see that the separation theorem for convex sets can ensure the existence of a supporting hyperplane without the assumption of differentiability and hence without resort to the calculus. It is not surprising then that the separation theorem provides the modern apparatus for the optimization of a function.

Our main topic in this section is the famous Kuhn–Tucker Theorem of nonlinear programming.

Definition 13. A real-valued function $f(\boldsymbol{x})$ defined over $\mathbb{R}^n$ is said to be *concave* if

$$f(\lambda \boldsymbol{x}_1 + (1-\lambda)\boldsymbol{x}_2) \geqq \lambda f(\boldsymbol{x}_1) + (1-\lambda) f(\boldsymbol{x}_2) \quad \text{for all } \lambda \in [0, 1]$$

and *convex* if the inequality is reversed.

Consider now the nonlinear programming problem

$$\text{(I)} \qquad \begin{aligned} &\max_{\boldsymbol{x}} f(\boldsymbol{x}) \\ &\text{subject to } g_i(\boldsymbol{x}) \geqq 0 \qquad (i = 1, \ldots, m) \\ &\qquad\qquad \boldsymbol{x} \in X \end{aligned}$$

where X is a given convex subset of $\mathbb{R}^n$ (X may be the nonnegative orthant in $\mathbb{R}^n$). The set $\Gamma = \{\boldsymbol{x} \in X : g_i(\boldsymbol{x}) \geqq 0, i = 1, \ldots, m\}$ is called the *feasible set for* (I). Corresponding to (I) we can form the Lagrangian function

$$L(\boldsymbol{x}, \boldsymbol{u}) = f(\boldsymbol{x}) + \boldsymbol{u}' \cdot \boldsymbol{g}(\boldsymbol{x}),$$

where $\boldsymbol{g}(\boldsymbol{x}) = (g_1(\boldsymbol{x}), \ldots, g_m(\boldsymbol{x}))'$.

In order to prove the so-called Kuhn–Tucker Theorem in nonlinear programming, it is convenient to introduce

Lemma 13. *Let f be a real-valued continuous function defined over a metric space X. If $f(\boldsymbol{x}) \geqq 0$ for every $\boldsymbol{x}$ of a subset K of X then $f(\boldsymbol{x}) \geqq 0$ for all $\boldsymbol{x} \in \partial K$, where ∂K is the set of all boundary points of K.*

PROOF. If $\partial K \subseteq K$ then the assertion is trivial. Hence it suffices to verify the assertion when $\partial K \nsubseteq K$. Suppose there exists $\boldsymbol{x}_0 \in \partial K$ such that $\boldsymbol{x}_0 \notin K$ and $f(\boldsymbol{x}_0) < 0$. Then, by the continuity of f, there exists $\delta > 0$ such that $f(\boldsymbol{x}) < 0$ for every $\boldsymbol{x} \in B(\boldsymbol{x}_0, \delta)$. Since $\boldsymbol{x}_0 \in \partial K$, $B(\boldsymbol{x}_0, \delta) \cap K \neq \varnothing$. Then from the definition of δ it follows that $f(\boldsymbol{x}) < 0$ for all $\boldsymbol{x} \in B(\boldsymbol{x}_0, \delta) \cap K$, which, however, contradicts the assumption that $f(\boldsymbol{x}) \geqq 0$ for all $\boldsymbol{x} \in K$. □

Theorem 21 (Kuhn and Tucker [1951]). *If $(\boldsymbol{x}^*, \boldsymbol{u}^*)$ is a saddle point of $L(\boldsymbol{x}, \boldsymbol{u})$ in $\boldsymbol{x} \in X$, $\boldsymbol{u} \geqq \boldsymbol{0}$, then $\boldsymbol{x}^*$ is an optimal vector for* (I). *Suppose that $f(\boldsymbol{x})$ and $g_i(\boldsymbol{x})$, $i = 1, \ldots, m$, are concave in $\boldsymbol{x}$ and that there exists $\boldsymbol{x}_0 \in X$ such that $\boldsymbol{g}(\boldsymbol{x}_0) > 0$. Then with an optimal vector $\boldsymbol{x}^*$ for* (I) *there is associated a nonnegative vector $\boldsymbol{u}^*$ such that $L(\boldsymbol{x}^*, \boldsymbol{u}^*)$ is a saddle point of $L(\boldsymbol{x}, \boldsymbol{u})$.*

PROOF. From the definition of a saddle point,

$$f(\boldsymbol{x}^*) + (\boldsymbol{u}^*)' \cdot \boldsymbol{g}(\boldsymbol{x}^*) \leqq f(\boldsymbol{x}^*) + \boldsymbol{u}' \cdot \boldsymbol{g}(\boldsymbol{x}^*) \quad \text{for any } \boldsymbol{u} \geqq \boldsymbol{0}.$$

Thus

$$(\boldsymbol{u} - \boldsymbol{u}^*)' \cdot \boldsymbol{g}(\boldsymbol{x}^*) \geqq 0 \quad \text{for any } \boldsymbol{u} \geqq \boldsymbol{0}.$$

Since the inequality must hold for $\boldsymbol{u} = \boldsymbol{u}^* + \boldsymbol{e}_i$, $i = 1, \ldots, m$, and $\boldsymbol{u} = \boldsymbol{0}$, we obtain

$$\boldsymbol{g}(\boldsymbol{x}^*) \geqq \boldsymbol{0}$$

and

$$(\boldsymbol{u}^*)' \cdot \boldsymbol{g}(\boldsymbol{x}^*) \leqq 0.$$

On the other hand, $\boldsymbol{u}^* \geqq \boldsymbol{0}$ and $\boldsymbol{g}(\boldsymbol{x}^*) \geqq \boldsymbol{0}$ imply $(\boldsymbol{u}^*)' \cdot \boldsymbol{g}(\boldsymbol{x}^*) \geqq 0$. Hence $(\boldsymbol{u}^*)' \cdot \boldsymbol{g}(\boldsymbol{x}^*) = 0$. Again, from the definition of a saddle point, and making use of the result just obtained,

$$f(\boldsymbol{x}) + (\boldsymbol{u}^*)' \cdot \boldsymbol{g}(\boldsymbol{x}) \leqq f(\boldsymbol{x}^*) \quad \text{for any } \boldsymbol{x} \in X.$$

It being clear that X contains Γ, $\boldsymbol{x}^*$ is surely an optimal vector for (I).

Turning to the converse proposition, let $\boldsymbol{x}^*$ be an optimal vector and let us define

$$A_{\boldsymbol{x}} = \left(\begin{pmatrix} z_0 \\ \boldsymbol{z} \end{pmatrix} : \begin{pmatrix} z_0 \\ \boldsymbol{z} \end{pmatrix} \leqq \begin{pmatrix} f(\boldsymbol{x}) \\ \boldsymbol{g}(\boldsymbol{x}) \end{pmatrix} \text{ for some } \boldsymbol{x} \in X \right),$$

$$A = \bigcup_{\boldsymbol{x} \in X} A_{\boldsymbol{x}},$$

and

$$B = \left(\begin{pmatrix} u_0 \\ \boldsymbol{u} \end{pmatrix} : \begin{pmatrix} u_0 \\ \boldsymbol{u} \end{pmatrix} > \begin{pmatrix} f(\boldsymbol{x}^*) \\ \boldsymbol{0} \end{pmatrix} \right).$$

The assumed concavity of $f(\boldsymbol{x})$ and $g_i(\boldsymbol{x})$, $i = 1, \ldots, m$, implies that A and B are convex sets. Corresponding to any two points

$$\begin{pmatrix} z_{0i} \\ \boldsymbol{z}_i \end{pmatrix} \in A \qquad (i = 1, 2),$$

there exist $\boldsymbol{x}_i \in X$, $i = 1, 2$, such that

$$\begin{pmatrix} z_{0i} \\ \boldsymbol{z}_i \end{pmatrix} \leqq \begin{pmatrix} f(\boldsymbol{x}_i) \\ \boldsymbol{g}(\boldsymbol{x}_i) \end{pmatrix} \qquad (i = 1, 2).$$

Then, for all $\lambda \in [0, 1]$, we have

$$\begin{pmatrix} z_0 \\ \boldsymbol{z} \end{pmatrix} \equiv \lambda \begin{pmatrix} z_{01} \\ \boldsymbol{z}_1 \end{pmatrix} + (1 - \lambda) \begin{pmatrix} z_{02} \\ \boldsymbol{z}_2 \end{pmatrix} \leqq \lambda \begin{pmatrix} f(\boldsymbol{x}_1) \\ \boldsymbol{g}(\boldsymbol{x}_1) \end{pmatrix} + (1 - \lambda) \begin{pmatrix} f(\boldsymbol{x}_2) \\ \boldsymbol{g}(\boldsymbol{x}_2) \end{pmatrix}$$
$$\leqq \begin{pmatrix} f(\lambda \boldsymbol{x}_1 + (1 - \lambda) \cdot \boldsymbol{x}_2) \\ \boldsymbol{g}(\lambda \boldsymbol{x}_1 + (1 - \lambda) \cdot \boldsymbol{x}_2) \end{pmatrix}.$$

Since X is convex, the point $\lambda \boldsymbol{x}_1 + (1 - \lambda)\boldsymbol{x}_2$ is in X; hence the point $(z_0, \boldsymbol{z})'$ is also contained in A. Similarly for all $\lambda \in [0, 1]$,

$$\lambda \begin{pmatrix} u_{01} \\ \boldsymbol{u}_1 \end{pmatrix} + (1 - \lambda) \begin{pmatrix} u_{02} \\ \boldsymbol{u}_2 \end{pmatrix} > \lambda \begin{pmatrix} f(\boldsymbol{x}^*) \\ \boldsymbol{0} \end{pmatrix} + (1 - \lambda) \begin{pmatrix} f(\boldsymbol{x}^*) \\ \boldsymbol{0} \end{pmatrix} = \begin{pmatrix} f(\boldsymbol{x}^*) \\ \boldsymbol{0} \end{pmatrix}.$$

Next we show that A and B have no points in common. Suppose the contrary. Then there exist

$$\boldsymbol{x} \in X \quad \text{and} \quad \begin{pmatrix} z_0 \\ \boldsymbol{z} \end{pmatrix}$$

such that

$$\begin{pmatrix} f(\boldsymbol{x}^*) \\ \boldsymbol{0} \end{pmatrix} < \begin{pmatrix} z_0 \\ \boldsymbol{z} \end{pmatrix} \leqq \begin{pmatrix} f(\boldsymbol{x}) \\ \boldsymbol{g}(\boldsymbol{x}) \end{pmatrix},$$

that is, $\boldsymbol{g}(\boldsymbol{x}) > \boldsymbol{0}$ and $f(\boldsymbol{x}^*) < f(\boldsymbol{x})$, contradicting the assumed optimality of $\boldsymbol{x}^*$.

Applying Theorem 20, there exists a nonzero vector $(\xi_0, \boldsymbol{\xi})$ such that

$$\xi_0 z_0 + \boldsymbol{\xi}'\boldsymbol{z} \leq \xi_0 u_0 + \boldsymbol{\xi}'\boldsymbol{u} \quad \text{for any} \begin{pmatrix} z_0 \\ \boldsymbol{z} \end{pmatrix} \in A, \begin{pmatrix} u_0 \\ \boldsymbol{u} \end{pmatrix} \in B. \tag{44}$$

Moreover, $\boldsymbol{\xi} \geqq \boldsymbol{0}$. For, from the definition of B, $\boldsymbol{u}$ can be arbitrarily large; hence $\boldsymbol{\xi} \ngeqq \boldsymbol{0}$ implies the existence of a vector in B which violates (44). Similarly $\xi_0 \geqq 0$.

From the definition of $A_{\boldsymbol{x}}$, $(f(\boldsymbol{x}), \boldsymbol{g}(\boldsymbol{x}))'$ is in $A_{\boldsymbol{x}}$ and hence in A. The point $(f(\boldsymbol{x}^*), \boldsymbol{0}')'$ being on the boundary of B, the application of Lemma 13 yields

$$\xi_0 f(\boldsymbol{x}^*) \geqq \xi_0 z_0 + \boldsymbol{\xi}'\boldsymbol{z} \quad \text{for any } (z_0, \boldsymbol{z}')' \in A.$$

Since $(f(\boldsymbol{x}), \boldsymbol{g}(\boldsymbol{x}))'$ belongs to A for any $\boldsymbol{x} \in X$, we obtain

$$\xi_0 f(\boldsymbol{x}) + \boldsymbol{\xi}' \cdot \boldsymbol{g}(\boldsymbol{x}) \leqq \xi_0 f(\boldsymbol{x}^*) \quad \text{for any } \boldsymbol{x} \in X.$$

Suppose that $\xi_0 = 0$. Then $\boldsymbol{\xi}' \cdot \boldsymbol{g}(\boldsymbol{x}) \leqq 0$ for any $\boldsymbol{x} \in X$ and $\boldsymbol{\xi} \neq \mathbf{0}$. It was shown earlier that $\boldsymbol{\xi} \geqq \mathbf{0}$; hence the assumed existence of $\boldsymbol{x}_0 \in X$ such that $\boldsymbol{g}(\boldsymbol{x}_0) > \mathbf{0}$ implies that $\boldsymbol{\xi}' \cdot \boldsymbol{g}(\boldsymbol{x}_0) > 0$, contradicting the fact that $\boldsymbol{\xi}' \cdot \boldsymbol{g}(\boldsymbol{x}) \leqq 0$ for any $\boldsymbol{x} \in X$. Hence $\xi_0 > 0$. Let us define $\boldsymbol{u}^* = \boldsymbol{\xi}/\xi_0$. Then

$$\boldsymbol{u}^* = \frac{\boldsymbol{\xi}}{\xi_0} \geqq \mathbf{0}$$

and

$$f(\boldsymbol{x}) + (\boldsymbol{u}^*)' \cdot (\boldsymbol{g}(\boldsymbol{x})) \leqq f(\boldsymbol{x}^*) \quad \text{for any } \boldsymbol{x} \in X.$$

Hence $(\boldsymbol{u}^*)' \cdot \boldsymbol{g}(\boldsymbol{x}^*) \leqq 0$. On the other hand $\boldsymbol{u}^* \geqq \mathbf{0}$; and optimality of $\boldsymbol{x}^*$ implies that $\boldsymbol{g}(\boldsymbol{x}^*) \geqq \mathbf{0}$. Thus $(\boldsymbol{u}^*)' \cdot \boldsymbol{g}(\boldsymbol{x}^*) = 0$ and, for any $\boldsymbol{x} \in X$ and $\boldsymbol{u} \geqq \mathbf{0}$,

$$f(\boldsymbol{x}) + (\boldsymbol{u}^*)' \cdot \boldsymbol{g}(\boldsymbol{x}) \leqq f(\boldsymbol{x}^*) = f(\boldsymbol{x}^*) + (\boldsymbol{u}^*)' \cdot \boldsymbol{g}(\boldsymbol{x}^*) \leqq f(\boldsymbol{x}^*) + \boldsymbol{u}' \cdot \boldsymbol{g}(\boldsymbol{x}^*)$$

□

Theorem 21 reveals a close relationship between a saddle point of the Lagrangian function corresponding to a nonlinear programming problem and an optimal vector for the problem. On the other hand, Theorem 13 enables us to identify a saddle point of the Lagrangian function associated with a linear programming problem with a pair of optimal vectors for the problem; and, by virtue of Theorem 17, these vectors can be further interpreted as a pair of optimal strategies of a matrix game. Therefore it is natural to ask whether we can characterize a saddle point of the Lagrangian function corresponding to a nonlinear programming problem as a min-max (or max-min) strategy. The following theorem partly clarifies that question.

Theorem 22. *Let* $(\boldsymbol{x}^*, \boldsymbol{u}^*)$ *be a saddle point of a real-valued function* $\psi(\boldsymbol{x}, \boldsymbol{u})$, *where* $\boldsymbol{x}$ *and* $\boldsymbol{u}$ *are vectors in* $\mathbb{R}^n$ *and* $\mathbb{R}^m$ *respectively. Then*

$$\psi(\boldsymbol{x}^*, \boldsymbol{u}^*) = \min_{\boldsymbol{u} \in \mathbb{R}^m} \max_{\boldsymbol{x} \in \mathbb{R}^n} \psi(\boldsymbol{x}, \boldsymbol{u}) = \max_{\boldsymbol{x} \in \mathbb{R}^n} \min_{\boldsymbol{u} \in \mathbb{R}^m} \psi(\boldsymbol{x}, \boldsymbol{u}) \tag{45}$$

Conversely, if there are $\boldsymbol{x}^* \in \mathbb{R}^n$ *and* $\boldsymbol{u}^* \in \mathbb{R}^m$ *satisfying* (45), *then* $(\boldsymbol{x}^*, \boldsymbol{u}^*)$ *is a saddle point of* $\psi(\boldsymbol{x}, \boldsymbol{u})$.

PROOF. From the definition of a saddle point, $\boldsymbol{x}^* \in \mathbb{R}^n$, $\boldsymbol{u}^* \in \mathbb{R}^m$, and $\psi(\boldsymbol{x}, \boldsymbol{u}^*) \leq \psi(\boldsymbol{x}^*, \boldsymbol{u}^*)$ for all $\boldsymbol{x} \in \mathbb{R}^n$. Hence it is clear that

$$\psi(\boldsymbol{x}^*, \boldsymbol{u}^*) = \max_{\boldsymbol{x} \in \mathbb{R}^n} \psi(\boldsymbol{x}, \boldsymbol{u}^*) \geqq \min_{\boldsymbol{u} \in \mathbb{R}^m} \max_{\boldsymbol{x} \in \mathbb{R}^n} \psi(\boldsymbol{x}, \boldsymbol{u}). \tag{46}$$

Similarly,

$$\psi(\boldsymbol{x}^*, \boldsymbol{u}^*) = \min_{\boldsymbol{u} \in \mathbb{R}^m} \psi(\boldsymbol{x}^*, \boldsymbol{u}) \leqq \max_{\boldsymbol{x} \in \mathbb{R}^n} \min_{\boldsymbol{u} \in \mathbb{R}^m} \psi(\boldsymbol{x}, \boldsymbol{u}). \tag{47}$$

From (46) and (47),

$$\min_{\boldsymbol{u} \in \mathbb{R}^m} \max_{\boldsymbol{x} \in \mathbb{R}^n} \psi(\boldsymbol{x}, \boldsymbol{u}) \leqq \psi(\boldsymbol{x}^*, \boldsymbol{u}^*) \leqq \max_{\boldsymbol{x} \in \mathbb{R}^n} \min_{\boldsymbol{u} \in \mathbb{R}^m} \psi(\boldsymbol{x}, \boldsymbol{u}). \tag{48}$$

Let us define

$$\min_{\boldsymbol{u} \in \mathbb{R}^m} \psi(\boldsymbol{x}, \boldsymbol{u}) = \psi(\boldsymbol{x}, \boldsymbol{u}_{\boldsymbol{x}}),$$

$$\max_{\boldsymbol{x} \in \mathbb{R}^n} \psi(\boldsymbol{x}, \boldsymbol{u}) = \psi(\boldsymbol{x}_{\boldsymbol{u}}, \boldsymbol{u}),$$

$$\min_{\boldsymbol{u} \in \mathbb{R}^m} \max_{\boldsymbol{x} \in \mathbb{R}^n} \psi(\boldsymbol{x}, \boldsymbol{u}) = \psi(\boldsymbol{x}_{\boldsymbol{u}_1}, \boldsymbol{u}_1),$$

and

$$\max_{\boldsymbol{x} \in \mathbb{R}^n} \min_{\boldsymbol{u} \in \mathbb{R}^m} \psi(\boldsymbol{x}, \boldsymbol{u}) = \psi(\boldsymbol{x}_1, \boldsymbol{u}_{\boldsymbol{x}_1}).$$

Then

$$\begin{aligned} \psi(\boldsymbol{x}_{\boldsymbol{u}_1}, \boldsymbol{u}_1) &\geqq \psi(\boldsymbol{x}_1, \boldsymbol{u}_1) \quad [\text{because } \psi(\boldsymbol{x}_{\boldsymbol{u}}, \boldsymbol{u}) \geqq \psi(\boldsymbol{x}, \boldsymbol{u}) \text{ for any } \boldsymbol{x} \in \mathbb{R}^n, \boldsymbol{u} \in \mathbb{R}^m] \\ &\geqq \psi(\boldsymbol{x}_1, \boldsymbol{u}_{\boldsymbol{x}_1}) \quad [\text{because } \psi(\boldsymbol{x}, \boldsymbol{u}_{\boldsymbol{x}}) \geqq \psi(\boldsymbol{x}, \boldsymbol{u}) \text{ for any } \boldsymbol{u} \in \mathbb{R}^m, \boldsymbol{x} \in \mathbb{R}^n]. \end{aligned}$$

Hence we can assert that, in general,

$$\min_{\boldsymbol{u} \in \mathbb{R}^m} \max_{\boldsymbol{x} \in \mathbb{R}^n} \psi(\boldsymbol{x}, \boldsymbol{u}) \geqq \max_{\boldsymbol{x} \in \mathbb{R}^n} \min_{\boldsymbol{u} \in \mathbb{R}^m} \psi(\boldsymbol{x}, \boldsymbol{u}). \tag{49}$$

Combining (48) with (49), it is quite clear that

$$\min_{\boldsymbol{u} \in \mathbb{R}^m} \max_{\boldsymbol{x} \in \mathbb{R}^n} \psi(\boldsymbol{x}, \boldsymbol{u}) = \psi(\boldsymbol{x}^*, \boldsymbol{u}^*) = \max_{\boldsymbol{x} \in \mathbb{R}^n} \min_{\boldsymbol{u} \in \mathbb{R}^m} \psi(\boldsymbol{x}, \boldsymbol{u}).$$

Conversely, notice that if

$$\psi(\boldsymbol{x}^*, \boldsymbol{u}^*) = \max_{\boldsymbol{x} \in \mathbb{R}^n} \min_{\boldsymbol{u} \in \mathbb{R}^m} \psi(\boldsymbol{x}, \boldsymbol{u})),$$

then

$$\psi(\boldsymbol{x}^*, \boldsymbol{u}^*) = \min_{\boldsymbol{u} \in \mathbb{R}^m} \psi(\boldsymbol{x}^*, \boldsymbol{u}),$$

from which it follows that

$$\psi(\boldsymbol{x}^*, \boldsymbol{u}^*) \leqq \psi(\boldsymbol{x}^*, \boldsymbol{u}) \quad \text{for all } \boldsymbol{u} \in \mathbb{R}^m.$$

In a similar manner, we obtain

$$\psi(\boldsymbol{x}^*, \boldsymbol{u}^*) \geqq \psi(\boldsymbol{x}, \boldsymbol{u}^*) \quad \text{for all } \boldsymbol{x} \in \mathbb{R}^n.$$ □

In stating the Kuhn–Tucker Theorem of nonlinear programming it was not found necessary to require that the functions $f(\boldsymbol{x})$ and $g_i(\boldsymbol{x})$, $i = 1, \ldots, m$, be differentiable. In economic analysis, however, differentiability is often assumed. One may wonder, therefore, whether the conclusions of the theorem can be sharpened by introducing differentiability. On the other hand, the assumption of concavity is stronger than is required in many branches of economic theory, and one may wonder if it can be relaxed. To these questions we now turn.

Definition 14. A function $\boldsymbol{f}$ of $\mathbb{R}^n$ into $\mathbb{R}^m$ is said to be *differentiable* at $\boldsymbol{x}_0$ of $\mathbb{R}^n$ if there exists a linear transformation $T_{\boldsymbol{x}_0}$ of $\mathbb{R}^n$ into $\mathbb{R}^m$ such that

$$\lim_{\boldsymbol{h}\to\boldsymbol{0}} \frac{\|\boldsymbol{f}(\boldsymbol{x}_0 + \boldsymbol{h}) - \boldsymbol{f}(\boldsymbol{x}_0) - T_{\boldsymbol{x}_0}(\boldsymbol{h})\|}{\|\boldsymbol{h}\|} = 0. \tag{50}$$

Alternatively, set $(\boldsymbol{f}(\boldsymbol{x}_0 + \boldsymbol{h}) - \boldsymbol{f}(\boldsymbol{x}_0) - T_{\boldsymbol{x}_0}(\boldsymbol{h}))/\|\boldsymbol{h}\| = \boldsymbol{\alpha}(\boldsymbol{x}_0, \boldsymbol{h})$. Then (50) is equivalent to

$$\lim_{\boldsymbol{h}\to\boldsymbol{0}} \boldsymbol{\alpha}(\boldsymbol{x}_0, \boldsymbol{h}) = \boldsymbol{0}. \tag{51}$$

In view of the well-known one-to-one correspondence between the family of linear transformations of an n-dimensional vector space into an m-dimensional vector space and the family of $m \times n$ matrices over the underlying field, the image $T_{\boldsymbol{x}_0}(\boldsymbol{h})$ of $\boldsymbol{h}$ under $T_{\boldsymbol{x}_0}$ can be expressed as $\boldsymbol{f}'(\boldsymbol{x}_0)\cdot\boldsymbol{h}$ where $\boldsymbol{f}'(\boldsymbol{x}_0)$ is an $m \times n$ matrix corresponding to $T_{\boldsymbol{x}_0}$. Therefore, if $\boldsymbol{f}(\boldsymbol{x})$ is differentiable at $\boldsymbol{x}_0$, then

$$\boldsymbol{f}(\boldsymbol{x}_0 + \boldsymbol{h}) = \boldsymbol{f}(\boldsymbol{x}_0) + \boldsymbol{f}'(\boldsymbol{x}_0)\cdot\boldsymbol{h} + \boldsymbol{\alpha}(\boldsymbol{x}_0, \boldsymbol{h})\cdot\|\boldsymbol{h}\| \tag{52}$$

where

$$\lim_{\boldsymbol{h}\to\boldsymbol{0}} \boldsymbol{\alpha}(\boldsymbol{x}_0, \boldsymbol{h}) = \boldsymbol{0}.$$

If $\boldsymbol{f}(\boldsymbol{x})$ is differentiable at every point $\boldsymbol{x}$ of $\mathbb{R}^n$, we simply say that $\boldsymbol{f}(\boldsymbol{x})$ *is differentiable on* $\mathbb{R}^n$.

Let $\boldsymbol{f}(\boldsymbol{x}) = (f_1(\boldsymbol{x}), \ldots, f_m(\boldsymbol{x}))'$, where $f_i(\boldsymbol{x})$, $i = 1, \ldots, m$, are real valued functions on $\mathbb{R}^n$. The vector $\boldsymbol{h}$ in Definition 14 can be chosen arbitrarily; hence, letting $\boldsymbol{h} = t\cdot\boldsymbol{e}_j$ $(j = 1, \ldots, n)$, differentiability implies that

$$\begin{aligned} 0 &= \lim_{t\boldsymbol{e}_j\to\boldsymbol{0}} \frac{\|\boldsymbol{f}(\boldsymbol{x}_0 + t\cdot\boldsymbol{e}_j) - \boldsymbol{f}(\boldsymbol{x}_0) - t\cdot\boldsymbol{f}'(\boldsymbol{x}_0)\boldsymbol{e}_j\|}{\|t\cdot\boldsymbol{e}_j\|} \\ &= \lim_{t\to 0} \left\| \frac{\boldsymbol{f}(\boldsymbol{x}_0 + t\cdot\boldsymbol{e}_j) - \boldsymbol{f}(\boldsymbol{x}_0) - t\cdot\boldsymbol{f}'(\boldsymbol{x}_0)\boldsymbol{e}_j}{t} \right\|, \end{aligned}$$

whence

$$\lim_{t\to 0} \frac{\boldsymbol{f}(\boldsymbol{x}_0 + t\cdot\boldsymbol{e}_j) - \boldsymbol{f}(\boldsymbol{x}_0)}{t} = \boldsymbol{f}'(\boldsymbol{x}_0)\boldsymbol{e}_j \equiv (f_{1j}(\boldsymbol{x}_0), \ldots, f_{mj}(\boldsymbol{x}_0))' \qquad (j = 1, \ldots, n). \tag{53}$$

The ith element of (53) gives us

$$\lim_{t\to 0} \frac{f_i(\boldsymbol{x}_0 + t\cdot\boldsymbol{e}_j) - f_i(\boldsymbol{x}_0)}{t} = f_{ij}(\boldsymbol{x}_0) \qquad (i = 1, \ldots, m). \tag{54}$$

The left-hand side of (54), if it exists, is called the partial derivative of $f_i(\boldsymbol{x})$ with respect to x_j, evaluated at $\boldsymbol{x} = \boldsymbol{x}_0$; it is usually denoted by

$$\left.\frac{\partial f_i}{\partial x_j}\right|_{\boldsymbol{x}_0}.$$

The matrix $\boldsymbol{f}'(\boldsymbol{x}_0)$ then consists of all partial derivatives evaluated at $\boldsymbol{x} = \boldsymbol{x}_0$. If $m = 1$, (52) reduces to

$$f(\boldsymbol{x}_0 + \boldsymbol{h}) = f(\boldsymbol{x}_0) + \nabla f(\boldsymbol{x}_0) \cdot \boldsymbol{h} + \alpha(\boldsymbol{x}_0, \boldsymbol{h})\|\boldsymbol{h}\| \tag{52'}$$

where

$$\lim_{\boldsymbol{h} \to \boldsymbol{0}} \alpha(\boldsymbol{x}_0, \boldsymbol{h}) = 0 \quad \text{and} \quad \nabla f(\boldsymbol{x}_0) = \left(\left.\frac{\partial f}{\partial x_1}\right|_{\boldsymbol{x}_0}, \ldots, \left.\frac{\partial f}{\partial x_n}\right|_{\boldsymbol{x}_0} \right).$$

A glance at (52) shows that $\boldsymbol{f}(\boldsymbol{x})$ is continuous at any point where it is differentiable. This is a straightforward generalization of the well-known proposition that, if $m = n = 1$, differentiability implies continuity.

We turn next to the assumption of concavity imposed on the functions in (I) and, in particular, consider the possibility of replacing it with the weaker requirement of quasi-concavity.

Definition 15. Let E be a convex set in $\mathbb{R}^n$. A function f from E into $\mathbb{R}$ is said to be *quasi-concave* if

$$f(\lambda \boldsymbol{x}_1 + (1 - \lambda)\boldsymbol{x}_2) \geqq \min(f(\boldsymbol{x}_1), f(\boldsymbol{x}_2)) \quad \text{for any } \lambda \in [0, 1] \text{ and } \boldsymbol{x}_1, \boldsymbol{x}_2 \in E.$$

If the inequality is reversed and min is replaced by max, f is called *quasi-convex*.

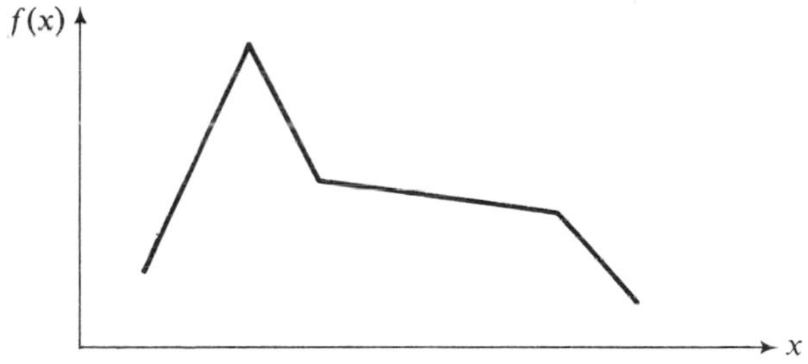

Figure 1.3a A quasi-concave function on $\mathbb{R}$.

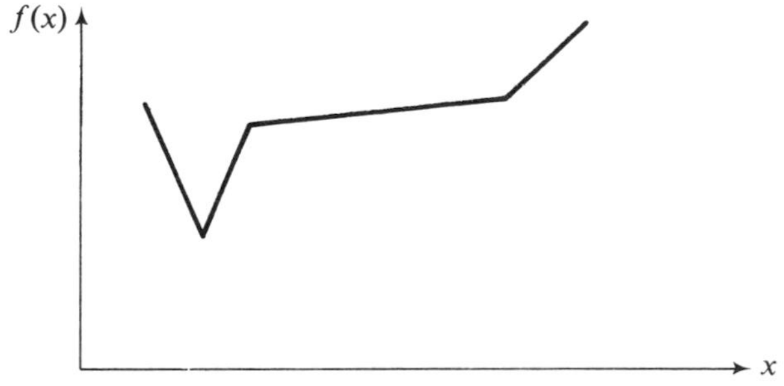

Figure 1.3b A quasi-convex function on $\mathbb{R}$.

Lemma 14. *Let E and f be as in Definition* 15. *Then the following assertions hold.*

1. *f is concave if and only if the set* $\Gamma_f = \{(x, \alpha): x \in E, \alpha \in \mathbb{R}, \text{ and } f(x) \geqq \alpha\}$ *is a convex set in* $\mathbb{R}^{n+1}$.
2. *Let f be differentiable on E. Then f is concave if and only if*

$$\nabla f(x_0)(x_1 - x_0) \geqq f(x_1) - f(x_0) \quad \text{for any } x_i \in E \qquad (i = 0, 1). \quad (55)$$

3. *If f is concave it is quasi-concave.*
4. *f is quasi-concave if and only if the set* $A_f(\alpha) = \{x \in E: f(x) \geqq \alpha\}$ *is convex for any* $\alpha \in \mathbb{R}$.
5. *Let f be differentiable on E and let* x_0 *and* x_1 *be any two points of E. Then for f to be quasi-concave on E it is necessary and sufficient that the inequality*

$$f(x_1) \geqq f(x_0).$$

implies that

$$\nabla f(x_0)(x_1 - x_0) \geqq 0. \quad (56)$$

6. *Suppose, in addition, that E is open and that f is twice continuously differentiable on* E^2. *Then f is concave on E if and only if the Hessian of f is nonpositive definite everywhere on E.*
7. *Suppose that f is differentiable on E. Let* x_0 *and* x_1 *be any distinct points of E, let* $[x_0, x_1] = \{(1 - \lambda)x_0 + \lambda x_1, \lambda \in [0, 1]\}$ *denote the line segment connecting* x_0 *and* x_1, *and let x be any point of* $[x_0, x_1]$ *satisfying the condition* $\nabla f(x) \neq \mathbf{0}$. *Considering the following nonlinear programming problem:*

$$\text{(qcp)} \qquad \begin{aligned} & f(y) \Rightarrow \max_{y \in E} \\ & \textit{subject to } \nabla f(x)y \leqq \nabla f(x)x. \end{aligned}$$

Suppose further that (qcp) *is solvable. Then f is quasi-concave on E if and only if x is optimal for* (qcp).

Proof (1) (*Necessity*). Let (x_i, α_i), $i = 1, 2$, be any two points in Γ_f. Since f is concave,

$$f(\lambda x_1 + (1 - \lambda)x_2) \geqq \lambda f(x_1) + (1 - \lambda)f(x_2) \geqq \lambda\alpha_1 + (1 - \lambda)\alpha_2 \quad \text{for every } \lambda \in [0, 1].$$

This implies that $\lambda(x_1, \alpha_1) + (1 - \lambda)(x_2, \alpha_2)$ is in Γ_f, for $\lambda x_1 + (1 - \lambda)x_2 \in E$ and $\lambda\alpha_1 + (1 - \lambda)\alpha_2 \in \mathbb{R}$.

(1) (*Sufficiency*). Let x_i, $i = 1, 2$, be any two points in E. Then, from the definition of Γ_f, the points $(x_i, f(x_i))$, $i = 1, 2$, are in Γ_f. Since Γ_f is convex, the point $\lambda(x_1, f(x_1)) + (1 - \lambda)(x_2, f(x_2))$ is in Γ_f for any $\lambda \in [0, 1]$, which, in turn, implies that

$$f(\lambda x_1 + (1 - \lambda)x_2) \geqq \lambda f(x_1) + (1 - \lambda)f(x_2) \quad \text{for any } \lambda \in [0, 1].$$

(2) (*Necessity*). Let μ be any real number in (0, 1] and suppose that $\boldsymbol{x}_i$, $i = 0, 1$, are any points in E. Then

$$\begin{aligned} f(\boldsymbol{x}_0 + \mu(\boldsymbol{x}_1 - \boldsymbol{x}_0)) &= f((1 - \mu)\boldsymbol{x}_0 + \mu \boldsymbol{x}_1) \\ &\geqq (1 - \mu) f(\boldsymbol{x}_0) + \mu f(\boldsymbol{x}_1) \qquad [\text{since } f(\boldsymbol{x}) \text{ is concave}] \\ &= f(\boldsymbol{x}_0) + \mu(f(\boldsymbol{x}_1) - f(\boldsymbol{x}_0)). \end{aligned}$$

Hence

$$f(\boldsymbol{x}_0 + \mu(\boldsymbol{x}_1 - \boldsymbol{x}_0)) - f(\boldsymbol{x}_0) \geqq \mu(f(\boldsymbol{x}_1) - f(\boldsymbol{x}_0)) \quad \text{for any } \mu \in (0, 1].$$

Since $f(\boldsymbol{x})$ is differentiable and $\mu > 0$, the above inequality implies that

$$\nabla f(\boldsymbol{x}_0) \cdot (\boldsymbol{x}_1 - \boldsymbol{x}_0) + \alpha(\boldsymbol{x}_0, \mu(\boldsymbol{x}_1 - \boldsymbol{x}_0)) \cdot \|\boldsymbol{x}_1 - \boldsymbol{x}_0\| \geqq f(\boldsymbol{x}_1) - f(\boldsymbol{x}_0) \quad \text{for any } \mu \in (0, 1].$$

Letting μ tend to 0, (55) follows at once.

(2) (*Sufficiency*). Suppose that $\boldsymbol{x}_0, \boldsymbol{x}_1 \in E$ and that $\lambda \in [0, 1]$. Since E is convex, it is clear that $\lambda \boldsymbol{x}_0 + (1 - \lambda)\boldsymbol{x}_1$ is in E. Then by assumption

$$f(\boldsymbol{x}_0) - f(\lambda \boldsymbol{x}_0 + (1 - \lambda)\boldsymbol{x}_1) \leqq (1 - \lambda) \cdot \nabla f(\lambda \boldsymbol{x}_0 + (1 - \lambda)\boldsymbol{x}_1) \cdot (\boldsymbol{x}_0 - \boldsymbol{x}_1)$$

and

$$f(\boldsymbol{x}_1) - f(\lambda \boldsymbol{x}_0 + (1 - \lambda)\boldsymbol{x}_1) \leqq -\lambda \cdot \nabla f(\lambda \boldsymbol{x}_0 + (1 - \lambda)\boldsymbol{x}_1) \cdot (\boldsymbol{x}_0 - \boldsymbol{x}_1).$$

Therefore

$$\lambda f(\boldsymbol{x}_0) + (1 - \lambda) f(\boldsymbol{x}_1) \leq f(\lambda \boldsymbol{x}_0 + (1 - \lambda)\boldsymbol{x}_1) \quad \text{for every } \lambda \in [0, 1].$$

(3) Since f is concave,

$$f(\lambda \boldsymbol{x}_1 + (1 - \lambda)\boldsymbol{x}_2) \geqq \lambda f(\boldsymbol{x}_1) + (1 - \lambda) f(\boldsymbol{x}_2) \quad \text{for any } \lambda \in [0, 1] \text{ and } \boldsymbol{x}_1, \boldsymbol{x}_2 \in E.$$

Let us define $\alpha = \min\{f(\boldsymbol{x}_1), f(\boldsymbol{x}_2)\}$. Then, from the above inequality,

$$f(\lambda \boldsymbol{x}_1 + (1 - \lambda)\boldsymbol{x}_2) \geqq \lambda\alpha + (1 - \lambda)\alpha = \min\{f(\boldsymbol{x}_1), f(\boldsymbol{x}_2)\}.$$

(4) (*Necessity*). Let $\boldsymbol{x}_1$ and $\boldsymbol{x}_2$ be any points in $A_f(\alpha)$. Then, from the definition of $A_f(\alpha)$,

$$\min\{f(\boldsymbol{x}_1), f(\boldsymbol{x}_2)\} \geqq \alpha.$$

Hence the assumed quasi-concavity of f ensures that

$$f(\lambda \boldsymbol{x}_1 + (1 - \lambda)\boldsymbol{x}_2) \geqq \alpha \quad \text{for any } \lambda \in [0, 1],$$

which establishes the convexity of $A_f(\alpha)$.

(4) (*Sufficiency*). Let $\boldsymbol{x}_i \in E$, $i = 1, 2$, and define $\beta = \min\{f(\boldsymbol{x}_1), f(\boldsymbol{x}_2)\}$. Since $A_f(\beta)$ is convex by assumption, $\lambda \in [0, 1]$ implies that $\lambda \boldsymbol{x}_1 + (1 - \lambda)\boldsymbol{x}_2$ is in $A_f(\beta)$. Hence

$$f(\lambda \boldsymbol{x}_1 + (1 - \lambda)\boldsymbol{x}_2) \geqq \min\{f(\boldsymbol{x}_1), f(\boldsymbol{x}_2)\} \quad \text{for any } \lambda \in [0, 1].$$

(5) (*Necessity*). Let $\boldsymbol{x}_j \in E, j = 0, 1$, and suppose that $f(\boldsymbol{x}_1) \geqq f(\boldsymbol{x}_0)$. Then the assumed quasi-concavity of f guarantees that

$$f(\boldsymbol{x}_0 + \lambda(\boldsymbol{x}_1 - \boldsymbol{x}_0)) = f((1 - \lambda)\boldsymbol{x}_0 + \lambda\boldsymbol{x}_1) \geqq f(\boldsymbol{x}_0) \quad \text{for any } \lambda \in [0, 1].$$

Since f is differentiable on E, it follows from the above inequality that

$$\begin{aligned} 0 &\leqq f(\boldsymbol{x}_0 + \lambda(\boldsymbol{x}_1 - \boldsymbol{x}_0)) - f(\boldsymbol{x}_0) \\ &= \lambda[\nabla f(\boldsymbol{x}_0) \cdot (\boldsymbol{x}_1 - \boldsymbol{x}_0) + \alpha(\boldsymbol{x}_0, \lambda(\boldsymbol{x}_1 - \boldsymbol{x}_0))\|\boldsymbol{x}_1 - \boldsymbol{x}_0\|] \quad \text{for any } \lambda \in [0, 1]. \end{aligned}$$

Since this inequality must hold for any $\lambda \in (0, 1]$,

$$\begin{aligned} \nabla f(\boldsymbol{x}_0) \cdot (\boldsymbol{x}_1 - \boldsymbol{x}_0) + \alpha(\boldsymbol{x}_0, \lambda(\boldsymbol{x}_1 - \boldsymbol{x}_0)) \cdot \|\boldsymbol{x}_1 - \boldsymbol{x}_0\| & \\ \geqq 0 \quad \text{for any } \lambda \in (0, 1]. & \quad (57) \end{aligned}$$

Suppose that $\nabla f(\boldsymbol{x}_0) \cdot (\boldsymbol{x}_1 - \boldsymbol{x}_0) = -\varepsilon < 0$. Then for

$$\varepsilon_0 = \frac{|\nabla f(\boldsymbol{x}_0) \cdot (\boldsymbol{x}_1 - \boldsymbol{x}_0)|}{\|\boldsymbol{x}_1 - \boldsymbol{x}_0\|}$$

there exists $\lambda_0 > 0$ such that if $\lambda < \lambda_0$ then

$$|\alpha(\boldsymbol{x}_0, \lambda(\boldsymbol{x}_1 - \boldsymbol{x}_0))| \cdot \|\boldsymbol{x}_1 - \boldsymbol{x}_0\| < \varepsilon;$$

otherwise, $\lim_{\lambda \to 0} \alpha(\boldsymbol{x}_0, \lambda(\boldsymbol{x}_1 - \boldsymbol{x}_0)) = 0$ would be false. Therefore, for λ such that $0 < \lambda < \min\{1, \lambda_0\}$,

$$\begin{aligned} 0 &\leqq \nabla f(\boldsymbol{x}_0) \cdot (\boldsymbol{x}_1 - \boldsymbol{x}_0) \\ &\quad + \alpha(\boldsymbol{x}_0, \lambda(\boldsymbol{x}_1 - \boldsymbol{x}_0)) \cdot \|\boldsymbol{x}_1 - \boldsymbol{x}_0\| \qquad \text{[since (57) holds]} \\ &\leqq \nabla f(\boldsymbol{x}_0) \cdot (\boldsymbol{x}_1 - \boldsymbol{x}_0) + |\alpha(\boldsymbol{x}_0, \lambda(\boldsymbol{x}_1 - \boldsymbol{x}_0))| \cdot \|\boldsymbol{x}_1 - \boldsymbol{x}_0\| < -\varepsilon + \varepsilon = 0, \end{aligned}$$

which is a contradiction.

(5) (*Sufficiency*). Suppose the contrary. Then there exist $\boldsymbol{x}_j \in E, j = 0, 1$, and $t \in (0, 1)$ such that

$$\min\{f(\boldsymbol{x}_0), f(\boldsymbol{x}_1)\} > f(\boldsymbol{x}_2),$$

where $\boldsymbol{x}_2 = t\boldsymbol{x}_0 + (1 - t)\boldsymbol{x}_1$. We further note that, without loss of generality, $\boldsymbol{x}_0$ and $\boldsymbol{x}_1$ can be labeled so that $f(\boldsymbol{x}_1) \leqq f(\boldsymbol{x}_0)$. Define

$$K = \{\boldsymbol{x} \in (\boldsymbol{x}_0, \boldsymbol{x}_1) : f(\boldsymbol{x}_1) > f(\boldsymbol{x})\}$$

where $(\boldsymbol{x}_0, \boldsymbol{x}_1) = \{\boldsymbol{x} = \lambda\boldsymbol{x}_0 + (1 - \lambda)\boldsymbol{x}_1 : \lambda \in (0, 1)\}$. Then from the assumed continuity of f as well as the definition of $\boldsymbol{x}_2$, it follows that the intersection of some neighborhoods of $\boldsymbol{x}_2$ and $(\boldsymbol{x}_0, \boldsymbol{x}_1)$ constitutes a nonempty subset of K, whence, in view of the assumed convexity of E, K is a nonempty subset of E that contains infinitely many points. Therefore, in view of the assumed implication, we have

$$\left.\begin{aligned} \nabla f(\boldsymbol{x})(\boldsymbol{x}_1 - \boldsymbol{x}) \geqq 0 \\ \nabla f(\boldsymbol{x})(\boldsymbol{x}_0 - \boldsymbol{x}) \geqq 0 \end{aligned}\right\} \quad \text{for any } \boldsymbol{x} \in K.$$

A direct calculation therefore yields

$$\left.\begin{aligned} \lambda \cdot \nabla f(\boldsymbol{x})(\boldsymbol{x}_1 - \boldsymbol{x}_0) &\geqq 0 \\ -(1-\lambda)\nabla f(\boldsymbol{x})(\boldsymbol{x}_1 - \boldsymbol{x}_0) &\geqq 0 \end{aligned}\right\} \quad \text{for any } \boldsymbol{x} \in K.$$

Since $\lambda \in (0, 1)$, these inequalities further imply that

$$\nabla f(\boldsymbol{x})(\boldsymbol{x}_1 - \boldsymbol{x}_0) = 0 \quad \text{for any } \boldsymbol{x} \in K.$$

On the other hand, from Lemma 13 it follows that

$$f(\boldsymbol{x}_1) \geqq f(\boldsymbol{x}) \quad \text{for all } \boldsymbol{x} \in \partial K.$$

Suppose that for some $\xi \in \partial K$, $f(\boldsymbol{x}_1) > f(\xi)$. Then, recalling that K is open relative to $(\boldsymbol{x}_0, \boldsymbol{x}_1)$,[3] there exists a neighborhood $B(\xi, \delta)$ of ξ such that

$$f(\boldsymbol{x}_1) > f(\boldsymbol{x}) \quad \text{for all } \boldsymbol{x} \in B(\xi, \delta) \cap (\boldsymbol{x}_0, \boldsymbol{x}_1).$$

However, this contradicts the fact that $\xi \in \partial K$. Hence we obtain

$$f(\boldsymbol{x}_1) = f(\boldsymbol{x}) \quad \text{for all } \boldsymbol{x} \in \partial K.$$

Moreover, every boundary point of K is on the line segment joining $\boldsymbol{x}_0$ and $\boldsymbol{x}_1$, for otherwise there is a boundary point of K which is, at the same time an exterior point of K, a selfcontradiction. Taking into account the possibility that $\boldsymbol{x}_1 \in \partial K$, the argument so far enables us to assert that there exists an $\boldsymbol{x}_3 \in \partial K$ with the property that

$$\boldsymbol{x}_3 = \lambda_0 \boldsymbol{x}_1 + (1 - \lambda_0)\boldsymbol{x}_2 \quad \text{for some } \lambda_0 \in (0, 1].$$

Then, in view of the well-known Mean Value Theorem[4], it is clear that for some $\theta \in (0, 1)$

$$\begin{aligned} 0 < f(\boldsymbol{x}_3) - f(\boldsymbol{x}_2) &= \nabla f(\boldsymbol{x}_2 + \theta(\boldsymbol{x}_3 - \boldsymbol{x}_2))(\boldsymbol{x}_3 - \boldsymbol{x}_2) \\ &= \lambda_0 \nabla f(\theta \boldsymbol{x}_3 + (1-\theta)\boldsymbol{x}_2)(\boldsymbol{x}_1 - \boldsymbol{x}_2) \\ &= \lambda_0 \cdot t \cdot \nabla f(\theta \boldsymbol{x}_3 + (1-\theta)\boldsymbol{x}_2)(\boldsymbol{x}_1 - \boldsymbol{x}_0). \end{aligned}$$

Since, by definition, $\lambda_0 t > 0$, $\nabla f(\theta \boldsymbol{x}_3 + (1-\theta)\boldsymbol{x}_2)(\boldsymbol{x}_1 - \boldsymbol{x}_0) > 0$. This contradicts our earlier observation that

$$\nabla f(\boldsymbol{x})(\boldsymbol{x}_1 - \boldsymbol{x}_0) = 0 \quad \text{for any } \boldsymbol{x} \in K.$$

because, by its construction, $\theta \boldsymbol{x}_3 + (1 - \theta)\boldsymbol{x}_2$ is in K.

(6) (*Necessity*). Let $\boldsymbol{x}_0$ and $\boldsymbol{x}$ be any two points of E and for $\lambda \in [0, 1]$ define $\boldsymbol{x}(\lambda) = \lambda \cdot \boldsymbol{x} + (1 - \lambda)\boldsymbol{x}_0$. Then, from the assumed convexity of E, $\boldsymbol{x}(\lambda) \in E$ for all $\lambda \in [0, 1]$. Hence, corresponding to $\boldsymbol{x}_0$ and $\boldsymbol{x}$, we can define

$$g_{\boldsymbol{x}}(\lambda) = f(\boldsymbol{x}(\lambda)) - f(\boldsymbol{x}_0) - \lambda \nabla f(\boldsymbol{x}_0)(\boldsymbol{x} - \boldsymbol{x}_0),$$

where λ is of course considered to vary on $[0, 1]$. Since f is concave, (2) of the lemma implies that, for any $\boldsymbol{x}_0$ and $\boldsymbol{x}$ of E, $g_{\boldsymbol{x}}(\lambda)$ attains its maximum ($g(0) = 0$) at $\lambda = 0$. Upon direct differentiation, we further obtain

$$g'_{\boldsymbol{x}}(\lambda) = (\nabla f(\boldsymbol{x}(\lambda)) - \nabla f(\boldsymbol{x}_0))(\boldsymbol{x} - \boldsymbol{x}_0)$$

and

$$g_x''(\lambda) = (\boldsymbol{x} - \boldsymbol{x}_0)'\boldsymbol{H}_f(\boldsymbol{x}(\lambda)) \cdot (\boldsymbol{x} - \boldsymbol{x}_0),$$

where $\boldsymbol{H}_f(\boldsymbol{x}(\lambda))$ denotes the Hessian of f evaluated at $\boldsymbol{x}(\lambda)$. It is thus straightforward that

$$\left.\begin{aligned} g_x'(0) &= 0 \\ g_x''(0) &= (\boldsymbol{x} - \boldsymbol{x}_0)'\boldsymbol{H}_f(\boldsymbol{x}_0)(\boldsymbol{x} - \boldsymbol{x}_0) \end{aligned}\right\} \quad \text{for all } \boldsymbol{x} \text{ and } \boldsymbol{x}_0 \text{ of } E.$$

Suppose now that for some $\boldsymbol{x}_0 \in E$, $\boldsymbol{H}_f(\boldsymbol{x}_0)$ is not nonpositive definite. Then there exists a $\boldsymbol{u} \neq \boldsymbol{0}$ such that $\boldsymbol{u}'\boldsymbol{H}_f(\boldsymbol{x}_0)\boldsymbol{u} > 0$. Since $\boldsymbol{x}_0$ is supposed to be in an open set E, there can be chosen a $t \in \mathbb{R}$ so small that

$$\boldsymbol{x}(t) = (\boldsymbol{x}_0 + t\boldsymbol{u}) \in E.$$

Then, from a direct calculation, it follows that

$$g_{x(t)}''(0) = (\boldsymbol{x}(t) - \boldsymbol{x}_0)'\boldsymbol{H}_f(\boldsymbol{x}_0) \cdot (\boldsymbol{x}(t) - \boldsymbol{x}_0) = t^2 \cdot \boldsymbol{u}'\boldsymbol{H}_f(\boldsymbol{x}_0) \cdot \boldsymbol{u} > 0.$$

Since, by assumption, $g_{x(t)}''(\lambda)$ is continuous, there is an open interval $(0, \delta)$ with the property that

$$g_{x(t)}''(\lambda) > 0 \quad \text{for all } \lambda \in (0, \delta).$$

Let λ_1 be in $(0, \delta)$. Then, resorting to Taylor's Theorem,[5] there exists a $\theta \in (0, 1)$ such that

$$g_{x(t)}(\lambda_1) = \frac{\lambda_1^2}{2}\, g_{x(t)}''(\theta\lambda_1) > 0 = g_{x(t)}(0).$$

This, however, contradicts our earlier observation that $g_x(\lambda)$ attains its maximum at $\lambda = 0$ for all $\boldsymbol{x}_0$ and $\boldsymbol{x}$ of E.

(6) (*Sufficiency*). Let $\boldsymbol{x}_0$, $\boldsymbol{x}$ and $g_x(\lambda)$ be as in the proof of necessity. Then, by hypothesis,

$$g_x''(\lambda) \leqq 0 \quad \text{for all } \boldsymbol{x}_0, \boldsymbol{x} \in E, \text{ and for all } \lambda \in [0, 1].$$

This, together with the fact that $g_x'(0) = 0$ for any $\boldsymbol{x}_0$ and $\boldsymbol{x}$ of E, implies that

$$g_x'(\lambda) \leqq 0 \quad \text{for all } \boldsymbol{x}_0, \boldsymbol{x} \in E, \text{ and for all } \lambda \in [0, 1].$$

Since, by definition, $g_x(0) = 0$ for all $\boldsymbol{x}_0$ and $\boldsymbol{x}$ of E, we can similarly assert that

$$g_x(\lambda) \leqq 0 \quad \text{for all } \boldsymbol{x}_0, \boldsymbol{x} \in E \text{ and for all } \lambda \in [0, 1].$$

Our proof is completed by setting $\lambda = 1$.

(7) (*Necessity*). Let $A_{f(\boldsymbol{x})} \equiv \{\boldsymbol{y} \in E : f(\boldsymbol{y}) \geqq f(\boldsymbol{x})\}$ and suppose that there exists a feasible $\boldsymbol{x}_2 \neq \boldsymbol{x}$ such that $f(\boldsymbol{x}_2) > f(\boldsymbol{x})$. Applying Lemma 14(5), $\nabla f(\boldsymbol{x})\boldsymbol{x}_2 \geqq \nabla f(\boldsymbol{x})\boldsymbol{x}$, whence $\nabla f(\boldsymbol{x})\boldsymbol{x}_2 = \nabla f(\boldsymbol{x})\boldsymbol{x}$. On the other hand, the inequality $f(\boldsymbol{x}_2) > f(\boldsymbol{x})$, together with the continuity of f, implies that there exists an ε-neighborhood $B(\boldsymbol{x}_2, \varepsilon)$ of $\boldsymbol{x}_2$ such that $f(\boldsymbol{y}) > f(\boldsymbol{x})$ for all $\boldsymbol{y}$ of $B(\boldsymbol{x}_2, \varepsilon)$. Hence, in view of the hypothesis that $\nabla f(\boldsymbol{x}) \neq \boldsymbol{0}$ and the fact that $\nabla f(\boldsymbol{x})\boldsymbol{x} = \nabla f(\boldsymbol{x})\boldsymbol{x}_2$, we can find an $\boldsymbol{x}_3$ of $B(\boldsymbol{x}_2, \varepsilon)$ for which $\nabla f(\boldsymbol{x})\boldsymbol{x} > \nabla f(\boldsymbol{x})\boldsymbol{x}_3$, contradicting the quasi-concavity of f and the fact that $f(\boldsymbol{x}_3) > f(\boldsymbol{x})$.

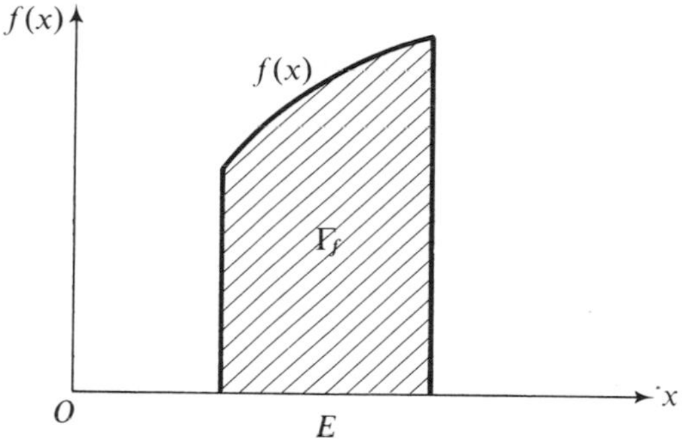

Figure 1.4

(7) (*If part*). Since, by definition, $\boldsymbol{x}$ can be expressed as a convex linear combination of $\boldsymbol{x}_0$ and $\boldsymbol{x}_1$,

$$\nabla f(\boldsymbol{x})\boldsymbol{x} \geqq \min\{\nabla f(\boldsymbol{x})\boldsymbol{x}_0, \nabla f(\boldsymbol{x})\boldsymbol{x}_1\}.$$

Therefore, either $\boldsymbol{x}_0$ or $\boldsymbol{x}_1$ is clearly a feasible vector of (qcp). Hence, by hypothesis, $f(\boldsymbol{x}) \geqq \min\{f(\boldsymbol{x}_0), f(\boldsymbol{x}_1)\}$. □

Remark Concerning Lemma 14. (i) If f is a quasi-concave (concave) then $-f$ is quasi-convex (convex); hence an analogous lemma trivially holds for quasi-convex (convex) functions.

(ii) The set Γ_f is said to be a *hypograph* of $f(\boldsymbol{x})$. Figure 1.4 above may help the reader to understand intuitively the geometric meaning of Lemma 14(1). Since the hypograph of a quasi-concave function is not always convex (see Figure 1.3a), the converse of Lemma 4(3) is not necessarily true.

(iii) Rearranging the terms of (55), we obtain $f(\boldsymbol{x}_1) \leq f(\boldsymbol{x}_0) + \nabla f(\boldsymbol{x}_0)\cdot(\boldsymbol{x}_1 - \boldsymbol{x}_0)$, the right-hand side of which is a linear approximation to $f(x_1)$. Hence we can say that if f is concave then any linear approximation gives us an overestimate of its true value. For $m = n = 1$, Figure 1.5 provides an illustration.

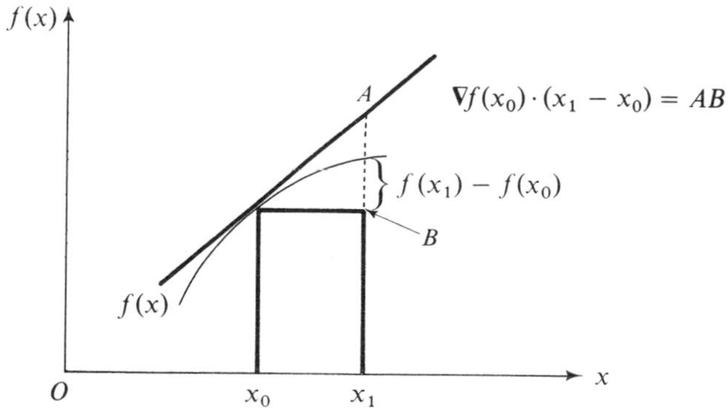

Figure 1.5

In economic analysis, frequent use is made of a sharpened version of quasi-concave and/or concave functions.

Definition 16. Let E and f be as in Definition 15. Then:

i. f is said to be *strictly concave on* E if, for any distinct $\boldsymbol{x}_0$ and $\boldsymbol{x}_1$ of E and for any $\lambda \in (0, 1)$,

$$f((1-\lambda)\boldsymbol{x}_0 + \lambda\boldsymbol{x}_1) > (1-\lambda)f(\boldsymbol{x}_0) + \lambda f(\boldsymbol{x}_1);$$

ii. f is said to be *strictly quasi-concave on* E if, for any distinct $\boldsymbol{x}_0$ and $\boldsymbol{x}_1$ of E and for any $\lambda \in (0, 1)$,

$$f((1-\lambda)\boldsymbol{x}_0 + \lambda\boldsymbol{x}_1) > \min\{f(\boldsymbol{x}_0), f(\boldsymbol{x}_1)\}.$$

Remark concerning Definition 16. (i) If the inequality ($>$) in Definition 16 is reversed and if min is replaced by max then f is said to be *strictly convex* and *strictly quasi-convex*, respectively. It is therefore evident that f is strictly convex (respectively, quasi-convex) if and only if $-f$ is strictly concave (respectively, quasi-concave). (ii) We should notice that mathematicians often say that f is strictly quasi-concave if, for any distinct $\boldsymbol{x}_0$ and $\boldsymbol{x}_1$ and for any $\lambda \in (0, 1)$,

$$f(\boldsymbol{x}_1) > f(\boldsymbol{x}_0) \qquad \text{implies } f((1-\lambda)\boldsymbol{x}_0 + \lambda\boldsymbol{x}_1) > f(\boldsymbol{x}_0).$$

Needless to say, strict quasi-concavity in this sense is broader than that defined in Definition 16.

Lemma 15. *Let E and f be as in Definition* 16 *and, if necessary, let f be differentiable on E.*

(1) *f is strictly concave on E if and only if*

$$\nabla f(\boldsymbol{x}_0)(\boldsymbol{x}_1 - \boldsymbol{x}_0) > f(\boldsymbol{x}_1) - f(\boldsymbol{x}_0) \quad \textit{for any distinct} \quad \boldsymbol{x}_0 \quad \textit{and} \quad \boldsymbol{x}_1 \textit{ of } E. \tag{55'}$$

(2) *If, in addition, f is twice continuously differentiable on E then for f to be strictly concave on E it suffices that the Hessian $\boldsymbol{H}_f(\boldsymbol{x})$ evaluated at any $\boldsymbol{x} \in E$ be negative definite.*

(3) *Suppose that f is continuous on E. Then for f to be strictly quasi-concave on E it is necessary that, for any real α, the set $A_f(\alpha) = \{\boldsymbol{x} \in E : f(\boldsymbol{x}) \geqq \alpha\}$ be strictly convex, that is, for any distinct points $\boldsymbol{x}_0$ and $\boldsymbol{x}_1$ of $A_f(\alpha)$, $(\boldsymbol{x}_0, \boldsymbol{x}_1) = \{(1-\lambda)\boldsymbol{x}_0 + \lambda\boldsymbol{x}_1, \lambda \in (0, 1)\}$ be contained in the set of all interior points of $A_f(\alpha)$.*

(4) *Let f, $\boldsymbol{x}_0$, and $\boldsymbol{x}_1$ be as in assertion* (7) *of Lemma* 14 *and let $\boldsymbol{x}$ be any point on the open line segment $(\boldsymbol{x}_0, \boldsymbol{x}_1)$ connecting $\boldsymbol{x}_0$ and $\boldsymbol{x}_1$. Assume further that $\nabla f(\boldsymbol{x}) \neq \boldsymbol{0}$. Then Lemma* 14(7) *is strengthened as follows: f is strictly quasi-concave on E if and only if $\boldsymbol{x}$ is the unique optimal vector for* (qcp).

PROOF. (1) In view of the assumed strict concavity of f we have, from (2) of Lemma 14,

$$f(\boldsymbol{x}_1) - f(\boldsymbol{x}_0) \leqq \nabla f(\boldsymbol{x}_0)(\boldsymbol{x}_1 - \boldsymbol{x}_0) \quad \text{for any } \boldsymbol{x}_j \in E, j = 0, 1.$$

Suppose that there exist two distinct points $\boldsymbol{x}_0$ and $\boldsymbol{x}_1$ of E such that

$$f(\boldsymbol{x}_1) - f(\boldsymbol{x}_0) = \nabla f(\boldsymbol{x}_0)(\boldsymbol{x}_1 - \boldsymbol{x}_0).$$

Then, by hypothesis,

$$f(\boldsymbol{x}(\lambda)) > \lambda f(\boldsymbol{x}_0) + (1 - \lambda) f(\boldsymbol{x}_1) \quad \text{for all } \lambda \in (0, 1),$$

where $\boldsymbol{x}(\lambda) = \lambda \boldsymbol{x}_0 + (1 - \lambda)\boldsymbol{x}_1$.

From the above two expressions it follows that

$$f(\boldsymbol{x}(\lambda)) - f(\boldsymbol{x}_0) > (1 - \lambda)\nabla f(\boldsymbol{x}_0)(\boldsymbol{x}_1 - \boldsymbol{x}_0).$$

On the other hand, in view of the convexity of E, $\boldsymbol{x}(\lambda)$ is in E for all $\lambda \in (0, 1)$. Therefore, by (2) of Lemma 14 again,

$$f(\boldsymbol{x}(\lambda)) - f(\boldsymbol{x}_0) \leqq \nabla f(\boldsymbol{x}_0)(\boldsymbol{x}(\lambda) - \boldsymbol{x}_0) = (1 - \lambda)\nabla f(\boldsymbol{x}_0)(\boldsymbol{x}_1 - \boldsymbol{x}_0),$$

contradicting the inequality established earlier.

The converse assertion is easily proved by replacing the loose inequalities ($\leqq$) and the closed interval $[0, 1]$ in the sufficiency proof of Lemma 14(2) by strict inequalities ($<$) and the open interval $(0, 1)$ respectively.

(2) Let $\boldsymbol{x}_0$, $\boldsymbol{x}_1$ and λ be as in the previous proof and let $\boldsymbol{x}(\lambda) = (1 - \lambda)\boldsymbol{x}_0 + \lambda \boldsymbol{x}_1$. Applying the Mean Value Theorem, there exists a $\delta \in (0, 1)$ such that

$$\begin{aligned} & f(\boldsymbol{x}(\lambda)) - \{(1 - \lambda) f(\boldsymbol{x}_0) + \lambda f(\boldsymbol{x}_1)\} \\ &= \lambda[\nabla f(\boldsymbol{x}_0)(\boldsymbol{x}_1 - \boldsymbol{x}_0) - (f(\boldsymbol{x}_1) - f(\boldsymbol{x}_0))] \\ &\quad + \left(\frac{\lambda^2}{2}\right)(\boldsymbol{x}_1 - \boldsymbol{x}_0)' \boldsymbol{H}_f(\boldsymbol{x}_{\delta\lambda})(\boldsymbol{x}_1 - \boldsymbol{x}_0), \end{aligned}$$

where $\boldsymbol{x}_{\delta\lambda} = (1 - \delta\lambda)\boldsymbol{x}_0 + (\delta\lambda)\boldsymbol{x}_1$.

Since E is convex and since $\boldsymbol{H}_f(\cdot)$ is negative definite everywhere on E, the final term of the above expression is negative. In view of Lemma 14(6), f is clearly concave. Hence

$$\begin{aligned} & \nabla f(\boldsymbol{x}_0)(\boldsymbol{x}_1 - \boldsymbol{x}_0) - (f(\boldsymbol{x}_1) - f(\boldsymbol{x}_0)) \geqq \\ & -\left(\frac{\lambda^2}{2}\right)(\boldsymbol{x}_1 - \boldsymbol{x}_0)' \boldsymbol{H}_f(\boldsymbol{x}_{\delta\lambda})(\boldsymbol{x}_1 - \boldsymbol{x}_0) > 0. \end{aligned}$$

Therefore, with the aid of Lemma 15(1), f is strictly concave.

(3) For any $\alpha \in \mathbb{R}$, choose distinct $\boldsymbol{x}_0$ and $\boldsymbol{x}_1$ of $A_f(\alpha)$. Then, by hypothesis,

$$f(\boldsymbol{x}) > \min\{f(\boldsymbol{x}_0), f(\boldsymbol{x}_1)\} \quad \text{for all } \boldsymbol{x} \in (\boldsymbol{x}_0, \boldsymbol{x}_1).$$

Since f is continuous, $\boldsymbol{x}$ is evidently an interior point of $A_f(\alpha)$.

(4) (*Only if part*). Since by hypothesis f is quasi-concave, Lemma 14(7) ensures that $\boldsymbol{x}$ is optimal for (qcp). Suppose the contrary. Then there is a feasible $\boldsymbol{x}_2$ such that $f(\boldsymbol{x}_2) = f(\boldsymbol{x})$ and such that $\boldsymbol{x} \neq \boldsymbol{x}_2$. Suppose further

that $\nabla f(\boldsymbol{x})\boldsymbol{x}_2 < \nabla f(\boldsymbol{x})\boldsymbol{x}$. Then, from Lemma 14(5), it follows that $f(\boldsymbol{x}_2) < f(\boldsymbol{x})$, contradicting the definition of $\boldsymbol{x}_2$. Hence $\nabla f(\boldsymbol{x})\boldsymbol{x}_2 = \nabla f(\boldsymbol{x})\boldsymbol{x}$. Since f is strictly quasi-concave, we see that

$$f(\boldsymbol{y}) > f(\boldsymbol{x}) \quad \text{for any } \boldsymbol{y} \in (\boldsymbol{x}, \boldsymbol{x}_2).$$

On the other hand, a direct calculation, together with the earlier observation that $\nabla f(\boldsymbol{x})\boldsymbol{x}_2 = \nabla f(\boldsymbol{x})\boldsymbol{x}$, yields

$$\nabla f(\boldsymbol{x})\boldsymbol{y} = \nabla f(\boldsymbol{x})\boldsymbol{x}.$$

Hence $\boldsymbol{x}$ can never be optimal for (qcp), a contradiction.

(4) (*If part*). From the proof of the if part of Lemma 14(7), it is obvious that either $\boldsymbol{x}_0$ or $\boldsymbol{x}_1$ is feasible. Assume, for simplicity, that $\boldsymbol{x}_0$ is feasible. Then, by hypothesis, $f(\boldsymbol{x}) > f(\boldsymbol{x}_0)$. Hence

$$f(\boldsymbol{x}) > \min\{f(\boldsymbol{x}_0), f(\boldsymbol{x}_1)\}. \qquad \square$$

Remark concerning Lemma 15. (i) In view of the remark concerning Definition 16, Lemma 15 can be easily converted to cover strictly convex and/or strictly quasi-convex functions.

(ii) It should be noted that the converse of assertion (2) of the lemma is not always true. For example, $f(x) = -x^4$ is strictly concave on $\mathbb{R}$. However, $H_f(0) \equiv 0$, which is not negative definite but only nonpositive definite.

It is nevertheless noteworthy that $f''(x) = H_f(x)$ is negative definite for all $x \neq 0$. This suggests the following theorem:

(Katzner [1970; Theorem 8.5-5, p. 202]) *If f is strictly concave on an open convex set E then, except possibly on a nowhere dense subset of E, its Hessian is negative definite on E, where a subset A of E is said to be dense if every $\boldsymbol{x} \in A^c$ is a limit point of A.*

(iii) The converse of Lemma 15(3) is not necessarily true. For example, let $f(x)$ be a real-valued function defined on $\mathbb{R}$ and let the graph of $f(x)$ be as depicted below. Then, for any $\alpha \in \mathbb{R}$ the set $A_f(\alpha)$ is strictly convex (note here

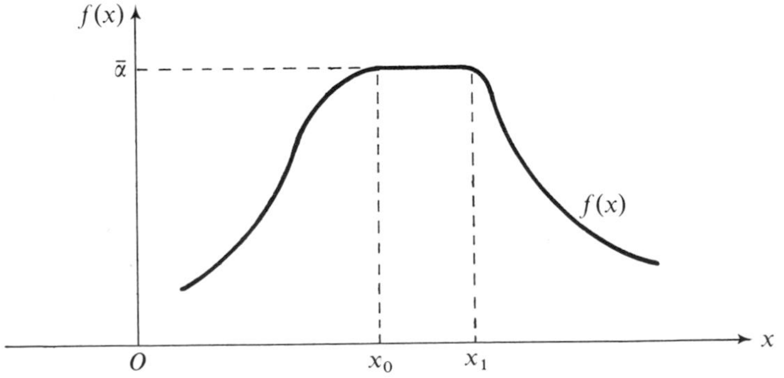

Figure 1.6

that for any $\alpha > \bar{\alpha}$, $A_f(\alpha) = \emptyset$, which is of course strictly convex). However, for all $\lambda \in (0, 1)$, $f((1-\lambda)x_0 + \lambda x_1) = \bar{\alpha}$. Hence, $f(x)$ is by no means strictly quasi-concave in the sense of Definition 16.

We can now proceed to a fundamental theorem of nonlinear programming in the differentiable case.

Theorem 23 (Modified Arrow–Enthoven [1961]). *Let $f(\boldsymbol{x})$ and $g_i(\boldsymbol{x})$, $i = 1, \ldots, m$, in the nonlinear programming problem* (I) *be differentiable on an open convex set X. Suppose, further, that $g_i(\boldsymbol{x})$, $i = 1, \ldots, m$, are quasi-concave on X. If $\boldsymbol{x}_0$ is optimal for* (I) *then there exists $\boldsymbol{u}_0 \geqq \boldsymbol{0}$ such that*

$$\nabla L_{\boldsymbol{x}}(\boldsymbol{x}_0, \boldsymbol{u}_0) = \boldsymbol{0}', \tag{58}$$

$$\nabla L_{\boldsymbol{u}}(\boldsymbol{x}_0, \boldsymbol{u}_0) \geqq \boldsymbol{0}, \tag{59}$$

$$\boldsymbol{u}_0' \cdot \nabla L_{\boldsymbol{u}}(\boldsymbol{x}_0, \boldsymbol{u}_0) = 0, \tag{60}$$

where

$$\nabla L_{\boldsymbol{x}}(\boldsymbol{x}, \boldsymbol{u}) \equiv \nabla f(\boldsymbol{x}) + \boldsymbol{u}' \cdot \begin{pmatrix} \nabla g_1(\boldsymbol{x}) \\ \vdots \\ \nabla g_m(\boldsymbol{x}) \end{pmatrix}$$

and

$$\nabla L_{\boldsymbol{u}}(\boldsymbol{x}, \boldsymbol{u}) \equiv \boldsymbol{g}(\boldsymbol{x}).$$

Conversely, let $\boldsymbol{x}_0 \in X$ and $\boldsymbol{u}_0 \geqq \boldsymbol{0}$ be the vectors satisfying (58) *through* (60). *If $f(\boldsymbol{x})$ is pseudoconcave at $\boldsymbol{x}_0$, that is, $f(\boldsymbol{x}) > f(\boldsymbol{x}_0)$ implies*

$$\nabla f(\boldsymbol{x}_0) \cdot (\boldsymbol{x} - \boldsymbol{x}_0) > 0,$$

then $\boldsymbol{x}_0$ is optimal for (I).

Proof. From the optimality of $\boldsymbol{x}_0$, (59) follows at once. If $\nabla f(\boldsymbol{x}_0) = \boldsymbol{0}$ then $\boldsymbol{x}_0$ and $\boldsymbol{u}_0 = \boldsymbol{0}$ clearly satisfy (58) through (60). Hence we can safely assume that $\nabla f(\boldsymbol{x}_0) \neq \boldsymbol{0}$. Let $\boldsymbol{x}_1$ and $\boldsymbol{x}_2$ be any elements of the feasible set $F = \{\boldsymbol{x} \in X : \boldsymbol{g}(\boldsymbol{x}) \geqq \boldsymbol{0}\}$. Then it is clear that

$$\min\{g_j(\boldsymbol{x}_1), g_j(\boldsymbol{x}_2)\} \geqq 0 \qquad (j = 1, \ldots, m).$$

Hence, from the assumed quasi-concavity of $g_j(\boldsymbol{x})$,

$$g_j(\lambda \boldsymbol{x}_1 + (1-\lambda)\boldsymbol{x}_2) \geqq 0 \qquad (j = 1, \ldots, m) \qquad \text{for any } \lambda \in [0, 1],$$

which shows the convexity of F. Since F is convex and $\boldsymbol{x}_0$ is optimal for (I), for any $\boldsymbol{x} \in F$ and every $\lambda \in [0, 1]$ we have

$$f(\boldsymbol{x}_0) \geqq f(\lambda \boldsymbol{x} + (1-\lambda)\boldsymbol{x}_0) = f(\boldsymbol{x}_0 + \lambda(\boldsymbol{x} - \boldsymbol{x}_0)).$$

In view of the differentiability of f and the optimality of $\boldsymbol{x}_0$, we have further

$$\nabla f(\boldsymbol{x}_0) \cdot (\boldsymbol{x} - \boldsymbol{x}_0) + \alpha(\boldsymbol{x}_0, \lambda(\boldsymbol{x} - \boldsymbol{x}_0)) \cdot \|\boldsymbol{x} - \boldsymbol{x}_0\| \leqq 0 \quad \text{for any } \lambda \in (0, 1].$$

Letting $\lambda \to 0$ in the above inequality we easily obtain

$$\nabla f(\boldsymbol{x}_0) \cdot (\boldsymbol{x} - \boldsymbol{x}_0) \leqq 0 \quad \text{for any } \boldsymbol{x} \in F. \tag{61}$$

Define $I_0 = \{i : 1 \leq i \leq m \text{ and } g_i(\boldsymbol{x}_0) = 0\}$. Suppose that $I_0 = \varnothing$. Then $\boldsymbol{g}(\boldsymbol{x}_0) > \mathbf{0}$, which implies that there exists $\delta > 0$ such that if $\boldsymbol{x} \in B(\boldsymbol{x}_0, \delta)$ then $\boldsymbol{g}(\boldsymbol{x}) > \mathbf{0}$, since the assumed differentiability of $g_i(\boldsymbol{x})$, $i = 1, \ldots, m$, implies their continuity. Now choose $\mu > 0$ so small that $\boldsymbol{x}_0 \pm \mu \boldsymbol{e}_j \in B(\boldsymbol{x}_0, \delta)$, $j = 1, \ldots, m$. Then from (61) we can assert that

$$\nabla f(\boldsymbol{x}_0) \leqq \mathbf{0} \tag{62}$$

and that

$$\nabla f(\boldsymbol{x}_0) \geqq \mathbf{0}, \tag{62'}$$

from which $\nabla f(\boldsymbol{x}_0) = \mathbf{0}$ follows immediately. This is absurd since we assume that $\nabla f(\boldsymbol{x}_0) \neq \mathbf{0}$, and hence $I_0 \neq \varnothing$. Then

$$g_i(\boldsymbol{x}) \geqq g_i(\boldsymbol{x}_0) = 0 \quad \text{for all } i \in I_0 \text{ and all } \boldsymbol{x} \in F.$$

Therefore, from Lemma 14(5),

$$\nabla g_i(\boldsymbol{x}_0) \cdot (\boldsymbol{x} - \boldsymbol{x}_0) \geqq 0 \quad \text{for all } i \in I_0 \text{ and all } \boldsymbol{x} \in F. \tag{63}$$

In view of Theorem 3, (61) and (63) are valid if and only if there exist $v_i \geqq 0$, $i \in I_0$, such that

$$\left.\frac{\partial f}{\partial x_j}\right|_{\boldsymbol{x}_0} + \sum_{i \in I_0} v_i \left.\frac{\partial g_i}{\partial x_j}\right|_{\boldsymbol{x}_0} = 0 \qquad (j = 1, \ldots, n).$$

Let $\boldsymbol{u}_0 = (u_{i0}, \ldots, u_{m0})$ be a vector obtained by putting $u_{i0} = v_i$ for $i \in I_0$ and $u_{i0} = 0$ for $i \notin I_0$. Then we can easily show that

$$\boldsymbol{u}_0' \cdot \nabla L_{\boldsymbol{u}}(\boldsymbol{x}_0, \boldsymbol{u}_0) = \boldsymbol{u}_0' \cdot \boldsymbol{g}(\boldsymbol{x}_0) = 0,$$

$$\nabla L_{\boldsymbol{x}}(\boldsymbol{x}_0, \boldsymbol{u}_0) = \mathbf{0}'.$$

From (59), (60) and the nonnegativity of $\boldsymbol{u}_0$,

$$u_{i0} = 0 \quad \text{for any } i \notin I_0. \tag{64}$$

Let $\boldsymbol{x}$ be any feasible vector. Then $g_i(\boldsymbol{x}) \geqq 0 = g_i(\boldsymbol{x}_0)$ for any $i \in I_0$. Hence, by Lemma 14(5),

$$\nabla g_i(\boldsymbol{x}_0) \cdot (\boldsymbol{x} - \boldsymbol{x}_0) \geqq 0 \quad \text{for any } i \in I_0 \text{ and any } \boldsymbol{x} \in F. \tag{65}$$

Multiplying both sides of (58) by $(\boldsymbol{x} - \boldsymbol{x}_0)$, and applying (64) and (65), we obtain

$$\nabla f(\boldsymbol{x}_0) \cdot (\boldsymbol{x} - \boldsymbol{x}_0) = - \sum_{i \in I_0} u_{i0} \cdot \nabla g_i(\boldsymbol{x}_0) \cdot (\boldsymbol{x} - \boldsymbol{x}_0) \leqq 0.$$

Therefore,

$$\nabla f(\boldsymbol{x}_0) \cdot (\boldsymbol{x} - \boldsymbol{x}_0) \leqq 0 \quad \text{for every } \boldsymbol{x} \in F. \tag{66}$$

Since $f(x)$ is pseudoconcave at x_0, (66) implies that

$$f(x_0) \geqq f(x) \quad \text{for every } x \in F,$$

which establishes the optimality of x_0. □

Our next task is to observe how Theorem 23 is modified if the domain X is a closed convex set in $\mathbb{R}^n$. (Recall that the nonnegative orthant $\mathbb{R}^n_+$, with which we are frequently concerned in economic analysis, is a special case of a closed convex domain. In fact $\mathbb{R}^n_+$ is a closed convex set in $\mathbb{R}^n$ with full dimensionality.)

Theorem 23′. *Let X be a closed convex set of $\mathbb{R}^n$ and let $f(x)$ and $g_i(x)$, $i = 1, \ldots, m$, be as in Theorem 23. If x_0 of X is optimal for* (I) *then there exists $u_0 \geqq 0$ such that*

$$\nabla L_x(x_0, u_0) \cdot (x - x_0) \leqq 0 \quad \text{for any } x \in F, \tag{58'}$$

$$\nabla L_u(x_0, u_0) \geqq 0, \tag{59'}$$

$$u_0' \cdot \nabla L_u(x_0, u_0) = 0. \tag{60'}$$

Conversely, let $x_0 \in X$ and $u_0 \geqq 0$ satisfy (58′)–(60′), *let f be quasi-concave on X, and let one of the following conditions be satisfied:*

i. $\nabla f(x_0) \cdot (x_1 - x_0) < 0$ *for some* $x_1 \subset X$;
ii. $\nabla f(x_0) \cdot (x_2 - x_0) \neq 0$ *for some* $x_2 \in X$, *and $f(x)$ is twice differentiable in some neighborhood of x_0;*
iii. *$f(x)$ is pseudoconcave at x_0.*

Then x_0 is optimal for (I).

PROOF. In view of the optimality of x_0, it suffices to verify the existence of $u_0 \geqq 0$ satisfying (58′) and (60′). Moreover, note that (61) remains valid in this case. Let I_0 be as in the proof of the previous theorem. If $I_0 = \varnothing$ then the choice of $u_0 = 0$ suffices to establish the assertion. Suppose therefore that I_0 is nonempty. Then, proceeding as in the proof of Theorem 23, there exists $u_0 \geqq 0$ for which the conditions (58)–(60) are satisfied. Noticing that (60′) is identical with (60) and that (58′) follows automatically from (58), our assertion is evident.

We turn to the proof of the converse assertion. From the nonnegativity of u_0, the quasi-concavity of each $g_i(x)$ and (60′), it follows that

$$u_0' \nabla g(x_0) \cdot (x - x_0) \geqq 0 \quad \text{for all } x \in F.$$

This, together with (58′), yields (66) again.

(i) For any $x \in F$ and x_1 in condition (i), define

$$x(\lambda) = (1 - \lambda)x + \lambda x_1.$$

Then a direct calculation and the fact that $(\boldsymbol{x}_1 - \boldsymbol{x}) = (\boldsymbol{x}_1 - \boldsymbol{x}_0) - (\boldsymbol{x} - \boldsymbol{x}_0)$ show that

$$\nabla f(\boldsymbol{x}_0)\cdot(\boldsymbol{x}(\lambda) - \boldsymbol{x}_0) = (1-\lambda)\cdot\nabla f(\boldsymbol{x}_0)\cdot(\boldsymbol{x} - \boldsymbol{x}_0) + \lambda\cdot\nabla f(\boldsymbol{x}_0)(\boldsymbol{x}_1 - \boldsymbol{x}_0) < 0 \quad \text{for any } \lambda \in (0, 1].$$

Taking Lemma 14(5) into account, the above is valid only if $f(\boldsymbol{x}(\lambda)) < f(\boldsymbol{x}_0)$. As λ tends to zero, $\boldsymbol{x}(\lambda)$ approaches $\boldsymbol{x}$. The optimality of $\boldsymbol{x}_0$ then follows from the continuity of $f(\boldsymbol{x})$.

(ii) Since under condition (i) there remains nothing to prove, we can safely assume that

$$\nabla f(\boldsymbol{x}_0)(\boldsymbol{x} - \boldsymbol{x}_0) \geqq 0 \quad \text{for all } \boldsymbol{x} \in X.$$

Hence condition (ii) should be altered to read:

(ii′) $\nabla f(\boldsymbol{x}_0)\cdot(\boldsymbol{x}_2 - \boldsymbol{x}_0) > 0$ for some $\boldsymbol{x}_2 \in X$.

Define the function

$$\psi(\alpha, \beta) = f(\boldsymbol{x}_0 + \alpha(\boldsymbol{x} - \boldsymbol{x}_0) + \beta(\boldsymbol{x}_2 - \boldsymbol{x}_0)) - f(\boldsymbol{x}_0)$$

for any $\boldsymbol{x} \in X$ and for any $(\alpha, \beta) \in \mathbb{R}^2$. For simplicity, denote $\boldsymbol{x}_0 + \alpha_i(\boldsymbol{x} - \boldsymbol{x}_0) + \beta_i(\boldsymbol{x}_2 - \boldsymbol{x}_0)$ by $\boldsymbol{\xi}_i$, $i = 1, 2$. Then, for any $\lambda \in [0, 1]$, we have

$$\begin{aligned} &\psi((1-\lambda)(\alpha_1, \beta_1) + \lambda(\alpha_2, \beta_2)) = \psi((1-\lambda)\alpha_1 + \lambda\alpha_2, (1-\lambda)\beta_1 + \lambda\beta_2) \\ &= f((1-\lambda)\boldsymbol{x}_0 + \lambda\boldsymbol{x}_0 + ((1-\lambda)\alpha_1 + \lambda\alpha_2)(\boldsymbol{x} - \boldsymbol{x}_0) \\ &\qquad\qquad + ((1-\lambda)\beta_1 + \lambda\beta_2)(\boldsymbol{x}_2 - \boldsymbol{x}_0)) - f(\boldsymbol{x}_0) \\ &= f((1-\lambda)\xi_1 + \lambda\xi_2) - f(\boldsymbol{x}_0) \\ &\geqq \min\{f(\xi_1), f(\xi_2)\} - f(\boldsymbol{x}_0) \qquad [\text{from the quasi-concavity of } f] \\ &= \min\{\psi(\alpha_1, \beta_1), \psi(\alpha_2, \beta_2)\} \qquad [\text{from the definition of } \psi(\cdot)] \end{aligned}$$

The function $\psi(\alpha, \beta)$ is thus quasi-concave. Furthermore, by direct differentiation,

$$\frac{\partial\psi(\alpha, \beta)}{\partial\alpha} \equiv \psi_\alpha(\alpha, \beta) = \nabla f(\xi)(\boldsymbol{x} - \boldsymbol{x}_0)$$

and

$$\frac{\partial\psi(\alpha, \beta)}{\partial\beta} \equiv \psi_\beta(\alpha, \beta) = \nabla f(\xi)(\boldsymbol{x}_2 - \boldsymbol{x}_0),$$

where $\xi = \boldsymbol{x}_0 + \alpha(\boldsymbol{x} - \boldsymbol{x}_0) + \beta(\boldsymbol{x}_2 - \boldsymbol{x}_0)$.

Inspection then yields

$$\psi(0, 0) = 0,$$

$$\psi(1, 0) = f(\boldsymbol{x}) - f(\boldsymbol{x}_0),$$

and

$$\psi_\beta(0, 0) = \nabla f(\boldsymbol{x}_0)(\boldsymbol{x}_2 - \boldsymbol{x}_0) > 0 \qquad [\text{because of (ii′)}].$$

Hence it suffices to verify that

$$\psi(1, 0) \leqq 0.$$

Suppose the contrary. Then, $\psi(1, 0) > 0 = \psi(0, 0)$. In view of the quasi-concavity of $\psi(\alpha, \beta)$ just proved, it is obvious that

$$\psi(\lambda, 0) \geqq \psi(0, 0) = 0 \quad \text{for any } \lambda \in [0, 1].$$

Since $\psi(\cdot)$ is continuous and since $\psi(1, 0)$ is assumed to be positive, the set $K = \{\varepsilon \in (0, 1]: \psi(\lambda, 0) > 0 \text{ for all } \lambda \in (\varepsilon, 1]\}$ is surely nonvacuous. Evidently K is bounded below. Hence $\varepsilon_0 = \inf_{\varepsilon \in K} \varepsilon$ does exist. Clearly, $\varepsilon_0 \geqq 0$ and $\psi(\varepsilon_0, 0) = 0$. Suppose that $\psi(\varepsilon_0, 0) > 0$. Then, there is an $\varepsilon_1 \in K$ such that $\varepsilon_1 < \varepsilon_0$, contradicting the definition of ε_0. If $\varepsilon_0 = 0$ then, taking into account the earlier observation that $\psi_\beta(0, 0) > 0$, there can be found a function $\alpha(\beta)$ defined on some neighborhood of zero, with the properties that $\psi(\alpha(\beta), 0) = \psi(0, \beta)$ and that $\lim_{\beta \to 0} \alpha(\beta) = 0$.[6] Applying Lemma 14(5), we obtain

$$\left(\frac{1}{\beta}\right) \cdot \psi_\alpha(0, \beta) \cdot \alpha(\beta) \geqq \psi_\beta(0, \beta).$$

Letting $\beta \to 0$, the right-hand side of the above tends to $\psi_\beta(0, 0)$ which is positive, while the left-hand side converges to zero because the assumed twice differentiability of $f(\mathbf{x})$ ensures that $(1/\beta) \cdot \psi_\alpha(0, \beta)$ converges to a finite limit as β approaches zero and because $\lim_{\beta \to 0} \alpha(\beta) = 0$. This is absurd. Hence $0 < \varepsilon_0 < 1$, which further implies that

$$\psi(\lambda, 0) = 0 \quad \text{for any } \lambda \in [0, \varepsilon_0]$$

and that

$$\psi(\lambda, 0) > 0 \quad \text{for any } \lambda \in (\varepsilon_0, 1].$$

Since the above inequality is clear from the definition of ε_0, it suffices to prove the first equality alone. Suppose the contrary. Then there exists a $\lambda_0 \in [0, \varepsilon_0)$ such that $\psi(\lambda_0, 0) > 0$, for we have already shown that $\psi(\lambda, 0)$ is nonnegative on the closed interval $[0, 1]$. By virtue of the quasi-concavity of $\psi(\alpha, \beta)$ again, we have

$$\psi((1 - \lambda)\lambda_0 + \lambda\varepsilon, 0) > 0 \quad \text{for any } \lambda \in [0, 1] \text{ and any } \varepsilon \in (\varepsilon_0, 1].$$

However, for $\lambda \in [0, 1]$ such that $\varepsilon_0 = (1 - \lambda)\lambda_0 + \lambda\varepsilon$, we arrive at a self-contradiction:

$$0 = \psi(\varepsilon_0, 0) = \psi((1 - \lambda)\lambda_0 + \lambda\varepsilon, 0) > 0.$$

Since $\psi_\beta(0, 0)$ is positive, we can again find a function $\alpha(\beta) \geqq \varepsilon_0$ defined on some neighborhood of zero such that $\psi(\alpha(\beta), 0) = \psi(0, \beta)$ and such that $\lim_{\beta \to 0} \alpha(\beta) = \varepsilon_0$.[7] Define $\theta = 1 - (\varepsilon_0/\alpha(\beta))$. Then clearly $\theta \in [0, 1]$. Hence the quasi-concavity of $\psi(\alpha, \beta)$, together with the fact that $\psi(\alpha(\beta), 0) = \psi(0, \beta)$, assures us that

$$\psi((1 - \theta) \cdot \alpha(\beta), \theta\beta) = \psi(\varepsilon_0, \theta\beta) \geqq \psi(0, \beta).$$

Recalling that $\psi(\varepsilon_0, 0) = 0$, it follows from the Mean Value Theorem that

$$\psi(\varepsilon_0, \theta\beta) = \psi(\varepsilon_0, \theta\beta) - \psi(\varepsilon_0, 0) = \psi_\beta(\varepsilon_0, \mu\theta\beta)\cdot(\theta\beta) \quad \text{for some } \mu \in (0, 1).$$

Therefore,

$$\psi_\beta(\varepsilon_0, \mu\theta\beta) \geqq \left(\frac{1}{\theta}\right)\cdot\left\{\frac{\psi(0, \beta)}{\beta}\right\}.$$

Note further than μ and θ approach zero as β approaches zero. Hence the right-hand side of the above inequality tends to infinity as β approaches zero, since $\lim_{\beta\to 0}(\psi(0, \beta)/\beta) = \psi_\beta(0, 0) > 0$. On the other hand, by hypothesis, the left-hand side approaches a finite value of $\psi_\beta(\varepsilon_0, 0)$. This is a contradiction, whence $\boldsymbol{x}_0$ is optimal.

(iii) The proof of the third assertion is completely similar to that of the corresponding assertion of Theorem 23 and hence is omitted. □

In Theorems 21, 23, and 23′ we have been interested in nonlinear programming in a convex domain X. In economic analysis, variables are often intrinsically nonnegative. As a special case of great interest, therefore, we may take X to be the n-dimensional nonnegative orthant $\mathbb{R}^n_+$. We are then led to the Kuhn–Tucker Theorem in the differentiable case, which we state as a corollary to Theorem 23.

Corollary 4. *Consider the nonlinear programming problem*

$$\text{(I}'\text{)} \qquad \max_{\boldsymbol{x}} f(\boldsymbol{x}) \quad \textit{subject to } \boldsymbol{g}(\boldsymbol{x}) \geqq \boldsymbol{0}, \quad \boldsymbol{x} \geqq \boldsymbol{0},$$

where $f(\boldsymbol{x})$ and $g_i(\boldsymbol{x})$, $i = 1, \ldots, m$, are real-valued differentiable functions defined on $\mathbb{R}^n_+$, and $g_i(\boldsymbol{x})$, $i = 1, \ldots, m$, are quasi-concave on $\mathbb{R}^n_+$. Let $\boldsymbol{x}_0$ be an optimal vector for (I′). *Then there exists $\boldsymbol{u}_0 \geqq \boldsymbol{0}$ such that*

$$\nabla L_{\boldsymbol{x}}(\boldsymbol{x}_0, \boldsymbol{u}_0) \leqq \boldsymbol{0}, \tag{67}$$

$$\nabla L_{\boldsymbol{x}}(\boldsymbol{x}_0, \boldsymbol{u}_0)\cdot\boldsymbol{x}_0 = 0, \tag{68}$$

$$\boldsymbol{g}(\boldsymbol{x}_0) \geqq \boldsymbol{0}, \tag{69}$$

$$\boldsymbol{u}_0' \cdot \boldsymbol{g}(\boldsymbol{x}_0) = 0. \tag{70}$$

Conversely, let $\boldsymbol{x}_0$ and $\boldsymbol{u}_0$ be nonnegative vectors satisfying (67)–(70), *let $f(\boldsymbol{x})$ be quasi-concave, and let one of the following conditions be satisfied:*

i. $\partial f/\partial x_k|_{\boldsymbol{x}_0} < 0$ *for some k;* (71)
ii. $\nabla f(\boldsymbol{x}_0)\boldsymbol{x}_0 > 0$;
iii. $\nabla f(\boldsymbol{x}_0) \neq 0$ *and $f(\boldsymbol{x})$ is twice differentiable in some neighborhood of $\boldsymbol{x}_0$;*
iv. *$f(\boldsymbol{x})$ is pseudoconcave at $\boldsymbol{x}_0$.*

Then $\boldsymbol{x}_0$ is optimal.

PROOF. Define $\Gamma = \{x \in \mathbb{R}^n_+ : g(x) \geqq 0\}$ and $J_0 = \{j \in N : x_{j_0} = 0\}$, and let I_0 be as in the proof of Theorem 23, where $N = \{1, \ldots, n\}$. If $\nabla f(x_0) = 0$ or $I_0 = \varnothing$, the choice $u_0 = 0$ will do for our present purpose. Moreover, it is immediate that

$$\nabla L_x(x_0, 0) = \nabla f(x_0) \tag{67'}$$

and that

$$\nabla L_x(x_0, 0)x_0 = \nabla f(x_0) \cdot x_0. \tag{68'}$$

Hence, if $\nabla f(x_0) = 0$ the corollary is rather trivial. We therefore suppose that $\nabla f(x_0) \neq 0$ and that $I_0 = \varnothing$. Then, by the reasoning employed in the proof of Theorem 23 and by (67′),

$$\nabla L_x(x_0, 0) = \nabla f(x_0) \leqq 0,$$

which, coupled with the assumed nonnegativity of x_0 and (68′), further implies that

$$\nabla L_x(x_0, 0) \cdot x_0 = \nabla f(x_0) \cdot x_0 \leqq 0.$$

If $\nabla f(x_0)x_0 = 0$, there remains nothing to prove. Suppose therefore that $\nabla f(x_0)x_0 < 0$. Then there exists an index i_0 such that $\nabla f_{i_0}(x_0) < 0$ and $x^0_{i_0} > 0$ where

$$\nabla f_{i_0}(x_0) = \left.\frac{\partial f}{\partial x_{i_0}}\right|_{x_0}$$

and $x^0_{i_0}$ is the i_0th element of x_0. From these inequalities, the hypothesis that $I_0 = \varnothing$, and the assumed continuity of $g_i(x_0)$ it follows that there exists $\mu > 0$ such that $x \equiv x_0 - \mu e_{i_0} \in \Gamma$. Taking into account (61) and the fact that $\mu > 0$, we can assert that

$$\nabla f_{i_0}(x_0) \geqq 0,$$

contradicting the hypothesis that $\nabla f_{i_0}(x_0) < 0$. It remains to prove the corollary on the assumption that $\nabla f(x_0) \neq 0$ and that $I_0 \neq \varnothing$. Then, from (63)

$$\nabla g_i(x_0) \cdot (x - x_0) \equiv \nabla g_i(x_0)y \geqq 0 \quad \text{for all } i \in I_0 \text{ and } x \in \Gamma. \tag{63'}$$

Assuming that J_0 has m_0 components, with m_0 possibly zero, we define an $m_0 \times n$ matrix D with elements d_{ij} such that if $j \in J_0$ then $d_{ij} = 1$ and such that, otherwise, $d_{ij} = 0$. In addition, we let A_1 be the coefficient matrix of the inequalities (63′). Since $y_j = x_j - x_{j0} \geqq 0$ for all $j \in J_0$, (61) and (63′) imply that $-\nabla f(x_0) \cdot y \geqq 0$ for all y satisfying

$$\begin{pmatrix} A_1 \\ D \end{pmatrix} \cdot y \geqq 0.$$

Applying Theorem 3 again, there exist $u_1 \geqq 0$ and $w_1 \geqq 0$ such that

$$A_1'u_1 + D'w_1 = -(\nabla f(x_0))'. \tag{72}$$

Let $\boldsymbol{u}_0$ be an m-dimensional vector obtained by adding suitable zeros to $\boldsymbol{u}_1$. Then, from (72),

$$\mathbf{0} = \nabla f(\boldsymbol{x}_0) + (\boldsymbol{u}_0)'\nabla \boldsymbol{g}(\boldsymbol{x}_0) + \boldsymbol{w}_1' \boldsymbol{D} \tag{73}$$

where

$$\nabla \boldsymbol{g}(\boldsymbol{x}_0) = \begin{pmatrix} \nabla g_1(\boldsymbol{x}_0) \\ \vdots \\ \nabla g_m(\boldsymbol{x}_0) \end{pmatrix}.$$

In view of the nonnegativity of $\boldsymbol{w}_1'\boldsymbol{D}$, (67) follows directly from (73). Postmultiplying (73) by $\boldsymbol{x}_0$, we obtain

$$0 = \boldsymbol{L}_{\boldsymbol{x}}(\boldsymbol{x}_0, \boldsymbol{u}_0) \cdot \boldsymbol{x}_0 + \boldsymbol{w}_1'\boldsymbol{D}\boldsymbol{x}_0 = \boldsymbol{L}_{\boldsymbol{x}}(\boldsymbol{x}_0, \boldsymbol{u}_0) \cdot \boldsymbol{x}_0,$$

since $\boldsymbol{D}\boldsymbol{x}_0 = \mathbf{0}$ from the definition of $\boldsymbol{D}$.

We now turn to the proof of the converse assertion. Let $\boldsymbol{x}$ be any feasible vector. Then (67) and the nonnegativity of $\boldsymbol{x}$ imply that

$$(\nabla f(\boldsymbol{x}_0) + \boldsymbol{u}_0' \cdot \nabla \boldsymbol{g}(\boldsymbol{x}_0)) \cdot \boldsymbol{x} \leqq 0 \quad \text{for any } \boldsymbol{x} \in \Gamma.$$

Taking (68) into account, the inequality just obtained further implies that

$$(\nabla f(\boldsymbol{x}_0) + \boldsymbol{u}_0' \cdot \nabla \boldsymbol{g}(\boldsymbol{x}_0)) \cdot (\boldsymbol{x} - \boldsymbol{x}_0) \leqq 0 \quad \text{for any } \boldsymbol{x} \in \Gamma. \tag{74}$$

On the other hand, (69) and (70), together with the nonnegativity of $\boldsymbol{u}_0$, imply that

$$u_{i0} = 0 \quad \text{for any } i \notin I_0,$$

Consequently, from (65) and (74),

$$\nabla f(\boldsymbol{x}_0) \cdot (\boldsymbol{x} - \boldsymbol{x}_0) \leqq -\sum_{i \in I_0} u_{i0} \cdot (\nabla g_i(\boldsymbol{x}_0) \cdot (\boldsymbol{x} - \boldsymbol{x}_0)) \leqq 0 \quad \text{for any } \boldsymbol{x} \in \Gamma.^8 \tag{75}$$

(i) For any $\boldsymbol{x} \in \Gamma$, let $\boldsymbol{x}(\lambda) = (1 - \lambda)\boldsymbol{x} + \lambda(\boldsymbol{x}_0 + \boldsymbol{e}_k)$. Then, from (71) and (75),

$$\begin{aligned}\nabla f(\boldsymbol{x}_0) \cdot (\boldsymbol{x}(\lambda) - \boldsymbol{x}_0) &= (1 - \lambda)\nabla f(\boldsymbol{x}_0) \cdot (\boldsymbol{x} - \boldsymbol{x}_0) \\ &\quad + \lambda \nabla f(\boldsymbol{x}_0)\boldsymbol{e}_k < 0 \quad \text{for any } \lambda \in (0, 1].\end{aligned} \tag{76}$$

Taking Lemma 14(5) into account, (76) holds only if $f(\boldsymbol{x}(\lambda)) < f(\boldsymbol{x}_0)$. As λ tends to zero, $\boldsymbol{x}(\lambda)$ approaches $\boldsymbol{x}$. Hence, from the continuity of $f(\boldsymbol{x})$,

$$f(\boldsymbol{x}) \leqq f(\boldsymbol{x}_0) \quad \text{for any } \boldsymbol{x} \in \Gamma.$$

(ii) For any $\boldsymbol{x} \in \Gamma$ and $\lambda \in (0, 1]$, we have

$$\nabla f(\boldsymbol{x}_0)((1 - \lambda)\boldsymbol{x} - \boldsymbol{x}_0) = (1 - \lambda)\nabla f(\boldsymbol{x}_0)(\boldsymbol{x} - \boldsymbol{x}_0) - \lambda \nabla f(\boldsymbol{x}_0)\boldsymbol{x}_0 < 0$$

[from (75), $\nabla f(\boldsymbol{x}_0)\boldsymbol{x}_0 > 0$ and $\lambda > 0$]. Again from the reasoning employed in the proof of previous assertion, the optimality of $\boldsymbol{x}_0$ follows at once.

(iii) Without loss, we can assume that neither condition (i) nor condition (ii) holds, for otherwise there would remain nothing to prove. In view of the hypothesis that $\nabla f(\boldsymbol{x}_0) \neq \boldsymbol{0}$, it suffices to verify the assertion under the condition

$$\nabla f(\boldsymbol{x}_0) \geq \boldsymbol{0} \quad \text{and} \quad \nabla f(\boldsymbol{x}_0)\boldsymbol{x}_0 = 0.$$

Let $J = \{i \in N : \nabla f_i(\boldsymbol{x}_0) > 0\}$. Then, from the above condition, J is surely nonempty. Note further that (75), coupled with the present hypothesis and the nonnegativity of $\boldsymbol{x} \in \Gamma$, implies that

$$\nabla f(\boldsymbol{x}_0)\boldsymbol{x} = 0 \quad \text{for any } \boldsymbol{x} \in \Gamma.$$

We can therefore assert that

$$x_i = 0 \quad \text{for all } \boldsymbol{x} \in \Gamma \text{ and for any } i \in J.$$

Let $J = N$. Then, from the above, it is evident that Γ consists of $\boldsymbol{0}$ alone. Hence the optimality of $\boldsymbol{x}_0$ (in fact, $\boldsymbol{x}_0 = \boldsymbol{0}$, in this case) is rather trivial. Henceforth, therefore, we assume that J is a proper subset of N. Then, without loss of generality, we can renumber so that $J = \{m_0 + 1, \ldots, n\}$, where m_0 is of course an integer such that $1 < m_0 < n$. Partition any $\boldsymbol{x}$ of Γ into $(\boldsymbol{y}, \boldsymbol{z})$, where $\boldsymbol{y}$ and $\boldsymbol{z}$ are respectively subvectors of $\boldsymbol{x}$, possessing the first m_0 elements of and the remaining elements of $\boldsymbol{x}$. Corresponding to this partition of $\boldsymbol{x}$, we decompose $\nabla f(\boldsymbol{x})$ so that

$$\nabla f_{\boldsymbol{y}}(\boldsymbol{x}) = (\nabla f_1(\boldsymbol{x}), \ldots, \nabla f_{m_0}(\boldsymbol{x}))$$

and

$$\nabla f_{\boldsymbol{z}}(\boldsymbol{x}) = (\nabla f_{m_0+1}(\boldsymbol{x}), \ldots, \nabla f_n(\boldsymbol{x})).$$

Then

$$\nabla f_{\boldsymbol{y}}(\boldsymbol{x}_0) - \boldsymbol{0}, \qquad \nabla f_{\boldsymbol{z}}(\boldsymbol{x}_0) > \boldsymbol{0}, \qquad \boldsymbol{z}_0 = \boldsymbol{0},$$

where $\boldsymbol{x}_0 = (\boldsymbol{y}_0, \boldsymbol{z}_0)$.

It is also clear that

$$\boldsymbol{z} = \boldsymbol{0} \quad \text{for all } \boldsymbol{x} = (\boldsymbol{y}, \boldsymbol{z}) \in \Gamma.$$

Associated with any $\boldsymbol{x}$ of Γ and any semipositive vector $\bar{\boldsymbol{z}}$ of order $n - m_0$, define the function

$$\kappa(\alpha, \beta) \equiv f(\boldsymbol{y}_0 + \alpha(\boldsymbol{y} - \boldsymbol{y}_0), \beta\bar{\boldsymbol{z}}) - f(\boldsymbol{x}_0)$$

for any $\alpha \in [0, 1]$ and $\beta \geqq 0$, where $\boldsymbol{y}$ is of course the subvector of $\boldsymbol{x}$ consisting of its first m_0 components.

Proceeding as in the proof of Theorem 23′(ii), the function $\kappa(\alpha, \beta)$ is seen to be quasi-concave and we easily obtain

$$\kappa(0, 0) = 0,$$
$$\kappa_\alpha(0, 0) = \nabla f_{\boldsymbol{y}}(\boldsymbol{x}_0) \cdot (\boldsymbol{y} - \boldsymbol{y}_0) = 0 \quad \text{for all } (\boldsymbol{y}, \boldsymbol{0}) \in \Gamma,$$

and

$$\kappa_\beta(0, 0) = \nabla f_{\boldsymbol{z}}(\boldsymbol{x}_0)\bar{\boldsymbol{z}} > 0.$$

The remaining proof is completely analogous to that of Theorem 23′(ii), and hence is omitted.

(iv) The proof of this assertion is exactly the same as that of Theorem 23 and hence is neglected. □

Remark Concerning Theorems 23 and 23′ and Corollary 4. (i) Suppose that f is concave and differentiable on E. Then, by Lemma 14(2),

$$\nabla f(\boldsymbol{x}_0)\cdot(\boldsymbol{x}-\boldsymbol{x}_0) \geqq f(\boldsymbol{x}) - f(\boldsymbol{x}_0).$$

Hence, if $f(\boldsymbol{x}) > f(\boldsymbol{x}_0)$ then $\nabla f(\boldsymbol{x}_0)\cdot(\boldsymbol{x}-\boldsymbol{x}_0) > 0$. In other words, the concavity of f implies its pseudoconcavity. Therefore, the condition that f is pseudoconcave at $\boldsymbol{x}_0$ is a tiny generalization of the condition that f is concave, which Arrow–Enthoven [1961; Theorems 1 and 3] employed.

(ii) It may be useful to clarify the relationship among the various kinds of concavity we have so far mentioned. In general, the following implications are obvious.

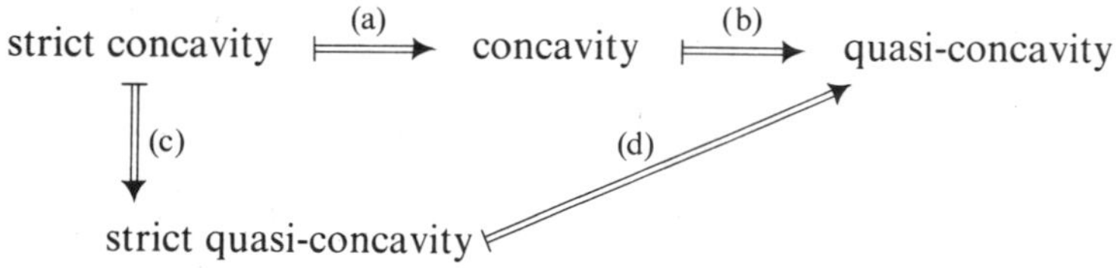

If, in addition, f is differentiable then we can add:

$$\text{concavity} \overset{(e)}{\Longrightarrow} \text{pseudoconcavity}.$$

As a straightforward application of the corollary just shown, we characterize a differentiable quasi-concave function in terms of its bordered Hessian. Let $f(\boldsymbol{x})$ be a real-valued function which is twice continuously differentiable on its domain. Then,

$$D = \det\begin{pmatrix} 0 & f_1 & f_2 & \cdots & f_n \\ f_1 & f_{11} & f_{12} & \cdots & f_{1n} \\ \vdots & \vdots & \vdots & & \vdots \\ f_n & f_{n1} & f_{n2} & \cdots & f_{nn} \end{pmatrix}$$

is said to be the bordered Hessian of f or simply the bordered Hessian. Here $f_i = \partial f/\partial x_i|_{\boldsymbol{x}}$ and $f_{ij} = \partial f_i/\partial x_j|_{\boldsymbol{x}}$, $i, j = 1, \ldots, n$. Moreover, by D_r we denote the principal minor of D consisting of its first $r + 1$ rows and columns, that is,

$$D_r = \det\begin{pmatrix} 0 & f_1 & \cdots & f_r \\ f_1 & f_{11} & \cdots & f_{1r} \\ \vdots & \vdots & & \vdots \\ f_r & f_{r1} & \cdots & f_{rr} \end{pmatrix} \qquad (r = 1, \ldots, n).$$

We now state and prove:

Theorem 24 (Arrow–Enthoven [1961: Theorem 5]). *Let $f(\boldsymbol{x})$ be a real-valued and twice continuously differentiable function defined on the nonnegative orthant $\mathbb{R}^n_+$ with $n \geqq 2$. Suppose that*

$$(-1)^r D_r > 0 \qquad (r = 1, \ldots, n) \quad \textit{everywhere on } \mathbb{R}^n_+.$$

Then, $f(\boldsymbol{x})$ is quasi-concave on its domain.

Conversely, the quasi-concavity of $f(\boldsymbol{x})$ implies that

$$(-1)^r D_r \geqq 0 \quad \textit{for } r = 1, \ldots, n \textit{ and for any } \boldsymbol{x} \in \mathbb{R}^n_+.$$

PROOF. Let $\boldsymbol{x}_0$ be any point of $\mathbb{R}^n_+$ and by f_i^0, f_{ij}^0, and D_r^0 respectively denote f_i, f_{ij}, and D_r evaluated at $\boldsymbol{x}_0$. Consider further the following problem:

$$\text{(P)} \qquad \begin{aligned} &\max_{\boldsymbol{x}} f(\boldsymbol{x}) \\ &\text{subject to } \nabla f(\boldsymbol{x}_0)\boldsymbol{x} = \nabla f(\boldsymbol{x}_0)\boldsymbol{x}_0 \\ &\text{and } \boldsymbol{x} \geqq \boldsymbol{0}. \end{aligned}$$

Since, by hypothesis, $D_1^0 = -(f_1^0)^2 < 0$, $f_1^0 \neq 0$. Hence,

$$x_1 = \left(\frac{1}{f_1^0}\right) \cdot \left(\nabla f(\boldsymbol{x}_0)\boldsymbol{x}_0 - \sum_{j \neq 1} f_i^0 x_j\right).$$

Eliminating x_1, the function $f(\boldsymbol{x})$ reduces to:

$$F(\boldsymbol{y}) = f\left(\left(\frac{1}{f_1^0}\right) \cdot \left(\nabla f(\boldsymbol{x}_0)\boldsymbol{x}_0 - \sum_{j \neq 1} f_j^0 x_j\right), x_2, \ldots, x_n\right),$$

where $\boldsymbol{y} = (x_2, \ldots, x_n)$. A direct differentiation then yields

$$\left.\frac{\partial F}{\partial x_i}\right|_{\boldsymbol{y}} \equiv F_i = f_1 \cdot \left(-\frac{f_i^0}{f_1^0}\right) + f_i \qquad (i = 2, \ldots, n),$$

$$\left.\frac{\partial F_i}{\partial x_j}\right|_{\boldsymbol{y}} \equiv F_{ij} = \left\{f_{ij} - \left(\frac{f_i^0}{f_1^0}\right) f_{1j}\right\} - \left(\frac{f_j^0}{f_1^0}\right) \cdot \left(f_{i1} - \left(\frac{f_i^0}{f_1^0}\right) f_{11}\right) \ (i, j = 2, \ldots, n).$$

From these and a direct but tedious calculation it follows that

$$D_r = -(f_1^0)^2 \det \begin{pmatrix} F_{22} & \cdots & F_{2r} \\ \vdots & & \vdots \\ F_{r2} & \cdots & F_{rr} \end{pmatrix} \qquad (r = 2, \ldots, n).$$

Therefore the present hypothesis is equivalent to:

$$(-1)^{r+1}(f_1^0)^2 \det \begin{pmatrix} F_{22} & \cdots & F_{2r} \\ \vdots & & \vdots \\ F_{r2} & \cdots & F_{rr} \end{pmatrix} > 0 \qquad (r = 2, \ldots, n).$$

This, together with Lemma 4 of Chapter 4, implies that the Hessian of $F(\mathbf{y})$ is negative definite. Hence, with the aid of Lemma 15(2), $F(\mathbf{y})$ is strictly concave. Moreover,

$$F_i^0 = f_1^0\left(-\frac{f_i^0}{f_1^0}\right) + f_i^0 = 0 \quad \text{for } i = 2, \ldots, n.$$

Applying Lemma 15(1), $\mathbf{x}_0$ is seen to be a global optimum for (P). Let (P′) be the constrained maximization problem obtained from (P) by replacing its equality constraints by $\nabla f(\mathbf{x}_0)\mathbf{x}_0 \geqq \nabla f(\mathbf{x}_0)\mathbf{x}$. We shall show that $\mathbf{x}_0$ is optimal also for the problem (P′). Suppose the contrary. Then there exists an $\mathbf{x}_1 \geqq \mathbf{0}$ such that $f(\mathbf{x}_1) > f(\mathbf{x}_0)$ and such that $\nabla f(\mathbf{x}_0)\mathbf{x}_0 > \nabla f(\mathbf{x}_0)\mathbf{x}_1$. Define the function

$$g(\lambda) = f((1 - \lambda)\mathbf{x}_0 + \lambda \mathbf{x}_1)$$

for any $\lambda \in [0, 1]$.

Then

$$g(1) = f(\mathbf{x}_1) > f(\mathbf{x}_0) = g(0),$$

$$g'(\lambda) = \nabla f(\mathbf{x}(\lambda))(\mathbf{x}_1 - \mathbf{x}_0) \quad \text{for all } \lambda \in [0, 1],$$

and in particular

$$g'(0) = \nabla f(\mathbf{x}_0)(\mathbf{x}_1 - \mathbf{x}_0) < 0,$$

where of course $\mathbf{x}(\lambda) = (1 - \lambda)\mathbf{x}_0 + \lambda\mathbf{x}_1$. Since $g'(0)$ is negative and $g(1) > g(0)$, and since $g(\lambda)$ is continuous on the closed interval $[0, 1]$, there exists a $\lambda_0 \in (0, 1)$ with the properties that $g'(\lambda_0) = 0$ and $g(\lambda_0) \leqq g(\lambda)$ for all $\lambda \in [0, 1]$. In view of the fact that $0 < \lambda_0 < 1$, we can choose an $h > 0$ so small that $(\lambda_0 + h) \in (0, 1)$. Consequently, $\mathbf{x}(\lambda_0 + h) - \mathbf{x}(\lambda_0) = h(\mathbf{x}_1 - \mathbf{x}_0)$. It therefore follows that

$$\nabla f(\mathbf{x}(\lambda_0)) \cdot (\mathbf{x}(\lambda_0 + h) - \mathbf{x}(\lambda_0)) = 0.$$

Hence, by the foregoing argument, $g(\lambda_0 + h) = f(\mathbf{x}(\lambda_0 + h)) < f(\mathbf{x}(\lambda_0)) = g(\lambda_0)$, contradicting our earlier observation that $g(\lambda)$ attains its minimum at λ_0. Thus the quasi-concavity of $f(\mathbf{x})$ follows directly from Lemma 14(7).

Conversely, let $\mathbf{x}_0$ be any vector of $\mathbb{R}^n_+$. If $\nabla f(\mathbf{x}_0) = \mathbf{0}$ then there remains nothing to prove. Suppose therefore that $\nabla f(\mathbf{x}_0) \neq \mathbf{0}$. Choosing $u_0 = 1$, $\mathbf{x}_0$ and u_0 satisfy conditions (67)–(70) for problem (P′). Since $f(\mathbf{x})$ is now supposed to be quasi-concave as well as twice continuously differentiable, condition (iii) of Corollary 4 is surely met, so that $\mathbf{x}_0$ is a global maximum for (P′), which in turn implies that it is also a local maximum for (P). The conclusion therefore follows from the necessary condition for classical constrained maximization. □

Remarks Concerning Theorem 24. Noting that the nonnegativity of $\mathbf{x}$ plays no essential role in the proof of the first half of the assertion, it is evident that the condition that sgn $D_r = (-1)^r$, $r = 1, \ldots, n$, ensures the quasi-concavity of f defined on a convex domain.

We bring this foundation-laying first chapter to an end by stating a duality theorem of nonlinear programming. From the considerable variety of available duality theorems we have chosen one of fairly simple character. Readers interested in more general duality theorems may consult Karlin [1959; Volume 1, Chapter 7] and Luenberger [1968: Sections 7.8, 7.12].

Theorem 25 (Wolfe [1961]). *Let $f(\boldsymbol{x})$ and $g_i(\boldsymbol{x})$, $i = 1, \ldots, m$, be real-valued differentiable functions defined on a convex domain X and let them be concave on X. Then, with an optimal vector $\boldsymbol{x}_0$ for the nonlinear programming problem* (I), *there is associated a $\boldsymbol{u}_0 \geqq \boldsymbol{0}$ which solves*

$$\text{(II)} \qquad \begin{aligned} &\min_{(\boldsymbol{x}, \boldsymbol{u})} L(\boldsymbol{x}, \boldsymbol{u}) \\ &\textit{subject to } \nabla L_{\boldsymbol{x}}(\boldsymbol{x}, \boldsymbol{u}) = \boldsymbol{0} \\ &\qquad\qquad \boldsymbol{x} \in X, \\ &\qquad\qquad \boldsymbol{u} \geqq \boldsymbol{0}. \end{aligned}$$

Moreover,

$$f(\boldsymbol{x}_0) = L(\boldsymbol{x}_0, \boldsymbol{u}_0). \tag{77}$$

Proof. Since, by assumption and Lemma 14(3), $f(\boldsymbol{x})$ and $\boldsymbol{g}(\boldsymbol{x})$ satisfy the conditions of Theorem 23, there exists $\boldsymbol{u}_0 \geqq \boldsymbol{0}$ fulfilling conditions (58) through (60). Hence the feasible set F_d of problem (II) is surely nonvacuous. Let $(\boldsymbol{x}, \boldsymbol{u})$ be any vector (or pair) of F_d. Then, we obtain

$$\begin{aligned} &L(\boldsymbol{x}_0, \boldsymbol{u}_0) - L(\boldsymbol{x}, \boldsymbol{u}) \\ &= f(\boldsymbol{x}_0) - f(\boldsymbol{x}) - \boldsymbol{u}'\boldsymbol{g}(\boldsymbol{x}) \qquad \text{[by (60)]} \\ &\leqq \nabla f(\boldsymbol{x})(\boldsymbol{x}_0 - \boldsymbol{x}) - \boldsymbol{u}'\boldsymbol{g}(\boldsymbol{x}) \qquad \text{[by Lemma 14(2)]} \\ &\leqq \nabla f(\boldsymbol{x})(\boldsymbol{x}_0 - \boldsymbol{x}) - \boldsymbol{u}'(\boldsymbol{g}(\boldsymbol{x}_0) - \nabla \boldsymbol{g}(\boldsymbol{x})(\boldsymbol{x}_0 - \boldsymbol{x})) \qquad \text{[by Lemma 14(2) and } \boldsymbol{u} \geqq \boldsymbol{0}] \\ &= -\boldsymbol{u}'\boldsymbol{g}(\boldsymbol{x}_0) \leqq 0 \qquad [\nabla L_{\boldsymbol{x}}(\boldsymbol{x}, \boldsymbol{u}) = \boldsymbol{0}, \boldsymbol{u} \geqq \boldsymbol{0} \text{ and } \boldsymbol{g}(\boldsymbol{x}_0) \geqq \boldsymbol{0}]. \end{aligned}$$

In view of (60), the final assertion is immediate. □

Remark Concerning Theorem 25. Problem (II) is called the dual problem of problem (I), which, as with linear programming problems, is said to be the primary problem. The dual problem corresponding to problem (I′) is

$$\text{(II}'\text{)} \qquad \begin{aligned} &\min_{(\boldsymbol{x}, \boldsymbol{u})} L(\boldsymbol{x}, \boldsymbol{u}) \\ &\text{subject to } \nabla L_{\boldsymbol{x}}(\boldsymbol{x}, \boldsymbol{u}) \leqq \boldsymbol{0}, \\ &\qquad\qquad \nabla L_{\boldsymbol{x}}(\boldsymbol{x}, \boldsymbol{u}) \cdot \boldsymbol{x} = 0, \\ &\qquad\qquad \boldsymbol{x} \in \mathbb{R}^n_+, \\ &\qquad\qquad \boldsymbol{u} \geqq \boldsymbol{0}. \end{aligned}$$

For the pair of problems (I′) and (II′) we can state

Corollary 5. *Suppose that $f(\boldsymbol{x})$ and $g_i(\boldsymbol{x})$, $i = 1, \ldots, m$, are real-valued differentiable concave functions defined on $\mathbb{R}^n_+$. Then, corresponding to an optimal vector $\boldsymbol{x}_0$ for problem (I′), there exists a $\boldsymbol{u}_0 \geqq \boldsymbol{0}$ such that $(\boldsymbol{x}_0, \boldsymbol{u}_0)$ is optimal for the problem (II′) and*

$$f(\boldsymbol{x}_0) = L(\boldsymbol{x}_0, \boldsymbol{u}).$$

PROOF. By the reasoning employed in the proof of Theorem 24, corresponding to $\boldsymbol{x}_0$ there exists $\boldsymbol{u}_0 \geqq \boldsymbol{0}$ satisfying (67) through (70). This, in turn, implies that the feasible set F'_d of (II′) is nonempty and that $f(\boldsymbol{x}_0) = L(\boldsymbol{x}_0, \boldsymbol{u}_0)$. Let $(\boldsymbol{x}, \boldsymbol{u})$ be any vector of F'_d. Then

$$L(\boldsymbol{x}_0, \boldsymbol{u}_0) - L(\boldsymbol{x}, \boldsymbol{u}) = f(\boldsymbol{x}_0) - f(\boldsymbol{x}) - \boldsymbol{u}'\boldsymbol{g}(\boldsymbol{x}) \quad \text{[by (70)]}.$$

Since $\boldsymbol{u} \geqq \boldsymbol{0}$, the repeated application of Lemma 14(2) yields

$$\begin{aligned} L(\boldsymbol{x}_0, \boldsymbol{u}_0) - L(\boldsymbol{x}, \boldsymbol{u}) &\leqq \nabla f(\boldsymbol{x})(\boldsymbol{x}_0 - \boldsymbol{x}) - \boldsymbol{u}'(\boldsymbol{g}(\boldsymbol{x}_0) - \nabla \boldsymbol{g}(\boldsymbol{x})(\boldsymbol{x}_0 - \boldsymbol{x})) \\ &= \nabla L_{\boldsymbol{x}}(\boldsymbol{x}, \boldsymbol{u}) \cdot \boldsymbol{x}_0 - \nabla L_{\boldsymbol{x}}(\boldsymbol{x}, \boldsymbol{u}) \cdot \boldsymbol{x} - \boldsymbol{u}'\boldsymbol{g}(\boldsymbol{x}_0) \leqq 0. \end{aligned}$$

The final inequality of course follows from the assumed optimality of $\boldsymbol{x}_0$ and the hypothesis that $(\boldsymbol{x}, \boldsymbol{u}) \in F'_d$. □

Before proceeding, we apply the corollary to the linear programming problem investigated in Section 2 of this chapter. Define $\boldsymbol{g}(\boldsymbol{x}) = \boldsymbol{b} - A\boldsymbol{x}$ and $f(\boldsymbol{x}) = \boldsymbol{c}'\boldsymbol{x}$. Then problem (II′) reduces to

$$\begin{aligned} &\min_{(\boldsymbol{x}, \boldsymbol{u})} \boldsymbol{c}'\boldsymbol{x} + \boldsymbol{u}'(\boldsymbol{b} - A\boldsymbol{x}) \\ &\text{subject to } \boldsymbol{c}' - \boldsymbol{u}'A \leqq \boldsymbol{0}, \\ &\qquad (\boldsymbol{c}' - \boldsymbol{u}'A) \cdot \boldsymbol{x} = 0, \\ &\qquad (\boldsymbol{x}, \boldsymbol{u}) \geqq \boldsymbol{0}. \end{aligned}$$

Taking the second constraint $((\boldsymbol{c}' - \boldsymbol{u}'A) \cdot \boldsymbol{x} = 0)$ into account, the above problem further reduces to

$$\begin{aligned} &\min_{\boldsymbol{u}} \boldsymbol{u}'\boldsymbol{b} \\ &\text{subject to } \boldsymbol{u}'A \geqq \boldsymbol{c}', \\ &\qquad \boldsymbol{u} \geqq \boldsymbol{0}. \end{aligned}$$

We have thus obtained linear programming problem (II), the standard dual of linear programming problem (I). Since $f(\boldsymbol{x})$ and $\boldsymbol{g}(\boldsymbol{x})$ are clearly concave and differentiable on $\mathbb{R}^n_+$, the "only if" assertion of Theorem 12 directly follows from Corollary 5. This suggests the possibility of generalizing the converse assertion, which is called the converse duality theorem for nonlinear programming.

Theorem 26 (Mangasarian and Ponstein [1965]).[9] *Let X, $f(\boldsymbol{x})$ and $g_i(\boldsymbol{x})$, $i = 1, \ldots, m$, be as in Theorem 25 and let $(\hat{\boldsymbol{x}}, \hat{\boldsymbol{u}})$ be an optimal vector for the dual problem* (II). *If, in addition, $L(\boldsymbol{x},\hat{\boldsymbol{u}})$ is strictly concave at $\boldsymbol{x} = \hat{\boldsymbol{x}}$*[10] *then $\hat{\boldsymbol{x}}$ solves the primary problem* (I).

PROOF. By Theorem 4.5(b) of Mangasarian and Ponstein [1965], the set S of the optimal vectors for the primary problem (I) is nonempty.[11] Assume that $\hat{\boldsymbol{x}} \notin S$ so that, for any $\bar{\boldsymbol{x}} \in S$, $\bar{\boldsymbol{x}} \neq \hat{\boldsymbol{x}}$. Then Theorem 25 ensures the existence of $\bar{\boldsymbol{u}} \geqq \boldsymbol{0}$ such that $(\bar{\boldsymbol{x}}, \bar{\boldsymbol{u}})$ solves the dual problem (II) and such that $f(\bar{\boldsymbol{x}}) = L(\bar{\boldsymbol{x}}, \bar{\boldsymbol{u}})$. By assumption, therefore,

$$f(\bar{\boldsymbol{x}}) = L(\bar{\boldsymbol{x}}, \bar{\boldsymbol{u}}) = L(\hat{\boldsymbol{x}}, \hat{\boldsymbol{u}}).$$

This, coupled with the optimality of $(\hat{\boldsymbol{x}}, \hat{\boldsymbol{u}})$ and Lemma 15(1), yields

$$0 = \nabla L_{\boldsymbol{x}}(\hat{\boldsymbol{x}}, \hat{\boldsymbol{u}}) \cdot (\bar{\boldsymbol{x}} - \hat{\boldsymbol{x}}) > L(\bar{\boldsymbol{x}}, \hat{\boldsymbol{u}}) - L(\hat{\boldsymbol{x}}, \hat{\boldsymbol{u}}) = (\hat{\boldsymbol{u}})'\boldsymbol{g}(\bar{\boldsymbol{x}}) \geqq 0,$$

a self-contradiction. □

Notes to Chapter 1

1. Let $\boldsymbol{x}$ and $\boldsymbol{y}$ be n-dimensional vectors and $\boldsymbol{e}_i$ the ith unit vector of order n, and consider a special norm defined by $h_\infty(\boldsymbol{x}) = \max_j |x_j|$, where x_j is the jth component of $\boldsymbol{x}$. Let $h(\boldsymbol{x})$ be any norm defined on the vector space under consideration. Then

$$\begin{aligned} |h(\boldsymbol{x}) - h(\boldsymbol{y})| &= \max\{h(\boldsymbol{x}) - h(\boldsymbol{y}), h(\boldsymbol{y}) - h(\boldsymbol{x})\} \\ &\leqq h(\boldsymbol{x} - \boldsymbol{y}) \\ &\qquad [\text{since } h(\boldsymbol{x} - \boldsymbol{y}) = h(\boldsymbol{y} - \boldsymbol{x}) \text{ and since } \boldsymbol{x} = (\boldsymbol{x} - \boldsymbol{y}) + \boldsymbol{y} \\ &\leqq \sum_{j=1}^{n} |x_j - y_j| \cdot h(\boldsymbol{e}_j) \\ &\qquad [\text{from (1)–(3) of the vector norm and the fact that } \boldsymbol{x} = \textstyle\sum_j x_j \boldsymbol{e}_j] \\ &\leqq n \cdot h_\infty(\boldsymbol{x} - \boldsymbol{y}) \sum_{j=1}^{n} h(\boldsymbol{e}_j) \qquad [\text{from the definition of } h_\infty(\boldsymbol{x})]. \end{aligned}$$

Moreover, in view of (1) of the vector norm, $\sum_{j=1}^{n} h(\boldsymbol{e}_j)$ is a positive constant, henceforth denoted by K, so that, for any $e > 0$, the choice of $d \in (0, e/K)$ enables us to assert the continuity of $h(\boldsymbol{x})$ with respect to the metric induced by the norm $h_\infty(\boldsymbol{x})$. In order to show that any vector norm is continuous with respect to the metric induced by any vector norm, let $h'(\boldsymbol{x})$ be another vector norm and, for the time being, assume that $\boldsymbol{x} \neq \boldsymbol{0}$. Then the change of variables $\boldsymbol{x} = h_\infty(\boldsymbol{x})\boldsymbol{y}$ transforms $h'(\boldsymbol{x})$ into $h_\infty(\boldsymbol{x})h'(\boldsymbol{y})$. Since, by definition, $h_\infty(\boldsymbol{y}) = 1$, and since $h'(\boldsymbol{y})$ is continuous with respect to the metric induced by $h_\infty(\boldsymbol{y})$, it surely attains its maximum (M) and minimum (m) on the compact set $S = \{\boldsymbol{y}: h_\infty(\boldsymbol{y}) = 1\}$. Clearly, $\boldsymbol{y} \in S$ implies that $y \neq \mathbf{0}$. Hence neither M nor m can be nonpositive. Recalling that $h(\mathbf{0}) = h'(\mathbf{0})$, we have established that

$$m \cdot h_\infty(x) \leqq h'(x) \leqq M \cdot h_\infty(x) \quad \text{for all } x.$$

This inequality, in conjunction with the one obtained earlier, ensures that, for any $e > 0$ and for any z fulfilling $h'(x - z) < d' \equiv m \cdot d/K$,

$$|h(x) - h(z)| < e,$$

that is, that $h(x)$ is continuous.

2. If $f'(x)$ in Definition 14 is continuous at $x = x_0$ then f is said to be *continuously differentiable at* x_0; and if $f'(x)$ is continuous at every point of the domain, say S, f is called *continuously differentiable on* S.

   If not only f but every partial derivative of f is continuously differentiable at x_0 then it is said to be *twice continuously differentiable at* x_0. That f is twice continuously differentiable simply means that f is so at every point of its domain.

   As is well known, from the assumed continuity of each partial derivative of f the symmetry of the Hessian follows. A neat proof of this proposition can be found in Takeuchi [1966; Theorem 8.24, p. 212] or in Buck [1965; Theorem 13 and Corollary, p. 248].
3. Let Y be a subset of a metric space. Then a subset E of Y is said to be *open relative to* Y if to each $x \in E$ there corresponds a neighborhood $B(x, \varepsilon)$ such that $(B(x, \varepsilon) \cap Y) \subseteq E$.

   A fairly detailed discussion of this concept can be found in Section 2 of Appendix B.
4. See Subsection 2.3 of Appendix B.

   A mean value theorem for real valued functions of a single variable, from which the multivariate version follows, is elegantly proved in Rudin [1964; Theorem 5.10, p. 93].
5. See Subsection 2.3 of Appendix B.
6. Since $\psi_\beta(0, 0) > 0$, there is a sufficiently small $\beta_0 > 0$ such that $\psi_\beta(0, \beta) > 0$ for all $\beta \in (0, \beta_0)$. The Mean Value Theorem, together with the fact that $\psi(0, 0) = 0$, asserts that, for any $\beta \in (0, \beta_0)$, we can find a $\theta \in (0, 1)$ such that

$$\psi(0, \beta) = \psi_\beta(0, \theta\beta) > 0 \qquad [\text{since } \theta\beta \in (0, \beta_0)].$$

Hence, from the diagram depicted below, there exists the function $\alpha(\beta)$ with the properties stated above.

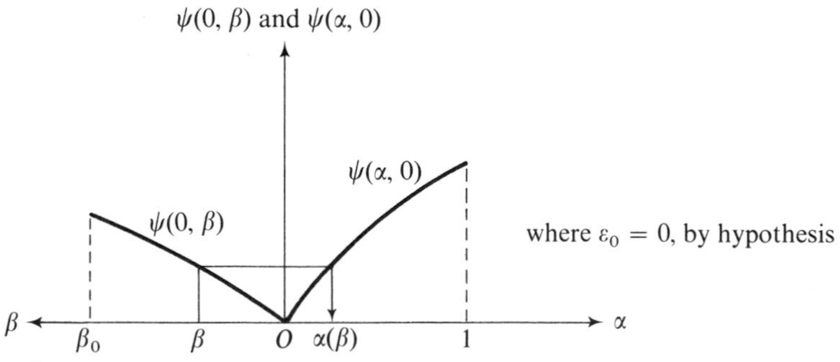

7. Proceeding as in Footnote 6, we can draw the following diagram, from which our assertion is obvious.

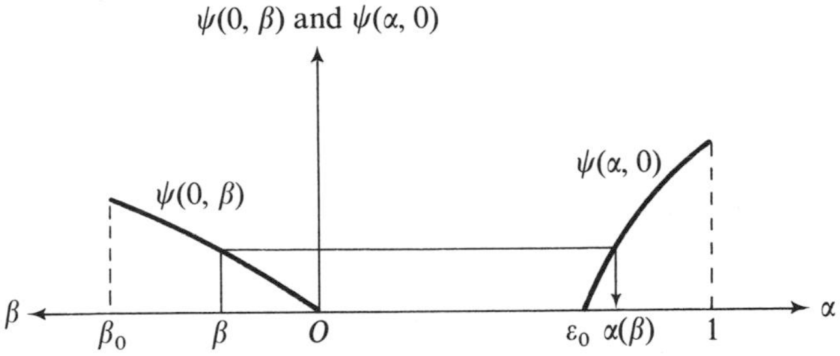

8. In deriving (75), we implicitly assume that $I_0 \neq \varnothing$. In the case $I_0 = \varnothing$, (69) and (70), together with the assumed nonnegativity of $\boldsymbol{u}_0$, imply that $\boldsymbol{u}_0 = \mathbf{0}$. Hence, from (74),

$$\nabla f(\boldsymbol{x}_0)(\boldsymbol{x} - \boldsymbol{x}_0) \leqq 0 \quad \text{for any } \boldsymbol{x} \in \Gamma.$$

9. Mangasarian and Ponstein [1965] calls this theorem the "strictly converse duality theorem" because the vector $\hat{\boldsymbol{x}}$ in the theorem becomes an optimal vector for the problem (I).
10. Since $\boldsymbol{u}$ is fixed at $\boldsymbol{u} = \hat{\boldsymbol{u}}$, $L(\boldsymbol{x}, \hat{\boldsymbol{u}})$ is a function of $\boldsymbol{x}$. Moreover, when $\boldsymbol{x}_0$ of Definition 16 is fixed, f is said to be *strictly concave at* $\boldsymbol{x} = \boldsymbol{x}_0$.
11. The proof of Theorem 4.5(b) of Mangasarian and Ponstein [1965] is fairly lengthy and hence is omitted.

2 Nonnegative Matrices

Nonnegative matrices play a central role in many branches of economic analysis. The uniqueness and stability of the path of balanced growth in a dynamic Leontief system, the global stability of some trading systems, and the Stolper–Samuelson Theorem of the neoclassical theories of international trade and distribution are a few examples. For detail the reader may refer to Dorfman, Samuelson and Solow [1958], Morishima [1964], and Nikaido [1968].

Our main purpose in this chapter is to describe the more important of those properties of nonnegative matrices of interest to economists. Section 1 is devoted to the classical theorem of Perron and Frobenius. For economists this theorem is perhaps the most fundamental of all propositions concerning nonnegative matrices. In Section 2 we consider an extended version of the Perron–Frobenius Theorem. Throughout the chapter our attention will be confined to *square* nonnegative matrices.

1 The Perron–Frobenius Theorem

In dealing with square matrices one inevitably encounters the concepts of decomposability and indecomposability. It is appropriate therefore that we begin with a definition.

Definition 1. Let A be an $n \times n$ matrix and N the set of indices $1, \ldots, n$, that is, $N = \{1, \ldots, n\}$. A is said to be *decomposable* if, for some nonempty proper subset J of N, $a_{ij} = 0$ for all indices $i \in J$ and $i \notin J$. A is *indecomposable* if A is not decomposable.

Notice that, by this definition, the real number zero is an indecomposable matrix of order one.

Lemma 1. *If an $n \times n$ matrix A is indecomposable then* (i) *A' also is indecomposable and* (ii) *for any $i \in N$ there exists $j_i \in N$ such that $a_{ij_i} \neq 0$ and $i \neq j_i$.*

Proof. (i) Let us denote the ijth element of A' by a'_{ij}, so that $a_{ji} = a'_{ij}$. If A' is decomposable, there exists a nonempty subset J of N such that $a'_{ij} = 0$ for any $i \in J$ and any $j \notin J$. However, $a_{ji} = 0$ for any $j \in J^c$ and $i \notin J^c$. Hence A also is decomposable, a contradiction.

(ii) From the definition of indecomposability, the nonempty subset $N(i) = \{1, \ldots, i-1, i+1, \ldots, n\}$ of N must contain an element j_i such that $a_{ij_i} \neq 0$. Obviously, $i \neq j_i$. □

Lemma 2. (a) *An $n \times n$ matrix A is decomposable if and only if there exists a permutation matrix P such that*

$$PAP' = \begin{pmatrix} A_{11} & A_{12} \\ 0 & A_{22} \end{pmatrix},$$

where A_{ii}, $i = 1, 2$, is a square matrix. (b) *Furthermore, if every off-diagonal element of A is nonnegative then A is decomposable if and only if there exists a real number λ such that the inequality $\lambda x \geqq Ax$ has a semipositive solution which is not positive.*

Proof. (a) Suppose that A is decomposable. Then, from Definition 1, there exists a nonempty proper subset J of N such that $a_{ij} = 0$ for any $i \subset J$ and any $j \notin J$. Let $J = \{k_1, \ldots, k_J\}$ and $J^c = \{k_{J+1}, \ldots, k_n\}$ and renumber k_{J+j} as $j, j = 1, \ldots, n - J$, and k_i as $n - J + i$, $i = 1, \ldots, J$. Then A can be transformed into

$$\begin{pmatrix} A_{11} & A_{12} \\ 0 & A_{22} \end{pmatrix}$$

by a permutation matrix P which corresponds to the renumbering of $k_i \in J$. The proof of sufficiency is easy and is omitted.

(b) (*Sufficiency*). Let us denote the set $\{i \in N : x_i = 0\}$ by I_x. By assumption, I_x is a nonempty proper subset of N. Since $x_i = 0$ for any $i \in I_x$ and since $\lambda x \geqq Ax$,

$$0 = (\lambda - a_{ii})x_i \geqq \sum_{j \neq i} a_{ij}x_j = \sum_{\substack{j \notin I_x \\ j \neq i}} a_{ij}x_j \geqq 0 \quad \text{for any } i \in I_x.$$

Hence $a_{ij} = 0$ for any $i \in I_x$ and any $j \notin I_x$.

(b) (*Necessity*). Since A is decomposable there exists a proper subset J of N such that $a_{ij} = 0$ for any $i \in J$ and any $j \notin J$. Let us define an n-dimensional vector $\boldsymbol{x} = (x_1, \ldots, x_n)'$, where $x_i = 0$ if $i \in J$ and $x_i > 0$ if $i \notin J$. Then, for any real λ and any $i \in J$,

$$0 = (\lambda - a_{ii})x_i = \sum_{j \neq i} a_{ij}x_j.$$

On the other hand, there clearly exists a real λ such that

$$(\lambda - a_{ii})x_i \geqq \sum_{j \neq i} a_{ij}x_j \quad \text{for any } i \notin J. \qquad \square$$

If in the decomposition of Lemma 2 one of the $\boldsymbol{A}_{ii}$ is itself decomposable, it can in turn be subjected to the procedure described in the proof of the lemma. Thus with any decomposable matrix there is associated a permutation matrix $\boldsymbol{P}$ such that

$$\boldsymbol{PAP}' = \begin{pmatrix} \boldsymbol{A}_{11} & \cdots\cdots & \boldsymbol{A}_{1m} \\ & \boldsymbol{A}_{22} \quad \cdots & \boldsymbol{A}_{2m} \\ & \ddots & \vdots \\ & \boldsymbol{0} \quad \ddots & \\ & & \boldsymbol{A}_{mm} \end{pmatrix} \tag{1}$$

where $\boldsymbol{A}_{ii}$, $i = 1, \ldots, m$, is an indecomposable square matrix. If $\boldsymbol{A}_{ij} = \boldsymbol{0}$ for every i and j such that $m \geqq j > i$, $i = 1, \ldots, m - 1$, then $\boldsymbol{A}$ is said to be *completely decomposable.*

Let us dwell for a moment on the economic sense of decomposability and indecomposability. Suppose, for example, that $\boldsymbol{A} = (a_{ij})$ is the matrix of technical coefficients of the Leontief type, where a_{ij} is the amount of the jth input per unit of the ith output. Suppose further that $\boldsymbol{A}$ is decomposable. Then industries belonging to the subset J of N such that $a_{ij} = 0$ for any $i \in J$ and any $j \notin J$ can increase their outputs without the benefit of additional inputs from outside the industrial group defined by J. Hence an increase in final demand for the product of any industry in J has no impact on industries outside the group. That is, if $\boldsymbol{A}$ is decomposable there exists in the economy at least one self-contained industrial group. If $\boldsymbol{A}$ is indecomposable, on the other hand, an increase in the final demand for the product of any industry must affect the input-demand for every other product.

We now state and prove a proposition which we shall need in later arguments.

Lemma 3. *Let* $f(x) = \sum_{j=0}^{n} a_j x^{n-j}$ *be a polynomial function of* x *of order* n. *If* $f(x) = 0$ *has a real root and if* a_j, $j = 0, 1, \ldots, n$, *is real then*

$$\operatorname{sign} f(x) = \operatorname{sign} a_0 \quad \textit{for any } x > \lambda^*,$$

where λ^* *is the largest real root of* $f(x) = 0$.

Proof. Renumbering the roots of $f(x) = 0$ we can, without loss of generality, list them as

$$\lambda_{2j-1} = \bar{\lambda}_{2j} \qquad (j = 1, \ldots, s),$$

$$\lambda_k = \bar{\lambda}_k \qquad (k = 2s + 1, \ldots, n).$$

The factoral decomposition of $f(x)$ yields

$$f(x) = a_0\left[\prod_{j=1}^{s}(x - \lambda_{2j-1})(x - \lambda_{2j})\right]\left[\prod_{k=2s+1}^{n}(x - \lambda_k)\right].$$

For real x, however,

$$x - \lambda_{2j} = \bar{x} - \bar{\lambda}_{2j-1} = \overline{x - \lambda_{2j-1}}.$$

Hence

$$f(x) = a_0\left[\prod_{j=1}^{s}|x - \lambda_{2j-1}|^2\right]\left[\prod_{k=2s+1}^{n}(x - \lambda_k)\right] \quad \text{for all } x \in \mathbb{R}.$$

Now, for every j,

$$|x - \lambda_{2j-1}|^2 = (x - \mathrm{Re}(\lambda_{2j-1}))^2 + (\mathrm{Im}(\lambda_{2j-1}))^2 \quad \text{and} \quad \mathrm{Im}(\lambda_{2j-1}) \neq 0,$$

where $\mathrm{Re}(\lambda_{2j-1})$ and $\mathrm{Im}(\lambda_{2j-1})$ denote the real and imaginary parts of λ_{2j-1}, respectively. Hence

$$\prod_{j=1}^{s}|x - \lambda_{2j-1}|^2 > 0 \quad \text{for all } x \in \mathbb{R}$$

Hence, $f(x)/a_0 > 0$ for all $x > \lambda^*$ or, equivalently,

$$\operatorname{sign} f(x) = \operatorname{sign} a_0 \quad \text{for all } x > \lambda^*.$$ □

That completes the preliminaries. We can now proceed to our main argument.

Lemma 4. *Let A be an $n \times n$ matrix with nonnegative off-diagonal elements. Then A has a real eigenvalue λ^* such that:*

i. $f_{ij}(\lambda) \geqq 0$ *for any* $\lambda \geqq \lambda^*$, $i, j = 1, \ldots, n$, *and* $f_{ii}(\lambda) > 0$ *for any* $\lambda > \lambda^*$, $i = 1, \ldots, n$;
ii. λ^* *is the largest real eigenvalue of* A;
iii. $f(\lambda) > 0$ *for any* $\lambda > \lambda^*$;
iv. *with* λ^* *there is associated a semipositive eigenvector* x^* *of* A;
v. *for any eigenvalue* λ *of* A, *not equal to* λ^*,

$$|\lambda| \leqq \lambda^* + \max_{j \in N}\{|a_{jj}| - a_{jj}\} \quad \textit{and} \quad \lambda^* > \mathrm{Re}(\lambda);$$

vi. $(\rho I - A)^{-1} \geq 0$ *if and only if* $\rho > \lambda^*$; *and*
vii. $\lambda^* \geqq 0$ *if and only if there exists a vector* $x \geq 0$ *such that* $Ax \geqq 0$ *where* $f(\lambda) = \det(\lambda I - A)$ *and* $f_{ij}(\lambda)$ *is the cofactor of the ijth element of* $f(\lambda)$.

PROOF. (i) We proceed by induction on n. Suppose that the proposition is valid for $n - 1$. Let $M_{kj}^i(\lambda)$ be the cofactor of $(\delta_{kj} - a_{kj})$ in $f_{ii}(\lambda)$, where δ_{kj} is the Kronecker delta. Then the cofactor expansion of $f_{ij}(\lambda)$ yields

$$f_{ij}(\lambda) = \sum_{k \neq i} a_{ki} M_{kj}^i(\lambda) \qquad (j \neq i; i = 1, \ldots, n). \tag{2}$$

On the other hand,

$$f(\lambda) = (\lambda - a_{ii})f_{ii}(\lambda) - \sum_{j \neq i} a_{ij} f_{ij}(\lambda). \tag{3}$$

Substituting (2) into (3),

$$f(\lambda) = (\lambda - a_{ii})f_{ii}(\lambda) - \sum_{k,\, j \neq i} a_{ij} a_{ki} M_{kj}^{i}(\lambda). \tag{4}$$

By virtue of the induction assumption, $f_{ii}(\lambda) = 0$ has a real root $\tilde{\mu}_i$ with the property (i). Hence

$$f(\tilde{\mu}_i) = -\sum_{k,\, j \neq i} a_{ij} a_{ki} M_{kj}^{i}(\tilde{\mu}_i) \leqq 0. \tag{5}$$

Suppose that $f(\lambda) = 0$ has no real root. Then

$$f(\lambda) = \prod_j |\lambda - \lambda_{2j-1}|^2 > 0 \quad \text{for any } \lambda \in \mathbb{R},$$

in contradiction of (5). Hence $\boldsymbol{A}$ has a real eigenvalue. Let λ^* be the largest of the real eigenvalues of $\boldsymbol{A}$. Then, for any i, $\lambda^* \geqq \tilde{\mu}_i$. Otherwise, from Lemma 3, $f(\tilde{\mu}_i)$ must be positive, which again contradicts (5). Hence the induction assumption, together with (2), enables us to infer that

$$f_{ij}(\lambda) = \sum_{k \neq i} a_{ki} M_{kj}^{i}(\lambda) \geqq 0 \quad \text{for all } \lambda \geqq \lambda^*, j \neq i.$$

Since $\tilde{\mu}_i$ is the largest real root of $f_{ii}(\lambda) = 0$, $i = 1, \ldots, n$, and since $\lambda^* \geqq \tilde{\mu}_i$, $i = 1, \ldots, n$, it follows from Lemma 3 that

$$f_{ii}(\lambda) \geqq 0 \quad \text{for all } \lambda \geqq \lambda^*$$

and

$$f_{ii}(\lambda) > 0 \quad \text{for all } \lambda > \lambda^*.$$

That is, λ^* has property (i).

It remains to consider the case $n = 2$. Let us again set $\lambda^* = \frac{1}{2}[(a_{11} + a_{22}) + \sqrt{(a_{11} - a_{22})^2 + 4a_{12}a_{21}}]$. Since $f_{ij}(\lambda) = a_{ji}$, the first half of (i) is valid. The second half follows from the easily verified fact that

$$\lambda^* - a_{jj} = \tfrac{1}{2}[(-1)^j(a_{11} + a_{22}) + \sqrt{(a_{11} - a_{22})^2 + 4a_{12}a_{21}}] \geqq 0 \quad (j = 1, 2).$$

(ii) We have shown this assertion in the proof of (i).

(iii) Since by definition λ^* is the largest real root of $f(\lambda) = 0$, property (iii) follows straightforwardly from Lemma 3.

(iv) Let $\boldsymbol{l}$ be an $n \times 1$ vector of ones. We have $(\lambda^*\boldsymbol{I} - \boldsymbol{A}) \cdot \operatorname{adj}(\lambda^*\boldsymbol{I} - \boldsymbol{A}) = \det(\lambda^*\boldsymbol{I} - \boldsymbol{A}) \cdot \boldsymbol{I} = \boldsymbol{0}$. Hence, if $\operatorname{adj}(\lambda^*\boldsymbol{I} - \boldsymbol{A})$ contains a positive element then $\operatorname{adj}(\lambda^*\boldsymbol{I} - \boldsymbol{A}) \cdot \boldsymbol{l}$ is a semipositive eigenvector corresponding to λ^*. Suppose that $\operatorname{adj}(\lambda^*\boldsymbol{I} - \boldsymbol{A}) = \boldsymbol{0}$. Then $\lambda^* = \tilde{\mu}_i$, $i = 1, \ldots, n$, where $\tilde{\mu}_i$ is the largest real eigenvalue of $\boldsymbol{A}\binom{i}{i}$; for, otherwise, there exists an index i_0 such that $\lambda^* > \tilde{\mu}_{i_0}$, which implies that the i_0th diagonal element of adj$(\lambda^* \boldsymbol{I} - \boldsymbol{A})$ is positive, contradicting the hypothesis. From the definition of λ^*, $\det(\lambda^* \boldsymbol{I} - \boldsymbol{A}) = 0$; hence at least one row of $(\lambda^* \boldsymbol{I} - \boldsymbol{A})$, say the kth row, can be represented as a linear combination of the remaining rows. Applying the induction assumption to $\boldsymbol{A}\binom{k}{k}$, there exists a semipositive $\boldsymbol{y} = (y_1, \ldots, y_{k-1}, y_{k+1}, \ldots, y_n)'$ such that $(\lambda^* \boldsymbol{I} - \boldsymbol{A}\binom{k}{k}) \cdot \boldsymbol{y} = \boldsymbol{0}$. Let us define $\hat{\boldsymbol{y}} = (y_1, \ldots, y_{k-1}, 0, y_{k+1}, \ldots, y_n)'$. Then $\hat{\boldsymbol{y}}$ is semipositive and $(\lambda^* \boldsymbol{I} - \boldsymbol{A}) \cdot \hat{\boldsymbol{y}} = \boldsymbol{0}$.

It remains to consider the case $n = 2$. From the definition of λ^*, there exists a nonzero t such that $(-a_{21}, \lambda^* - a_{22}) = t \cdot (\lambda^* - a_{11}, -a_{12})$. It therefore suffices to show the existence of a semipositive $\boldsymbol{x}$ such that $(\lambda^* - a_{11})x_1 = a_{12}x_2$. Suppose that a_{12} is positive. Then $\boldsymbol{x} = (1, (\lambda^* - a_{11})/a_{12})'$ satisfies the requirement. Otherwise, for $\boldsymbol{x}' = (0, 1)$, we have $(\lambda^* - a_{11})x_1 = a_{12}x_2$.

(v) Again proof is by induction on n. Suppose that $n = 2$. Since $\lambda_1 = \frac{1}{2}[(a_{11} + a_{22}) + \sqrt{(a_{11} - a_{22})^2 + 4a_{12}a_{21}}]$ is the larger of the two eigenvalues of $\boldsymbol{A}$, the assertion can be verified by showing that $\lambda_1 + \max_i\{|a_{ii}| - a_{ii}\} - |\lambda_2| \geqq 0$. This can be done by direct calculation.

Let $\boldsymbol{x}^*$ be a semipositive eigenvector of $\boldsymbol{A}$ corresponding to λ^*. Then the set $J_{x^*} = \{j \in N : x_j^* > 0\}$ is a nonempty subset of N; without loss of generality we may assume that J_{x^*} consists of $1, 2, \ldots, k$, where $k \leqq n$. In the case $k = n$, there is an index $h \in N$ such that $|\lambda - a_{hh}| \leq \lambda^* - a_{hh}$; for, otherwise, $(\lambda \boldsymbol{I} - \boldsymbol{A})$ has a row-dominant diagonal and, therefore, is nonsingular, contradicting the definition of λ. Taking into account the well-known inequality $|\lambda - a_{jj}| \geqq |\lambda| - |a_{jj}|$, it is obvious that

$$\lambda^* + \max_{i \in N}\{|a_{ii}| - a_{ii}\} \geq \lambda^* + |a_{jj}| - a_{jj} \geqq |\lambda|.$$

It may be noted that, in this case, we have no need of induction but can prove the desired result directly.

Suppose that $k < n$. Then, for any $j \notin J_{x^*}$,

$$0 = (\lambda^* - a_{jj})x_j = \sum_{l \in Jx^*} a_{jl}x_l.$$

Hence, $a_{jl} = 0$ for any $j \in J_{x^*}^c$ and any $l \in J_{x^*}$. Thus we are able to rewrite $\boldsymbol{A}$ as

$$\begin{pmatrix} \boldsymbol{A}_{11} & \boldsymbol{A}_{12} \\ \boldsymbol{0} & \boldsymbol{A}_{22} \end{pmatrix}.$$

Bearing this in mind, it can be seen that any eigenvalue of $\boldsymbol{A}$ satisfies either $\det(\lambda \boldsymbol{I} - \boldsymbol{A}_{11}) = 0$ or $\det(\lambda \boldsymbol{I} - \boldsymbol{A}_{22}) = 0$. Let λ be any eigenvalue of $\boldsymbol{A}_{11}$ different from λ^*. Then, by the same reasoning as in the case $k = n$,

$$\lambda^* + \max_{i \in Jx^*}\{|a_{ii}| - a_{ii}\} \geqq \lambda^* + (|a_{jj}| - a_{jj}) \geqq |\lambda|.$$

Let us denote the largest real eigenvalue of $\boldsymbol{A}_{22}$ by μ^*. Then the induction assumption applied to $\boldsymbol{A}_{22}$ yields $\mu^* + \max_{i \notin Jx^*}\{|a_{ii}| - a_{ii}\} \geqq |\mu|$ for any

μ such that $\det(\mu \boldsymbol{I} - \boldsymbol{A}_{22}) = 0$ and $\mu^* \neq \mu$. Recalling that $\lambda^* \geqq \mu^*$, the assertion is obvious.

For sufficiently large $\alpha > 0$, the matrix $\boldsymbol{B} = \boldsymbol{A} + \alpha \boldsymbol{I}$ is clearly nonnegative. Let β^* be the largest real eigenvalue of $\boldsymbol{B}$. Then, applying the first half of (v) to $\boldsymbol{B}$, it is trivially true that $\beta^* \geqq |\beta|$ for any eigenvalue β of $\boldsymbol{B}$. Suppose that there exists an eigenvalue λ_0 of $\boldsymbol{A}$ such that $\lambda^* \neq \lambda_0$ and $\text{Re}(\lambda_0) \geqq \lambda^*$. Then, defining $\beta_0 = \lambda_0 + \alpha$, we obtain

$$|\beta_0| \leqq \beta^* \equiv \lambda^* + \alpha \leqq \text{Re}(\lambda_0) + \alpha \equiv \text{Re}(\beta_0).$$

From this inequality and the fact that $\text{Re}(\beta_0) \leqq |\beta_0|$, $\beta^* = |\beta_0| = \text{Re}(\beta_0)$. Hence $\beta^* = \beta_0$, contradicting the hypothesis that $\lambda_0 \neq \lambda^*$.

(vi) If $\rho > \lambda^*$ then, from (i) and (iii) of this lemma, $(\rho \boldsymbol{I} - \boldsymbol{A})^{-1} \geq \boldsymbol{0}$.

To establish necessity, it suffices to prove that $\rho \not< \lambda^*$; for, by assumption, $\rho \neq \lambda^*$. Let $\hat{\boldsymbol{y}}$ be $(\rho \boldsymbol{I} - \boldsymbol{A})^{-1}\boldsymbol{x}^*$. Then $(\rho - \lambda^*)\hat{\boldsymbol{y}} = \boldsymbol{x}^*$ and $\hat{\boldsymbol{y}}$ is clearly nonnegative. If $\rho < \lambda^*$ then, for any $i \in N$ satisfying $x_i^* > 0$, we have

$$0 \geqq (\rho - \lambda^*)\hat{y}_i = x_i^* > 0,$$

which is a contradiction. Hence $\rho > \lambda^*$.

(vii) (*Necessity*). Since $\lambda^* \geqq 0$ and since $\boldsymbol{x}^* \geq \boldsymbol{0}$, $\boldsymbol{A}\boldsymbol{x}^* = \lambda^* \boldsymbol{x}^* \geqq \boldsymbol{0}$.

(vii) (*Sufficiency*). Suppose the contrary. Then, from (vi), $-\boldsymbol{A}^{-1}$ is nonnegative. Premultiplying both sides of $\boldsymbol{A}\boldsymbol{x} \geqq \boldsymbol{0}$ by $-\boldsymbol{A}^{-1}$, we see that $-\boldsymbol{x} \geqq \boldsymbol{0}$, contradicting the assumed semipositivity of $\boldsymbol{x}$. □

Remark Concerning Lemma 4. The second half of part (v) and part (vii) are Lemmas 1 and 2, respectively, of Arrow and Nerlove [1958]. The eigenvalue λ^* of Lemma 4 will be called the *Frobenius root of* $\boldsymbol{A}$.

If it is assumed that $\boldsymbol{A}$ is indecomposable, Lemma 4 can be strengthened.

Lemma 5. *Let* $\boldsymbol{A}$ *be an indecomposable* $n \times n$ *matrix which satisfies the conditions of Lemma* 4. *Then*:

i. $\boldsymbol{x}^*$ *is positive*;
ii. $\text{adj}(\lambda \boldsymbol{I} - \boldsymbol{A}) > \boldsymbol{0}$ *for any* $\lambda \geqq \lambda^*$;
iii. $\lambda^* > \tilde{\mu}_i$, $i = 1, 2, \ldots, n$;
iv. $N_{\lambda^* \boldsymbol{I} - \boldsymbol{A}} = \{\boldsymbol{x} : (\lambda^* \boldsymbol{I} - \boldsymbol{A})\boldsymbol{x} = \boldsymbol{0}\}$, *the annihilator space of* $\lambda^* \boldsymbol{I} - \boldsymbol{A}$ *is a linear subspace of dimension* 1, *that is, for any* $\boldsymbol{x}$ *in the space there exists a real number* $t_{\boldsymbol{x}}$ *such that* $\boldsymbol{x} = t_{\boldsymbol{x}} \cdot \boldsymbol{x}^*$;
v. *no eigenvector* $\boldsymbol{x}$ *of* $\boldsymbol{A}$ *which is associated with an eigenvalue* λ *different from* λ^* *is semipositive*;
vi. $(\rho \boldsymbol{I} - \boldsymbol{A})^{-1} > \boldsymbol{0}$ *if and only if* $\rho > \lambda^*$;
vii. λ^* *is a simple root of* $f(\lambda) = 0$; *and*
viii. $\lambda^* > 0$ *if and only if the inequality* $\boldsymbol{A}\boldsymbol{x} > \boldsymbol{0}$ *has a solution* $\boldsymbol{x} \geq \boldsymbol{0}$.

Proof. (i) In view of the second half of Lemma 2, the inequality $\lambda^* \boldsymbol{x} \geqq \boldsymbol{A}\boldsymbol{x}$ can have no solution which is semipositive but not positive. On the other hand, $\boldsymbol{x}^*$ is a semipositive solution of the above inequality. Hence $\boldsymbol{x}^*$ must be positive.

(ii) When $\lambda > \lambda^*$, from Lemma 4(i) and Lemma 4(iii), $f(\lambda)$ and $f_{ii}(\lambda)$, $i = 1, \ldots, n$, are positive. Hence every column of $\operatorname{adj}(\lambda \boldsymbol{I} - \boldsymbol{A})$ is a semipositive solution of $\lambda \boldsymbol{x} \geqq \boldsymbol{A}\boldsymbol{x}$, which, in view of Lemma 2(b), implies that $\operatorname{adj}(\lambda \boldsymbol{I} - \boldsymbol{A}) > \boldsymbol{0}$.

Suppose that $\lambda = \lambda^*$. Since the matrix $(\lambda^* \boldsymbol{I} - \boldsymbol{A})$ is singular, there exists a row of $(\lambda^* \boldsymbol{I} - \boldsymbol{A})$, say the kth row, which can be represented as a linear combination of the remaining rows. If $f_{kk}(\lambda^*) = 0$, then λ^* is the Frobenius root of $\boldsymbol{A}\binom{k}{k}$; hence there exists a semipositive vector

$$\boldsymbol{y} = (y_1, \ldots, y_{k-1}, y_{k+1}, \ldots, y_n)'$$

such that $\lambda^* \boldsymbol{y} = \boldsymbol{A}\binom{k}{k}\boldsymbol{y}$. Define

$$\hat{\boldsymbol{y}} = (y_1, \ldots, y_{k-1}, 0, y_{k+1}, \ldots, y_n)'.$$

Then $\hat{\boldsymbol{y}}$ is an eigenvector of $\boldsymbol{A}$ associated with λ^*. From the definition of $\hat{\boldsymbol{y}}$, it is obviously semipositive but not positive. However, this again contradicts Lemma 2(b), since $\boldsymbol{A}$ is an indecomposable matrix with off-diagonal elements nonnegative. Hence $f_{kk}(\lambda^*) > 0$. Let the kth row of $\operatorname{adj}(\lambda^* \boldsymbol{I} - \boldsymbol{A})$ be $\boldsymbol{z}_k^*$. Then $\lambda^*(\boldsymbol{z}_k^*)' = \boldsymbol{A}'(\boldsymbol{z}_k^*)'$. Since $\boldsymbol{A}'$ is indecomposable and nonnegative save for its diagonal elements and since $\boldsymbol{z}_k^*$ is semipositive, the contrapositive of Lemma (2b) asserts that $\boldsymbol{z}_k^*$ is positive. Therefore, every column of $\operatorname{adj}(\lambda^* \boldsymbol{I} - \boldsymbol{A})$ is semipositive, which, by the same reasoning as above, leads us to the conclusion that $\operatorname{adj}(\lambda^* \boldsymbol{I} - \boldsymbol{A}) > \boldsymbol{0}$.

(iii) Suppose the contrary. Then there exists an index $i_0 \in N$ such that $\lambda^* = \mu_{i_0}$, which implies that $f_{i_0 i_0}(\lambda^*) = 0$, contradicting (ii) above.

(iv) From (ii) the rank of the matrix $(\lambda^* \boldsymbol{I} - \boldsymbol{A})$ is $n - 1$. The dimension of the annihilator space of $(\lambda^* \boldsymbol{I} - \boldsymbol{A})$ is therefore 1 $(= n - (n - 1))$.

(v) Suppose the contrary. Then there exist an eigenvalue λ of $\boldsymbol{A}$ and a semipositive eigenvector $\boldsymbol{x}$ corresponding to λ. Needless to say, $\lambda \neq \lambda^*$. Notice that λ must be real; otherwise, $\boldsymbol{x}$ could not be semipositive. Suppose that $\boldsymbol{x}$ contains a zero element. Then $J_{\boldsymbol{x}} = \{j : x_j = 0, j \in N\}$ is a nonempty proper subset of N. In view of the assumed indecomposability of $\boldsymbol{A}$, there exist indices $k \in J_{\boldsymbol{x}}$ and $s \notin J_{\boldsymbol{x}}$ with the property that $a_{ks} > 0$. Then

$$0 = \sum_{j=1}^{n} (\lambda \delta_{kj} - a_{kj})x_j = \sum_{j \notin J_{\boldsymbol{x}}} (\lambda \delta_{kj} - a_{kj})x_j \leqq -a_{ks}x_s < 0,$$

which is a contradiction. Thus $\boldsymbol{x}$ must be positive. Since λ is a real eigenvalue of $\boldsymbol{A}$ and different from λ^*, we have $\lambda^* > \lambda$ and, therefore,

$$(\lambda^* - a_{ii})x_i > (\lambda - a_{ii})x_i = \sum_{j \neq 1} a_{ij}x_j > 0 \qquad (i = 1, \ldots, n).$$

This implies that $(\lambda^*\boldsymbol{I} - \boldsymbol{A})$ possesses a dominant diagonal, whence $(\lambda^*\boldsymbol{I} - \boldsymbol{A})$ is nonsingular. This is inconsistent with the definition of λ^*.

(vi) Sufficiency follows at once from (ii), and necessity has been verified in Lemma 4(vi).

(vii) Differentiation of $f(\lambda)$ yields $f'(\lambda) = \sum_{i=1}^n f_{ii}(\lambda)$. Recalling (ii) again, $f'(\lambda^*) = \sum_{i=1}^n f_{ii}(\lambda^*) > 0$. Hence λ^* is simple.

(viii) (*Necessity*). From (i) and the hypothesis, $\boldsymbol{A}\boldsymbol{x}^* = \lambda^*\boldsymbol{x}^* > \boldsymbol{0}$.

(viii) (*Sufficiency*). Let $\boldsymbol{y}^*$ be a row eigenvector of $\boldsymbol{A}$ corresponding to λ^*. Then, applying (i) to $\boldsymbol{A}'$, $\boldsymbol{y}^*$ is positive, which, together with the hypothesis, implies that both $\boldsymbol{y}^*\boldsymbol{A}\boldsymbol{x}$ and $\boldsymbol{y}^*\boldsymbol{x}$ are positive. Hence $\lambda^* > 0$, for $\lambda^*(\boldsymbol{y}^*\boldsymbol{x}) = (\boldsymbol{y}^*\boldsymbol{A}\boldsymbol{x}) > 0$. □

Remark Concerning Lemmas 4 and 5. In verifying these lemmas we have implicitly assumed that $n > 2$. Let $n = 1$. Then $\boldsymbol{A}$ reduces to a real number and the validity of all propositions except those concerning the cofactors of $f(\lambda)$ is self-evident.

We now turn to the main proposition of this section, the Perron–Frobenius Theorem. There exist many proofs of this famous theorem. Our proof is a modified version of that offered recently by Murata [1972].

Theorem 1 (Perron [1907]; Frobenius [1908], [1909], [1912]). *Let $\boldsymbol{A}$ be an $n \times n$ indecomposable nonnegative matrix and λ^* its Frobenius root. Then*:

i. *there is associated with λ^* a positive eigenvector $\boldsymbol{x}^*$ of $\boldsymbol{A}^*$ with $\lambda^* \geqq a_{jj}$ for $j = 1, \ldots, n$ and an exact equality only when $n = 1$;*
ii. *$\lambda^* \geqq |\lambda|$ for any eigenvalue λ of $\boldsymbol{A}$, with a strict inequality if $\lambda \neq \lambda^*$ and $a_{kk} > 0$ for some k;*
iii. *$(\rho\boldsymbol{I} - \boldsymbol{A})^{-1} > \boldsymbol{0}$ if and only if $\rho > \lambda^*$;*
iv. *λ^* is larger than the Frobenius root of any principal submatrix of $\boldsymbol{A}$ of order less than n;*
v. *no eigenvector $\boldsymbol{x}$ of $\boldsymbol{A}$ which is associated with an eigenvalue λ different from λ^* is semipositive;*
vi. *λ^* is a simple root of $f(\lambda) = 0$, whence $d[N_{(\lambda^*\boldsymbol{I}-\boldsymbol{A})}] = 1$. Therefore, for any eigenvector $\boldsymbol{x}$ of $\boldsymbol{A}$ associated with λ^* there exists a nonzero real number $t_{\boldsymbol{x}}$ such that $\boldsymbol{x} = t_{\boldsymbol{x}}\boldsymbol{x}^*$.*

PROOF. (i) All parts of this assertion except the last follow from Lemma 5(i). The final part will be verified under (iv).

(ii) Since $\boldsymbol{A}$ is nonnegative, $|a_{jj}| = a_{jj}$ for $j = 1, \ldots, n$. Hence, from Lemma 4(v), $\lambda^* \geqq |\lambda|$.

Suppose that $\lambda \neq \lambda^*$, that $a_{kk} > 0$ for some k and that, nevertheless, $\lambda^* = |\lambda|$. Let $\boldsymbol{x} = (x_1, \ldots, x_n)'$ be an eigenvector associated with λ and let $\boldsymbol{y}' = (y_1, \ldots, y_n) > \boldsymbol{0}'$ be the row eigenvector corresponding to λ^*. Then

$$\lambda^*|x_i| = |\lambda x_i| = \left|\sum_{j=1}^n a_{ij}x_j\right| \leqq \sum_{j=1}^n a_{ij}|x_j| \qquad (i = 1, \ldots, n).$$

Suppose that there exists an r such that

$$\left|\sum_{j=1}^{n} a_{rj}x_j\right| < \sum_{j=1}^{n} a_{rj}|x_j|.$$

Then, recalling that $\boldsymbol{y} > \mathbf{0}$, we arrive at a contradiction:

$$\sum_{i=1}^{n} \lambda^* y_i|x_i| < \sum_{i=1}^{n} y_i\left(\sum_{j=1}^{n} a_{ij}|x_j|\right) = \sum_{j=1}^{n} \lambda^* y_j|x_j|.$$

It follows that

$$\lambda^*|x_i| = \left|\sum_{j=1}^{n} a_{ij}x_j\right| = \sum_{j=1}^{n} a_{ij}|x_j| \qquad (i = 1, \ldots, n),$$

whence, applying (iv) of Lemma 5,

$$x_i \neq 0 \qquad (i = 1, \ldots, n).$$

Let $\pi_k = \{j : a_{kj} > 0\}$ and let θ_j be the argument of x_j. Then

$$\lambda^*|x_k| = \left|\sum_{j\in\pi_k} a_{kj}x_j\right| = \sum_{j\in\pi_k} a_{kj}|x_j|.$$

Since the modulus of a sum of complex numbers is equal to the sum of their moduli if and only if they share a common argument, we can write

$$\theta_j = \theta_k \quad \text{for any } j \in \pi_k.$$

(Notice that it is at this point that the assumption that $a_{kk} > 0$ for some k is needed. Without it, we could not replace θ_j $(j \in \pi_k)$ with θ_k.) A small calculation then yields

$$\begin{aligned}
\lambda^* x_k &= \lambda^*|x_k|e^{i\theta_k} \\
&= \sum_{j\in\pi_k} a_{kj}|x_j|e^{i\theta_k} \\
&= \sum_{j\in\pi_k} a_{kj}x_j \qquad [\text{since } x_j = |x_j|e^{i\theta_k} \text{ for all } j \in \pi_k] \\
&= \lambda x_k.
\end{aligned}$$

Recalling that $x_k \neq 0$, it follows that $\lambda = \lambda^*$, contradicting the hypothesis.

(iii) This assertion follows trivially from Lemma 5(vi).

(iv) In view of Lemma 5(iii), it is clear that $\lambda^* > \tilde{\mu}_i$, $i = 1, \ldots, n$. The rest of the assertion is easily verified by successively reducing n by 1. Moreover, if $n \geqq 2$ then $\lambda^* > a_{jj}$, $j = 1, \ldots, n$, since a_{jj} $(j = 1, \ldots, n)$ are the principal submatrices of order 1. Needless to say, $n = 1$ implies that $a_{11} = \lambda^*$.

(v)–(vi) These assertions may be viewed as special cases of (iv) and (vii) of Lemma 5. □

Remark Concerning Theorem 1. Our statement of the theorem is slightly unconventional. Usually, it is asserted that an indecomposable nonnegative matrix $\boldsymbol{A}$ has a positive Frobenius root provided that $\boldsymbol{A} \neq \mathbf{0}$. Without this restriction, zero (an indecomposable nonnegative matrix of order 1) would fall within the scope of the statement and invalidate it. However, our statement (Theorem 1(i), especially the final part of the assertion) is free from the restriction to which we have referred. Of course, zero could be excluded from the class of indecomposable matrices by definition; but we would then run into the difficulty that the principal matrices in decomposition (1) would not always be indecomposable, since some of them might happen to be zero of order 1.

Without the assumption of indecomposability, Theorem 1 takes a slightly weakened form.

Theorem 2. *Let* A, A_1 *and* A_2 *be* $n \times n$ *nonnegative matrices and let* λ^*, λ_1^*, *and* λ_2^* *be the Frobenius roots of* A, A_1, *and* A_2 *respectively. Then*:

i. $\lambda^* \geqq |\lambda|$ *for any eigenvalue* λ *of* A; $\lambda^* \geqq a_{jj}$, $j = 1, \ldots, n$; *and* A *has a semipositive eigenvector* x^* *corresponding to* λ^*;
ii. $(\rho I - A)^{-1} \geq 0$ *if and only if* $\rho > \lambda^*$;
iii. *if there exists a semipositive vector* x *such that* $\rho x \leqq Ax$, *where* ρ *is a given real number, then* $\rho \leqq \lambda^*$; *and*
iv. *if* $A_1 \geqq A_2 \geqq 0$ *then* $\lambda_1^* \geqq \lambda_2^*$.

Proof. (i)–(ii) It has been shown that $\lambda^* \geqq \tilde{\mu}_i$, $i = 1, \ldots, n$. The assertions then follow from Lemma 4(iv)–(vi), together with the assumed nonnegativity of A.

(iii) Suppose the contrary. Then, from (ii), $(\rho I - A)^{-1} \geqq 0$. On the other hand, by assumption, $(\rho I - A)x \leqq 0$. Hence

$$x = (\rho I - A)^{-1}(\rho I - A)x \leqq (\rho I - A)^{-1} \cdot 0 = 0,$$

which contradicts the assumption that $x \geq 0$.

(iv) Let x_2^* be the semipositive eigenvector of A_2 associated with λ_2^*. Then, by assumption, $A_1 x_2^* \geqq A_2 x_2^* = \lambda_2^* x_2^*$. By virtue of (iii) it can be seen that $\lambda_1^* \geqq \lambda_2^*$. □

Remarks Concerning Theorem 2. (1) Since (vi) of Lemma 4 is identical with (ii) of the theorem, (iii) and (iv) of the theorem hold not only for nonnegative square matrices but also for square matrices with nonnegative off-diagonal elements.

(2) Let ρ be a positive number. Then the following three statements are mutually equivalent:

a. the power series $I + (A/\rho) + (A/\rho)^2 + \cdots$ converges to $(I - (A/\rho))^{-1}$;
b. $\rho > \lambda^*$; and
c. all leading principal minors of $I - (A/\rho)$ are positive.

Proof (a) ⇒ (b). The (ij)th element of the power series is nonnegative and converges to the corresponding element of $(I - (A/\rho))^{-1}$; hence $(I - (A/\rho))^{-1} \geq 0$. From (ii) of the theorem, therefore, $\rho > \lambda^*$.

(b) ⇒ (c). The implication follows immediately from (ii) of the theorem and Theorem 7 of Chapter 1.

(c) ⇒ (a). Let us define $S_m = \sum_{\tau=0}^{m-1} (A/\rho)^\tau$. Then a simple manipulation yields $S_{m+1} = I + S_m(A/\rho)$. Since $(A/\rho) \geqq 0$, $S_{m+1} \geqq S_m$ for any positive integer m. Hence

$$I \geqq S_m\left(I - \left(\frac{A}{\rho}\right)\right).$$

Recalling Theorem 7 of Chapter 1 again, (c) implies the existence of an $x > 0$ such that

$$0 < \left(I - \left(\frac{A}{\rho}\right)\right)x.$$

From these inequalities and the positivity of x,

$$x \geqq S_m\left(I - \left(\frac{A}{\rho}\right)\right)x.$$

From this inequality, and from the fact that x and $(I - (A/\rho))x$ are given positive vectors, it follows that S_m is bounded for any positive integer m. Thus S_m is a nondecreasing bounded sequence and therefore converges to a limit, say B: $\lim_{m\to\infty} S_m = B$. A further simple manipulation yields $(I - (A/\rho))S_m = I - (A/\rho)^m$ whence, in view of the above property of convergence, $\lim_{m\to\infty} (A/\rho)^m = 0$. Thus, in the limit, $(I - (A/\rho))B = I$ which, in turn, implies that $B = (I - (A/\rho))^{-1}$, as desired. □

Theorem 2 alerts us to the fact that the Frobenius root of a decomposable matrix may be zero. We now provide a necessary and sufficient condition for the Frobenius root of a decomposable matrix to be zero.

Corollary 1. *The Frobenius root λ^* of an $n \times n$ decomposable nonnegative matrix A is zero if and only if $A^n = 0$.*

PROOF. We begin by noting that $\lambda^* = 0$ if and only if $A_{jj} = 0, j = 1, \ldots, n$, where A_{jj} denotes the typical principal matrix in the decomposition (1).

(*Sufficiency*). Since $A^n = 0$, $(QAQ')^n - QA^nQ' = 0$ for any permutation matrix Q. Hence the same holds for the permutation matrix P of (1). Thus

$$0 = PA^nP' = (PAP')^n = \begin{pmatrix} A_{11}^n & & & 0 \\ & A_{22}^n & & \\ & & \ddots & \\ 0 & & & A_{nn}^n \end{pmatrix},$$

which, in turn, means that $A_{jj} = 0, j = 1, \ldots, n$.

(*Necessity*). If $\lambda^* = 0$ then $A_{jj} = 0$, $j = 1, \ldots, n$. Hence $0 = (PAP')^n = PA^nP'$, whence $A^n = 0$. □

2 Power-Nonnegative Matrices

In some branches of economic theory one encounters matrices with elements some of which may be negative. In the analysis of the stability of competitive economies, for example, the matrix of partial derivatives of the excess demand functions plays a central role; and one hesitates to rule out gross complements. Similarly, in (matrix) multiplier analysis one hesitates to rule out inferior goods. It is useful therefore to be able to fall back on a considerably weaker notion of nonnegativity, *viz.*, power-nonnegativity. In the present section we derive some of the properties of power-nonnegative matrices.

Brauer [1961] has provided a mathematical analysis of power-*positive* matrices. However, power-positive matrices are necessarily indecomposable, and the strong assumption of indecomposability is one that economists do not make lightly. For this reason we have chosen to concentrate on the more general power-nonnegative matrices. For some economic applications of power-positive matrices, the reader may consult Sato [1972].

Definition 2. An $n \times n$ real matrix $\boldsymbol{A}$ is called *power-nonnegative of the* kth *degree* if there exist positive integers k_i such that $\boldsymbol{A}^{k_i} \geqq \mathbf{0}$ and if k is the smallest such integer.

Lemma 6. *Let $\boldsymbol{A}$ be a power-nonnegative matrix of the* kth *degree and suppose that $\boldsymbol{A}^k$ is indecomposable. Then the eigenvalue of $\boldsymbol{A}$ of greatest magnitude is real and simple, and the eigenvector associated with that eigenvalue is positive.*

PROOF Let us denote the eigenvalues of $\boldsymbol{A}$ by α_j, $j = 1, \ldots, n$. Then the eigenvalues of $\boldsymbol{A}^k$ are α_j^k, $j = 1, \ldots, n$. Since $\boldsymbol{A}^k$ is nonnegative and indecomposable, the Frobenius root of $\boldsymbol{A}^k$, say α_1^k, is positive and simple, and $\alpha_1^k \geqq |\alpha_j^k|, j = 2, \ldots, n$. Hence

$$|\alpha_1| \geqq |\alpha_j| \qquad (j \neq 1). \tag{6}$$

If the absolute value of α_1 is greater than that of any other root, it is obviously real and simple. Suppose that $\alpha_2, \ldots, \alpha_p$ have the same absolute value as α_1. As is well known, α_j^k $(j = 2, \ldots, p)$ can be expressed as

$$\alpha_j^k = \alpha_1^k \omega^{j-1} \qquad (j = 2, \ldots, p), \tag{7}$$

where ω is the primitive pth root of 1. Suppose that α_1 is complex. Then there exists an index j_0 such that

$$\bar{\alpha}_1 = \alpha_{j_0} \qquad (j_0 \in \{2, \ldots, p\}). \tag{8}$$

From (7) and (8),

$$\alpha_1^k = (\bar{\alpha}_1)^k = \alpha_1^k \omega^{j_0 - 1}, \tag{9}$$

a contradiction since $j_0 - 1$ is less than p. Thus α_1 cannot be both complex and of the same magnitude as other roots. Nor can α_1 be real and not simple. For then there would exist a root α_j such that

$$\alpha_j = \alpha_1 \qquad (j \in \{2, \ldots, p\}), \tag{10}$$

which implies that $\alpha_1^k = \alpha_1^k \omega^{j-1}$, a further contradiction. Hence α_1 is real and simple.

The second part of the proposition is trivially true. □

Theorem 3. *Let* A *be a power-nonnegative matrix of the* kth *degree. Then the eigenvalue of* A *with the greatest absolute value is real and the associated eigenvector is semipositive.*

PROOF. It suffices to verify the proposition for A^k decomposable. Let P be the permutation matrix of (1), and let α_j^* be the eigenvalue of A_{jj} with greatest absolute value. Then the eigenvalue of A with the greatest absolute value is found by calculating $\max_{1 \leqq j \leqq m} |\alpha_j^*|$. The conclusion follows from Lemma 6. □

Corollary 2. *Let* A *be a power-nonnegative matrix of the* kth *degree.*

i. *If* k *is odd then the eigenvalue of* A *of greatest magnitude is also positive.*
ii. *If* A *contains a nonnegative row or column and if* A^k *is indecomposable then the eigenvalue of* A *of greatest magnitude is also positive.*
iii. *If* $l'A < l'$ *then the eigenvalue of* A *of greatest magnitude is less than* 1.

PROOF. Suppose that α_1 is the eigenvalue of A of greatest magnitude.

(i) Since $\alpha_1^k > 0$, k odd, and since α_1 is real, α_1 must be positive.

(ii) Let x be the eigenvector corresponding to α_1. Then, from Lemma 6, x is positive. Let a^i be a nonnegative row of A. In fact, $a^i \geq 0$; for, otherwise, the ith row of any power of A is zero and A^k is decomposable. Thus

$$a^i \cdot x = \alpha_1 x_i$$

and therefore

$$\alpha_1 = \frac{a^i \cdot x}{x_i} > 0.$$

(iii) Employing the same notation as in (ii) above, we have

$$Ax = \alpha_1 x$$

where x is, at worst, semipositive. Premultiplying by l' and recalling the assumption, we obtain

$$\sum_i x_i > (l'A)x = \alpha_1 l'x = \alpha_1 \left(\sum_i x_i \right).$$

The conclusion then follows from the fact that $\sum_i x_i > 0$. □

3 Some Special Matrices

In the present chapter we examine the properties of several special matrices. Each of these matrices plays a central role in one or more branches of the neoclassical theories of international trade, public finance and income distribution. Indeed, it may be said that the properties of these matrices are the essential analytical content of the theories. For applications of the mathematical results developed in this chapter, the reader may refer to Gale and Nikaido [1965], Kuhn [1968], Nikaido [1968], Chipman [1969], Kemp and Wegge [1969], Wegge and Kemp [1969], Inada [1971], Uekawa [1971], Uekawa, Kemp and Wegge [1973], and Ethier [1974].

1 P-, NP-, N-, PN-, and Semi-PN Matrices

Definition 1 (Gale and Nikaido [1965]). An $n \times n$ real matrix is said to be a P-*matrix* if all of its principal minors are positive.

Definition 2 (Inada [1971]). An $n \times n$ real matrix is said to be an N-*matrix* if all of its principal minors are negative. An N-matrix is said to be *of the first kind* if it contains at least one positive element; otherwise, it is said to be *of the second kind*.

Definition 3. An $n \times n$ real matrix is said to be an NP-*matrix* if its principal minors of order r have the sign of $(-1)^r$, $r = 1, \ldots, n$. An NP-matrix is sometimes called a *Hicksian matrix*.

Definition 4. An $n \times n$ real matrix is said to be a PN-*matrix* if its principal minors of order r have the sign of $(-1)^{r+1}$, $r = 1, \ldots, n$.

Definition 5 (Nikaido [1968]). An $n \times n$ real matrix is said to be a *weak* P-*matrix* if all of its principal minors are nonnegative and if its determinant is positive. A weak NP-matrix is defined similarly.

Definition 6 (Arrow [1974]). An $n \times n$ real matrix is said to be a P_0-*matrix* if all of its principal minors are nonnegative.

Definition 7 (Uekawa, Kemp, and Wegge [1972]). An $n \times n$ real matrix is to be a *semi*-PN *matrix* if every principal minor of order r $(2 \leqq r \leqq n)$ has the sign of $(-1)^{r-1}$.

Remark. If a nonnegative square matrix A is a semi-PN matrix then the off-diagonal elements of A are all positive. To see this, suppose the contrary, that for some i and some j, $i \neq j$, $a_{ij} = 0$. Then

$$\det\begin{pmatrix} a_{ii} & a_{ij} \\ a_{ji} & a_{jj} \end{pmatrix} \geqq 0,$$

a contradiction.

Definition 8. An $n \times n$ real matrix is said to have the *Minkowski property* if it possesses an inverse with nonnegative diagonal elements and non positive off-diagonal elements.

Definition 9. An $n \times n$ real matrix is said to have the *Metzler property* if it possesses an inverse with nonpositive diagonal elements and nonnegative off-diagonal elements.

Lemma 1. *An $n \times n$ real matrix is a* P-*matrix* (*respectively,* N-, PN-, NP-*matrix, matrix with the Minkowski or Metzler property*) *if and only if its transpose is a* P-*matrix* (*respectively,* N-, PN-, NP-*matrix, matrix with the Minkowski or Metzler property*).

PROOF. The proof is easy and is left to the reader.

Lemma 2. *An $n \times n$ real matrix A is a* P-*matrix if and only if $-A$ is an* NP-*matrix, and A is an* N-*matrix if and only if $-A$ is a* PN-*matrix.*

PROOF. The proposition follows from Definitions 1–4 and from the elementary properties of determinants. □

Lemma 3. *If an $n \times n$ real matrix A is a* P-*matrix* (*respectively,* N-, PN-, *or* NP-*matrix*) *then so is the matrix obtained from A by a simultaneous and identical interchange of rows and columns.*

PROOF. The proposition follows from the fact that the value of any principal minor is not affected by such an interchange of its rows and columns. □

Lemma 4. *If an $n \times n$ real matrix $\boldsymbol{A}$ is a* P-*matrix (respectively,* N-, PN-, *or* NP-*matrix) then so is $\boldsymbol{B}_J \equiv \boldsymbol{I}_J \boldsymbol{A} \boldsymbol{I}_J$, where J is any subset of $N = \{1, \ldots, n\}$ and $\boldsymbol{I}_J$ is a matrix obtained from the identity matrix $\boldsymbol{I}$ by replacing each* jth *row $\boldsymbol{e}^j$ by $-\boldsymbol{e}^j$, $j \in J$.*

PROOF. Let us denote by

$$\det \boldsymbol{I}_J \boldsymbol{A} \boldsymbol{I}_J \begin{pmatrix} i_1, i_2, \ldots, i_r \\ i_1, i_2, \ldots, i_r \end{pmatrix} = \det \boldsymbol{B}_J^r$$

any principal submatrix of $\boldsymbol{I}_J \boldsymbol{A} \boldsymbol{I}_J$. Noting that $(\boldsymbol{B}_J^r)_{jk} = a_{i_j i_k}$ if both i_j and i_k are in J or if both i_j and i_k are not in J and that $(\boldsymbol{B}_J^r) = -a_{i_j i_k}$ otherwise, we have

$$\det \boldsymbol{B}_J^r = (-1)^{2h} \det \boldsymbol{A} \begin{pmatrix} i_1, i_2, \ldots, i_r \\ i_1, i_2, \ldots, i_r \end{pmatrix},$$

where h is the number of indices both in J and in $\{i_1, \ldots, i_r\}$. Needless to say, if $J \cap \{i_1, \ldots, i_r\} = \varnothing$ then $h = 0$. □

Remark. While $\boldsymbol{I}_J \boldsymbol{A} \boldsymbol{I}_J$ must be an N-matrix if $\boldsymbol{A}$ is an N-matrix, $\boldsymbol{I}_J \boldsymbol{A} \boldsymbol{I}_J$ and $\boldsymbol{A}$ need not be N-matrices of the same kind.

Lemma 5. *Let $\boldsymbol{A}$ be an $n \times n$ real matrix. If $\boldsymbol{A}$ is a* P-*matrix (respectively,* N-, NP-, *or* PN-*matrix) then so is every principal submatrix of $\boldsymbol{A}$.*

PROOF. The proof follows directly from Definitions 1–4. □

2 Fundamental Properties of P-, N-, PN-, and NP-Matrices

Theorem 1 (Gale and Nikaido [1965]). *If an $n \times n$ real matrix $\boldsymbol{A}$ is a* P-*matrix then the system of linear inequalities*

$$\boldsymbol{A}\boldsymbol{x} \leqq \boldsymbol{0}, \tag{1}$$

$$\boldsymbol{x} \geqq \boldsymbol{0}, \tag{2}$$

has only the trivial solution $\boldsymbol{x} = \boldsymbol{0}$.

PROOF. It suffices to show that if the system (1) has a nonnegative solution it cannot be semipositive. The proof is by induction. The case $n = 1$ is trivial. Suppose that the proposition is true for $n - 1$. Let $\tilde{\boldsymbol{x}}$ be a nonnegative solution of (1). If $\boldsymbol{A}\tilde{\boldsymbol{x}} = \boldsymbol{0}$ then $\tilde{\boldsymbol{x}}$ must be $\boldsymbol{0}$ since $\boldsymbol{A}$ is nonsingular by assumption. This leaves two possibilities: that $\boldsymbol{A}\tilde{\boldsymbol{x}} < \boldsymbol{0}$; and that $\boldsymbol{A}\tilde{\boldsymbol{x}} \leq \boldsymbol{0}$ with at least one strict equality. However, the first of these possibilities can be reduced to the second. To see this, suppose that $\boldsymbol{A}\tilde{\boldsymbol{x}} < \boldsymbol{0}$, $\tilde{\boldsymbol{x}} \geqq \boldsymbol{0}$. Then, since the diagonal elements of $\boldsymbol{A}$ are positive, there exists $\boldsymbol{y} \geq \tilde{\boldsymbol{x}} \geqq \boldsymbol{0}$ such that $\boldsymbol{A}\boldsymbol{y} \leq \boldsymbol{0}$ with at least one strict equality. In what follows, therefore, we suppose that there exists $y \geq 0$ such that $Ay \leq 0$ with at least one strict equality. For concreteness we suppose that $(Ay)_{i_0} = 0$.

Applying the usual method of elimination, with $a_{i_0 i_0} > 0$ as the pivotal element, we obtain the transformed system

$$\sum_{j \neq i_0} a_{ij}^* y_j \leqq 0 \qquad (i \neq i_0), \tag{3}$$

$$\sum_{j=1}^{n} a_{i_0 j}^* y_j = 0, \tag{4}$$

where

$$a_{i_0 j}^* = \frac{a_{i_0 j}}{a_{i_0 i_0}} \quad \text{and} \quad a_{ij}^* = a_{ij} - a_{i_0 j}^* \cdot a_{i i_0} \quad (j = 1, \ldots, n;\, i \neq i_0).$$

Let us denote by A^* a matrix consisting of $a_{ij}^*(i,j = 1, \ldots, i_0 - 1, i_0 + 1, \ldots, n)$. Then we easily see that

$$\det A^* = \det(a_{ij}^*) = \frac{1}{a_{i_0 i_0}} \det A > 0. \tag{5}$$

Moreover, any principal minor of det A^* can be expressed as

$$\det A^* \begin{pmatrix} i_1, i_2, \ldots, i_q \\ i_1, i_2, \ldots, i_q \end{pmatrix}$$
$$= \frac{1}{a_{i_0 i_0}} \det A \begin{pmatrix} i_0, i_1, \ldots, i_q \\ i_0, i_1, \ldots, i_q \end{pmatrix} > 0 \qquad (i_j \neq i_0;\, i_j = i_1, \ldots, i_q). \tag{6}$$

From (5) and (6), A^* is a P-matrix of order $n - 1$. From (3) and the induction assumption, therefore, $y_j = 0$ for all $j \neq i_0$; and, in view of (4), this in turn implies that $y_{i_0} = 0$. Thus the proposition assumed to hold for the system of order $n - 1$ holds also for the system of order n. □

Corollary 1 (Inada [1971]). (i) *If A is an* N-*matrix of the second kind then the inequalities $Ax \geqq 0$, $x \geqq 0$ have only the zero solution and the inequalities $Ax \leqq 0$, $x \geqq 0$ have a nontrivial (semipositive) solution.* (ii) *If A is an* N-*matrix of the first kind then the inequalities $Ax \leq 0$, $x \geq 0$ have only the zero solution and the inequalities $Ax \geqq 0$, $x \geqq 0$ have a nontrivial (semipositive) solution.*

PROOF. (i) Since A is nonpositive and $a_{ii} < 0$, $i = 1, \ldots, n$, $Ax \leq 0$ for any semipositive x. It follows that the inequalities $Ax \geqq 0$, $x \geqq 0$ have only the trivial zero solution.

(ii) The first half of proposition (ii) can be established in much the same way as Theorem 1. Indeed, if one bears in mind (a) that the diagonal elements of A are now negative instead of positive and (b) that A must now be taken to be of at least order 2 (a real matrix of order 1 cannot be an N-matrix of the first kind since it cannot contain both a positive element and a negative diagonal), then the earlier proof suffices.

Turning to the second half of proposition (ii), let us suppose that the system $\boldsymbol{A}\boldsymbol{x} \geqq \boldsymbol{0}$, $\boldsymbol{x} \geqq \boldsymbol{0}$ has no nontrivial solution. Then, from (i) of Corollary 1 in Chapter 1, there exists $\boldsymbol{y} \geq \boldsymbol{0}$ such that $\boldsymbol{y}'\boldsymbol{A} < \boldsymbol{0}$. Since $\boldsymbol{A}'$ is also an N-matrix of the first kind and $\boldsymbol{A}'\boldsymbol{y} < \boldsymbol{0}$, the inequalities $\boldsymbol{A}'\boldsymbol{x} \leqq \boldsymbol{0}$, $\boldsymbol{x} \geqq \boldsymbol{0}$ possess a nontrivial solution, in contradiction of the first part of (ii), already established. □

Remark. The assertion just verified (that is, the second part of (ii) in Corollary 1) can be further strengthened. Let us consider the system of linear inequalities

$$\boldsymbol{A}\boldsymbol{x} > \boldsymbol{0}, \qquad \boldsymbol{x} > \boldsymbol{0}, \tag{7}$$

where $\boldsymbol{A}$ is an N-matrix of the first kind. This system may be rewritten as

$$(\boldsymbol{I}, -\boldsymbol{A})\begin{pmatrix}\boldsymbol{u}\\ \boldsymbol{x}\end{pmatrix} = \boldsymbol{0}, \qquad \begin{pmatrix}\boldsymbol{u}\\ \boldsymbol{x}\end{pmatrix} > \boldsymbol{0}, \tag{8}$$

where $\boldsymbol{u}$ is an n-dimensional vector. From Theorem 2 in Chapter 1, (8) has a solution if and only if the system

$$\begin{pmatrix}\boldsymbol{I}\\ -\boldsymbol{A}\end{pmatrix}\boldsymbol{p} \geq \boldsymbol{0}$$

has no solution, that is, if and only if the system

$$\begin{pmatrix}\boldsymbol{I}\\ -\boldsymbol{A}'\end{pmatrix}\boldsymbol{p} \geqq \boldsymbol{0} \quad \text{or} \quad \boldsymbol{A}'\boldsymbol{p} \leqq \boldsymbol{0}, \boldsymbol{p} \geqq \boldsymbol{0} \tag{9}$$

has only the trivial solution. Since (9) is now known to have only the trivial solution, we may conclude that (7) possesses a solution.

Definition 10. Let $\boldsymbol{A}$ be an $n \times n$ real matrix and $\boldsymbol{x}$ a given real n-dimensional vector. Then $\boldsymbol{A}$ is said to *reverse the sign of* $\boldsymbol{x}$ if

$$x_i(\boldsymbol{A}\boldsymbol{x})_i \leqq 0 \qquad (i = 1, \ldots, n). \tag{10}$$

Theorem 2 (Gale and Nikaido [1965]). *An $n \times n$ real matrix $\boldsymbol{A}$ is a* P-*matrix if and only if $\boldsymbol{A}$ does not reverse the sign of any nonzero vector.*

Before proceeding to the proof of Theorem 2 it will be convenient to verify the following lemma.

Lemma 6. *Let $\boldsymbol{A}$ be an $n \times n$ real matrix and let $f(\mu) = \det(\mu\boldsymbol{I} - \boldsymbol{A})$ be its characteristic function. Then $\boldsymbol{A}$ has a nonpositive eigenvalue if* $\det \boldsymbol{A} \leqq 0$.

PROOF. Let $\lambda_j, j = 1, \ldots, n$, be the eigenvalues of $\boldsymbol{A}$, the first $2k$ of them being complex, with

$$\lambda_j = \bar{\lambda}_{j+1} \qquad (j = 2s - 1;\, s = 1, \ldots, k),$$

and the remainder real, with

$$\lambda_j = \bar{\lambda}_j \qquad (j = 2k + 1, \ldots, n),$$

where $\bar{\lambda}_j$ is the complex conjugate of λ_j. Since, as is well known, $f(\mu) = \prod_{j=1}^n (\mu - \lambda_j)$, we have

$$\det(0 \cdot \boldsymbol{I} - \boldsymbol{A}) = (-1)^n \det \boldsymbol{A} = f(0) = (-1)^n \prod_{j=1}^n \lambda_j,$$

whence

$$0 \geqq \det \boldsymbol{A} = \left(\prod_{j=1}^k |\lambda_{2j-1}|^2\right) \cdot \left(\prod_{j=2k+1}^n \lambda_j\right).$$

Hence $\boldsymbol{A}$ must possess a real nonpositive eigenvalue. □

Proof of Theorem 2 (*Necessity*). Suppose that there exists a nonzero vector $\boldsymbol{x}$ the sign of which is reversed by the P-matrix $\boldsymbol{A}$. We take first the case in which $\boldsymbol{x}$ is semipositive and define a subset of N,

$$M = \{i : x_i > 0, i \in N\}.$$

Clearly M is nonempty. Then, from (10),

$$(\boldsymbol{A}\boldsymbol{x})_i = \sum_{j \in M} a_{ij} x_j \leqq 0 \quad \text{for any } i \in M \tag{11}$$

However, there is associated with M a principal submatrix of $\boldsymbol{A}$, and this submatrix is also a P-matrix; hence, from Theorem 1, (11) has no semipositive solution. Thus we arrive at a contradiction.

Let us suppose now that $\boldsymbol{x}$ is not semipositive, so that at least one component of $\boldsymbol{x}$ is negative. Let us define the (nonempty) subset of N,

$$J = \{i : x_i < 0, i \in N\}.$$

Then $\boldsymbol{y} = \boldsymbol{I}_J \boldsymbol{x} \geq \boldsymbol{0}$ and, by virtue of Lemma 4, $\boldsymbol{I}_J \boldsymbol{A} \boldsymbol{I}_J$ is a P-matrix. Moreover, it is easy to see that $y_i \boldsymbol{e}^i = x_i(\boldsymbol{e}^i \boldsymbol{I}_J)$ and that $x_i(\boldsymbol{A}\boldsymbol{x})_i \leq 0$, $i = 1, \ldots, n$, if and only if $(x_i \boldsymbol{e}^i)(\boldsymbol{A}\boldsymbol{x}) \leqq 0$, $i = 1, \ldots, n$. Hence

$$\begin{aligned} y_i(\boldsymbol{I}_J \boldsymbol{A} \boldsymbol{I}_J \boldsymbol{y})_i &= (y_i \boldsymbol{e}^i)(\boldsymbol{I}_J \boldsymbol{A} \boldsymbol{I}_J \boldsymbol{y}) \\ &= (x_i \boldsymbol{e}^i \boldsymbol{I}_J)(\boldsymbol{I}_J \boldsymbol{A} \boldsymbol{I}_J \boldsymbol{I}_J \boldsymbol{x}) \\ &= (x_i \boldsymbol{e}^i)(\boldsymbol{A}\boldsymbol{x}) \\ &= x_i(\boldsymbol{A}\boldsymbol{x})_i \leqq 0. \end{aligned}$$

From this point the argument proceeds as in the preceding paragraph, and ends in the same contradiction.

(*Sufficiency*). It suffices to show that, under the assumed condition, any principal minor of $\boldsymbol{A}$ is positive. Suppose that there exists a principal submatrix

$$\hat{\boldsymbol{A}} = \boldsymbol{A}\begin{pmatrix} i_1, i_2, \ldots, i_r \\ i_1, i_2, \ldots, i_r \end{pmatrix}$$

of A such that $\det \hat{A} \leqq 0$. Then $\hat{A}$ has a real nonpositive eigenvalue λ and a real eigenvector $x \neq 0$ corresponding to λ. Hence

$$x_i(\hat{A}x)_i = \lambda x_i^2 \qquad (i \in K = \{i_1, i_2, \ldots, i_r\}).$$

Let us now define y_i, $i = 1, \ldots, n$, by

$$y_i = x_i \qquad (i \in K),$$
$$y_i = 0 \qquad (i \notin K).$$

Then

$$y_i \sum_{j=1}^{n} a_{ij} y_j = y_i\Big(\sum_{j \in k} a_{ij} y_j\Big) = 0 \qquad (i \notin K)$$

and

$$y_i \sum_{j=1}^{n} a_{ij} y_j = x_i(\hat{A}x)_i = \lambda x_i^2 \leqq 0 \qquad (i \in K).$$

Thus we have found a nonzero vector y of which the sign is reversed by A. This is a contradiction. □

Remark. For N-matrices there is nothing comparable to Theorem 2. For, as appears from the above proof of sufficiency, an N-matrix has an eigenvector which is reversed by the matrix.

Theorem 3 (Arrow [1974]). *Let A be an $n \times n$ matrix. The following assertions are equivalent to each other.*

i. *A is a P_0-matrix, that is, all principal minors of A are nonnegative;*
ii. *for any $n \times n$ positive diagonal matrix D (hereafter,* pdm D*), $A + D$ is a P-matrix;*
iii. *for any nonzero n-vector x, there is an index $i \in N$ such that $x_i \neq 0$ and $x_i(Ax)_i \geqq 0$;*
iv. *for any $n \times n$* pdm *D, every real eigenvalue of DA is nonnegative.*

Proof. We begin by showing that (i) ⇒ (ii). The proof is by induction on n. Differentiating $\det(A + D)$ with respect to d_{ii}, the ith diagonal element of D, we obtain

$$\frac{\partial}{\partial d_{ii}} \det(A + D) = \det(A + D)\binom{i}{i} = \det\left(A\binom{i}{i} + D\binom{i}{i}\right)$$

where, for example, $A\binom{i}{i}$ is the submatrix obtained from A by deleting the ith row and ith column. From Lemma 5 and the induction assumption,

$$A\binom{i}{i} + D\binom{i}{i}$$

is a P-matrix. Hence

$$\det(A + D)\begin{pmatrix} i \\ i \end{pmatrix} > 0,$$

that is, $\det(A + D)$ is a strictly increasing function of d_{ii}, which implies that $(A + D)$ is a P-matrix. It remains to note that the implication holds for $n = 1$.

Consider now the implication (ii) $\Rightarrow$ (iii). Suppose that $(A + D)$ is a P-matrix. Then, from Theorem 2, there exists an index $i \in N$ such that $x_i \neq 0$ and $x_i((A + D)x)_i > 0$, whence

$$x_i(Ax)_i + d_{ii}x_i^2 > 0.$$

If there is an index i of $I = \{i \in N : x_i((A + D)x)_i > 0\}$ such that $x_i(Ax)_i \geqq 0$, there remains nothing to be proved. Accordingly, let us suppose that $x_i(Ax)_i < 0$ for every $i \in I$. Then we can choose a new pdm D' with $0 < d'_{ii} < -x_i(Ax)_i/x_i^2$. Evidently $A + D'$ reverses the sign of the vector x under consideration. Hence $A + D'$ is not a P-matrix, which is a contradiction.

We turn to the implication (iii) $\Rightarrow$ (iv). Let λ and x be respectively a real eigenvalue and the associated eigenvector of DA, so that

$$\lambda x_j = d_{jj}(Ax)_j \qquad (j = 1, \ldots, n).$$

In view of (iii), there is an index $i \in N$ such that $x_i \neq 0$ and $x_i(Ax)_i \geqq 0$. Hence

$$\lambda x_i^2 = d_{ii}x_i(Ax)_i \geqq 0,$$

from which the nonnegativity of λ follows directly.

Finally, we consider the implication (iv) $\Rightarrow$ (i). Let J and K be any subsets of N. Then A_{JK} is the submatrix of A with elements a_{jk} $(j \in J,\ k \in K)$. It suffices to show that $\det A_{JJ}$ is nonnegative for any subset J of N. Suppose that, for some J, $\det A_{JJ} < 0$. Then any eigenvalue μ of A_{JJ} is nonzero. Now for any positive t we may define an $n \times n$ diagonal matrix

$$D(t) = \begin{pmatrix} I_{JJ} & 0 \\ 0 & tI_{\bar{J}\bar{J}} \end{pmatrix}$$

where $\bar{J} = \{j : j \in N, j \notin J\}$. The eigenvalues of $D(t)A$ are continuous in t.[1] Moreover

$$D(0)A = \begin{pmatrix} A_{JJ} & A_{J\bar{J}} \\ 0 & 0 \end{pmatrix},$$

implying that the eigenvalues of $D(0)A$ consist of the eigenvalues of A_{JJ} and zeros. It follows that as t goes to zero the eigenvalues of $D(t)A$ approach zero or the eigenvalues of A_{JJ}. Let $\Lambda_1 = \{\lambda(t) : \lambda(t) \to \mu \text{ as } t \to 0\}$ and let $\Lambda_2 = \{\lambda(t) : \lambda(t) \to 0 \text{ as } t \to 0\}$, where $\lambda(t)$ is an eigenvalue of $D(t)A$. Let us denote by μ_1 the eigenvalue of A_{JJ} with the smallest absolute value and choose an ε such that $0 < \varepsilon < \frac{1}{2}|\mu_1|$. Then there exists t_0 such that

$$|\lambda(t)| < \varepsilon \quad \text{for } \lambda(t) \in \Lambda_2 \text{ and } 0 < t < t_0 \tag{12a}$$

and

$$|\mu_1 - \lambda(t)| < \varepsilon \quad \text{for } \lambda(t) \in \Lambda_1 \text{ and } 0 < t < t_0. \tag{13}$$

From (13), and in view of the choice of ε,

$$|\lambda(t)| > \varepsilon \quad \text{for } \lambda(t) \in \Lambda_1 \text{ and } 0 < t < t_0. \tag{12b}$$

It follows from (12a) and (12b) that, for $t < t_0$, the eigenvalue conjugate to $\lambda(t) \in \Lambda_1$ also lies in Λ_1 and that, as t goes to zero,

$$\prod_{\lambda(t) \in \Lambda_1} \lambda(t) \geqq 0$$

goes to the product of the eigenvalues of A_{JJ}, which equals det A_{JJ} and is negative by hypothesis. On the other hand, for $t > 0$ any real eigenvalue of $D(t)A$ is nonnegative by assumption, so that

$$\lim_{t \to 0} \prod_{\lambda(t) \in \Lambda_1} \lambda(t) \geqq 0,$$

Hence det A_{JJ} is nonnegative. □

Theorem 4 (Uekawa [1971]). *An $n \times n$ real matrix A is a* P*-matrix if and only if, for any J, $\varnothing \subseteq J \subseteq N$, the inequality $x'(I_J A I_J) > 0'$ has a positive solution.*

In proving necessity we shall need the following proposition.

Lemma 7. *Let A be a* P*-matrix of order n. If the system of linear inequalities $x'A > 0'$ has no positive solution then it has no semipositive solution.*

Proof. It suffices to verify the contrapositive of the assertion. Suppose that there exists $x \geq 0$ such that $x'A > 0'$. If x is positive, there is nothing to be shown. Suppose then that x contains at least one zero element and let $I_x = \{i : x_i = 0, i \in N\}$. By assumption, I_x is a nonempty proper subset of N. Consider an arbitrary element, say i, of I_x. Then we can define a subset $K_i = \{k : a_{ik} < 0, i \neq k\}$ of N. If $K_i \neq \varnothing$ we define the positive number

$$\varepsilon_i = \tfrac{1}{2} \min_{k \in K_i} \sum_{j \neq i} x_j a_{jk} (-a_{ik})^{-1}.$$

Otherwise, ε_i is set equal to any positive number. Then

$$0 < \sum_{j \neq i} x_j a_{jk} + \varepsilon_i a_{ik} \quad \text{for any } k \in K_i$$

and, since a_{ii}, $i = 1, \ldots, n$, is positive,

$$0 < \sum_{j \neq i} x_j a_{ji} + \varepsilon_i a_{ii}.$$

In vector notation, there exists an

$$\boldsymbol{x}^i = (x_1, \ldots, x_{i-1}, \varepsilon_i, x_{i+1}, \ldots, x_n) \geq \boldsymbol{0}' \quad \text{such that} \quad \boldsymbol{x}^i\boldsymbol{A} > \boldsymbol{0}'.$$

Since $I_{\boldsymbol{x}}$ is a proper subset of N, after a finite number of repetitions of this procedure we obtain $\boldsymbol{y} > \boldsymbol{0}$ such that $\boldsymbol{y}'\boldsymbol{A} > \boldsymbol{0}'$. □

Proof of Theorem 4 (*Necessity*). $\boldsymbol{A}$ is a P-matrix; hence, from Lemma 4, so is $\boldsymbol{I}_J\boldsymbol{A}\boldsymbol{I}_J$. Suppose that necessity is false, so that there exists a subset J of N such that $\boldsymbol{x}'(\boldsymbol{I}_J\boldsymbol{A}\boldsymbol{I}_J) = \boldsymbol{x}'\boldsymbol{B} > \boldsymbol{0}'$ has no positive solution and, by virtue of Lemma 7, $\boldsymbol{x}'\boldsymbol{B} > \boldsymbol{0}'$ has no semipositive solution. Then, from (ii) of Corollary 1 in Chapter 1, there exists $\boldsymbol{u} \geq \boldsymbol{0}$ such that $\boldsymbol{B}\boldsymbol{u} \leqq \boldsymbol{0}$, contradicting Theorem 1.

(*Sufficiency*). Suppose that $\boldsymbol{A}$ is not a P-matrix. Then, from Theorem 2, there exists a nonzero vector $\boldsymbol{x}$ such that $x_i(\boldsymbol{A}\boldsymbol{x})_i \leq 0$, $i = 1, \ldots, n$. Let us define the subset of N

$$J_{\boldsymbol{x}} = \{j : x_j < 0\} \cup \{j : x_j = 0, (\boldsymbol{A}\boldsymbol{x})_j > 0\}.$$

We consider separately the following three cases: (i) $J_{\boldsymbol{x}} = \varnothing$; (ii) $J_{\boldsymbol{x}} = N$; (iii) $\varnothing \neq J_{\boldsymbol{x}} \subset N$.

(i) Since $J_{\boldsymbol{x}} = \varnothing$ we have

$$\{j : x_j < 0\} = \varnothing$$

and

$$\{j : x_j = 0, (\boldsymbol{A}\boldsymbol{x})_j > 0\} = \varnothing.$$

Hence, for any $j \in N$, both $x_j \geqq 0$ and $x_j \neq 0$ or $(\boldsymbol{A}\boldsymbol{x})_j \leqq 0$. It follows that $\boldsymbol{x} > \boldsymbol{0}$ or $\boldsymbol{A}\boldsymbol{x} \leqq \boldsymbol{0}$ has a semipositive solution ($\boldsymbol{x} \neq \boldsymbol{0}$, by assumption). Since $\boldsymbol{A}$ reverses the sign of $\boldsymbol{x}$, if $\boldsymbol{x} > \boldsymbol{0}$ then $\boldsymbol{A}\boldsymbol{x} \leqq \boldsymbol{0}$. Thus, if $J_{\boldsymbol{x}} = \varnothing$ then $\boldsymbol{A}\boldsymbol{x} \leqq \boldsymbol{0}$ has a semipositive solution. Again applying (ii) of Corollary 1 in Chapter 1, the system of inequalities $\boldsymbol{y}'\boldsymbol{A} > \boldsymbol{0}$ has no semipositive solution, contradicting the assumption for $J = N$.

(ii) If $J_{\boldsymbol{x}} = N$ then $\boldsymbol{x} \leqq \boldsymbol{0}$. Moreover, the set $\{j : x_j < 0\} \neq \varnothing$, for $\boldsymbol{x} \neq \boldsymbol{0}$. Since $\boldsymbol{A}$ reverses the sign of $\boldsymbol{x}$,

$$(\boldsymbol{A}\boldsymbol{x})_j \geqq 0 \quad \text{for any } j \in \{j : x_j < 0\}.$$

Hence the system of linear inequalities

$$\boldsymbol{A}\boldsymbol{y} \leqq \boldsymbol{0},$$
$$\boldsymbol{y} \geqq \boldsymbol{0},$$

has a solution $-\boldsymbol{x} \geq \boldsymbol{0}$. In view of Theorem 4 in Chapter 1, therefore, the system $\boldsymbol{z}'\boldsymbol{A} > \boldsymbol{0}'$ has no semipositive solution, which is a contradiction.

(iii) If $J_{\boldsymbol{x}}$ is a nonempty proper subset of N then $\bar{J}_{\boldsymbol{x}}$ is nonempty (here $\bar{J}_{\boldsymbol{x}}$ denotes the complement of J). It follows that, for any $j \in \bar{J}_{\boldsymbol{x}}$, $x_j \geqq 0$; for otherwise $\bar{J}_{\boldsymbol{x}}$ would intersect $J_{\boldsymbol{x}}$, a contradiction. Since $\boldsymbol{A}$ reverses the sign of $\boldsymbol{x}$ we have $(\boldsymbol{A}\boldsymbol{x})_j \leqq 0$ for any $j \notin J_{\boldsymbol{x}}$ and $(\boldsymbol{A}\boldsymbol{x})_j \geqq 0$ for any $j \in J_{\boldsymbol{x}}$ that is,

$$\begin{aligned} \boldsymbol{0} \geqq \boldsymbol{I}_{J_x}(\boldsymbol{A}\boldsymbol{x}) &= \boldsymbol{I}_{J_x}\boldsymbol{A}\boldsymbol{I}_{J_x}\boldsymbol{I}_{J_x}\boldsymbol{x} \\ &= (\boldsymbol{I}_{J_x}\boldsymbol{A}\boldsymbol{I}_{J_x})(\boldsymbol{I}_{J_x}\boldsymbol{x}). \end{aligned}$$

This means that the system of linear inequalities

$$(\boldsymbol{I}_{J_x}\, \boldsymbol{A}\boldsymbol{I}_{J_x})\boldsymbol{y} \leqq \boldsymbol{0}$$

has a semipositive solution

$$\boldsymbol{I}_{J_x}\boldsymbol{x},$$

since $\boldsymbol{x} \neq \boldsymbol{0}$ and $x_j \leqq 0$ for $j \in J_{\boldsymbol{x}}$. From Theorem 4 in Chapter 1, the system

$$\boldsymbol{z}'(\boldsymbol{I}_{J_x}\, \boldsymbol{A}\boldsymbol{I}_{J_x}) > \boldsymbol{0}'$$

has no semipositive solution, contradicting the assumption. □

For the remainder of this chapter $\boldsymbol{A}$ will be taken to be a nonnegative $n \times n$ matrix. In addition, several alternative conditions will be imposed on $\boldsymbol{A}$. Economic interpretations of some of these conditions may be found in Uekawa [1971], and in Uekawa, Kemp, and Wegge [1973].

We turn now to a series of corollaries to Theorem 4.

Corollary 2. *A nonnegative $n \times n$ matrix $\boldsymbol{A}$ is a* P-*matrix if and only if, for any nonempty proper subset J of N, there exists $\boldsymbol{x} > \boldsymbol{0}$ such that*

$$\boldsymbol{x}'(\boldsymbol{I}_J \boldsymbol{A}\boldsymbol{I}_J) > \boldsymbol{0}'.$$

Proof (*Sufficiency*). $\boldsymbol{A}$ is a nonnegative matrix and, by assumption, $\boldsymbol{x}'(\boldsymbol{I}_J \boldsymbol{A}\boldsymbol{I}_J) > \boldsymbol{0}'$ has a positive solution for any J, $\varnothing \neq J \subset N$; hence $a_{ii} > 0$, $i = 1, \ldots, n$, and, for any $\boldsymbol{x} > \boldsymbol{0}$, $\boldsymbol{x}'\boldsymbol{A} > \boldsymbol{0}'$. It follows that the system

$$\boldsymbol{x}'(\boldsymbol{I}_J \boldsymbol{A}\boldsymbol{I}_J) > \boldsymbol{0}'$$

has a positive solution for $J = \varnothing$ or $J = N$. Hence, from Theorem 4, $\boldsymbol{A}$ is a P-matrix.

Necessity is obvious. □

Corollary 3. *An $n \times n$ real matrix $\boldsymbol{A}$ is a* P-*matrix if and only if, for any subset J of N, the system of linear inequalities*

$$(\boldsymbol{I}_J \boldsymbol{A}\boldsymbol{I}_J)\boldsymbol{x} > \boldsymbol{0}$$

has a positive solution.

Proof. The conclusion follows from Lemma 2 and Theorem 4. □

Corollary 4. *An $n \times n$ nonnegative matrix $\boldsymbol{A}$ is a* P-*matrix if and only for any nonempty proper subset J of N, the system of linear inequalities*

$$(\boldsymbol{I}_J \boldsymbol{A}\boldsymbol{I}_J)\boldsymbol{x} > \boldsymbol{0}$$

has a positive solution.

Proof. The proposition follows from Corollary 2 and Lemma 2. □

We shall find it convenient to refer to the condition of Corollary 4 as Condition [1].

3 Fundamental Properties of Semi-PN Matrices

Lemma 8. *Let* $\boldsymbol{A}$ *be an* $n \times n$ *nonnegative matrix. If for any nonempty proper subset* J *of* N *the system of linear inequalities*

$$(\boldsymbol{I}_J \boldsymbol{A} \boldsymbol{I}_J)\boldsymbol{x} < \boldsymbol{0} \tag{14}$$

has a positive solution then the same is true of any principal submatrix of $\boldsymbol{A}$ *of order* $m \geqq 2$.

We shall refer to the condition of the lemma as Condition [2].

PROOF. Suppose the contrary. Then there exists a principal submatrix which does not satisfy Condition [2]. Without loss of generality we may suppose that this submatrix is of order $n - 1$. That is, we suppose that there exists $\boldsymbol{A}\binom{i}{i}$ which for some proper subset $J(i)$ of $N(i)$ does not satisfy Condition [2]. Now by assumption the linear equations

$$\begin{pmatrix} \boldsymbol{I}_{(n-1)} & \boldsymbol{I}_{J(i)} \boldsymbol{A}\begin{pmatrix} i \\ i \end{pmatrix} \boldsymbol{I}_{J(i)} \end{pmatrix} \begin{pmatrix} \boldsymbol{y}(i) \\ \boldsymbol{x}(i) \end{pmatrix} = \boldsymbol{0}$$

have no positive solution. ($\boldsymbol{I}_{(n-1)}$ is the $(n-1)$-dimensional identity matrix.) By virtue of Theorem 2 in Chapter 1, therefore, the inequalities

$$\boldsymbol{y}(i)' \begin{pmatrix} \boldsymbol{I}_{(n-1)} & \boldsymbol{I}_{J(i)} \boldsymbol{A}\begin{pmatrix} i \\ i \end{pmatrix} \boldsymbol{I}_{J(i)} \end{pmatrix} \geq \boldsymbol{0}'$$

have a semipositive solution. Let $J \equiv J(i)$ if $\boldsymbol{y}(i)'\boldsymbol{I}_{J(i)}\boldsymbol{a}_i(i) \geqq 0$ and let $J = J(i) \cup \{i\}$ if $\boldsymbol{y}(i)'\boldsymbol{I}_{J(i)}\boldsymbol{a}_i(i) < 0$. Then the system

$$\boldsymbol{y}'(\boldsymbol{I}_J \boldsymbol{A} \boldsymbol{I}_J) \geqq \boldsymbol{0}' \tag{15}$$

has a semipositive solution with $y_i = 0$. In view of (i) of Corollary 1 in Chapter 1, there is therefore no semipositive solution to the inequalities $(\boldsymbol{I}_J \boldsymbol{A} \boldsymbol{I}_J)\boldsymbol{x} < \boldsymbol{0}$, which is a contradiction. □

Lemma 9. *If a nonnegative* $n \times n$ *matrix* $\boldsymbol{A}$ *satisfies Condition* [2] *then* $\boldsymbol{A}$ *is nonsingular.*

PROOF. Suppose that $\boldsymbol{A}$ is singular. Then there exists a nonzero $\boldsymbol{y}$ such that $\boldsymbol{y}'\boldsymbol{A} = \boldsymbol{0}'$. By assumption, $\boldsymbol{A}$ is nonnegative and satisfies Condition [2] for $J = N(j)$, $j = 1, \ldots, n$; hence $a_{ij} > 0$ for $i \neq j$, $j = 1, \ldots, n$. Therefore $\boldsymbol{y}$ contains both negative and nonnegative elements. Let us define $J = \{j : y_j < 0, j \in N\}$ and $\bar{J} = \{j : y_j \geqq 0, j \in N\}$. It is obvious that both J and $\bar{J}$ are nonempty, whence the system $(\boldsymbol{y}'\boldsymbol{I}_J)(\boldsymbol{I}_J \boldsymbol{A} \boldsymbol{I}_J) = \boldsymbol{0}'$ and therefore the system $\boldsymbol{z}'(\boldsymbol{I}_J \boldsymbol{A} \boldsymbol{I}_J) \geqq \boldsymbol{0}'$ has a semipositive solution. Applying (i) of Corollary 1 in Chapter 1, there is therefore no semipositive solution to the inequalities $(\boldsymbol{I}_J \boldsymbol{A} \boldsymbol{I}_J)\boldsymbol{x} < \boldsymbol{0}$, which is a contradiction. □

Theorem 5. *A nonnegative* $n \times n$ *matrix* $\boldsymbol{A}$ *is a semi-*PN *matrix if and only if* $\boldsymbol{A}$ *satisfies Condition* [2].

PROOF (*Sufficiency*). Evidently the proposition is valid for $n = 2$. Suppose that $n \geqq 3$. Then $\boldsymbol{A}\binom{i}{i}$, $i = 1, \ldots, n$, is a matrix of order not less than 2. By virtue of Lemma 8 and Lemma 9, $\boldsymbol{A}\binom{i}{i}$ is nonsingular. Hence the equation

$$\boldsymbol{a}^i(i) = \boldsymbol{y}'(i)\boldsymbol{A}\begin{pmatrix} i \\ i \end{pmatrix} \tag{16}$$

has a unique solution $\boldsymbol{y}(i)$. Let $J(i) = \{j : y_j > 0, j \neq i\}$, so that $\bar{J}(i) = \{j : y_j \leqq 0, j \neq i\}$. Since the off-diagonal elements of $\boldsymbol{A}$ are all positive, $J(i)$ is nonempty and the nonsingularity of $\boldsymbol{A}$ implies that $a_{ii} \neq \boldsymbol{y}'(i)\boldsymbol{a}_i(i)$. Suppose that $a_{ii} > \boldsymbol{y}'(i)\boldsymbol{a}_i(i)$. Then

$$\left.\begin{aligned} a_{ii} - \sum_{k \neq i} y_k a_{ki} &> 0, \\ a_{ij} - \sum_{k \neq i} y_k a_{kj} &= 0 \qquad (j \neq i). \end{aligned}\right\} \tag{17}$$

Let $J_i = \bar{J}(i) \cup \{i\}$. Then, from the definitions, $J_i \cup J(i) = N$ and $J_i \cap J(i) = \varnothing$. Let $u_k = y_k$ when $k \neq i$ and $u_i = -1$, so that

$$\boldsymbol{u}'\boldsymbol{I}_{J_i} \geq \boldsymbol{0}' \quad \text{and} \quad -a_{ij} + \sum_{k \neq i} y_k a_{k_j} = 0 \qquad (j \neq i)$$

Then (17) may be rewritten as

$$\begin{aligned} \sum_{k \in J_i} (-u_k)a_{kj} + \sum_{k \notin J_i} u_k(-a_{kj}) &\geqq 0 \qquad (j \in J_i), \\ \sum_{k \in J_i} (-u_k)(-a_{kj}) + \sum_{k \notin J_i} u_k a_{kj} &= 0 \qquad (j \notin J_i). \end{aligned}$$

Equivalently, the inequalities

$$\boldsymbol{z}'(\boldsymbol{I}_{J_i}\boldsymbol{A}\boldsymbol{I}_{J_i}) \geqq \boldsymbol{0}' \tag{18}$$

have a semipositive solution $\boldsymbol{u}'\boldsymbol{I}_{J_i} \geq \boldsymbol{0}'$. Recalling (i) of Corollary 1 in Chapter 1, $(\boldsymbol{I}_{J_i}\boldsymbol{A}\boldsymbol{I}_{J_i})\boldsymbol{x} < \boldsymbol{0}$ has no semipositive solution, a contradiction. Thus

$$a_{ii} - \boldsymbol{y}'(i)\boldsymbol{a}_i(i) = \alpha < 0. \tag{19}$$

From (16) and (19) it follows that

$$\det \boldsymbol{A} = \alpha \cdot \det \boldsymbol{A}\begin{pmatrix} i \\ i \end{pmatrix}$$

so that

$$\operatorname{sign}(\det \boldsymbol{A}) = \operatorname{sign}\left(-\det \boldsymbol{A}\begin{pmatrix} i \\ i \end{pmatrix}\right) \quad \text{for any } i. \tag{20}$$

In view of Lemma 8, (20) holds for any principal minors of order $r + 1$ and r, $r = 2, \ldots, n - 1$; moreover, all principal minors of order 2 are negative. Hence $\boldsymbol{A}$ is a semi-PN matrix.

(*Necessity*). We shall proceed by induction on n, the order of the matrix. Let us consider first the case in which $n = 2$, the lowest possible order of a semi-PN matrix. Suppose that $a_{11} = a_{22} = 0$. Since $\boldsymbol{A}$ is a semi-PN matrix, $a_{12} > 0$ and $a_{21} > 0$. It is then easy to see that Condition [2] is satisfied for all positive $\boldsymbol{x}$. Suppose alternatively that some a_{ii}, say a_{11}, is positive. Since $\boldsymbol{A}$ is a semi-PN matrix,

$$\frac{a_{12}}{a_{11}} > \frac{a_{22}}{a_{21}}, \tag{21}$$

whence Condition [2] is satisfied by all positive $\boldsymbol{x}$ such that

$$\frac{a_{12}}{a_{11}} > \frac{x_1}{x_2} > \frac{a_{22}}{a_{21}}. \tag{22}$$

It therefore suffices to show that a nonnegative semi-PN matrix of order n satisfies Condition [2] if a nonnegative semi-PN matrix of order $n - 1$ satisfies the condition. Suppose the contrary. Then there exists a nonempty proper subset J of N such that

$$(\boldsymbol{I}_J \boldsymbol{A} \boldsymbol{I}_J)\boldsymbol{x} < \boldsymbol{0}$$

has no positive solution. Equivalently, the system of linear equations

$$(\boldsymbol{I}, \boldsymbol{I}_J \boldsymbol{A} \boldsymbol{I}_J)\begin{pmatrix}\boldsymbol{u}\\ \boldsymbol{x}\end{pmatrix} = \boldsymbol{0}$$

has no positive solution. (Here $\boldsymbol{I}$ is the identity matrix of order n.) Recalling Theorem 2 in Chapter 1, there exists a semipositive vector $\boldsymbol{y} \geq \boldsymbol{0}$ such that

$$\boldsymbol{y}'(\boldsymbol{I}_J \boldsymbol{A} \boldsymbol{I}_J) \geqq \boldsymbol{0}'. \tag{23}$$

Let $P_{\boldsymbol{y}} = \{j : y_j > 0, j \in N\}$. Evidently $P_{\boldsymbol{y}}$ is nonempty. We consider separately the two cases, $\bar{P}_{\boldsymbol{y}} = \{j : y_j = 0, j \in N\} = \varnothing$ and $\bar{P}_{\boldsymbol{y}} \neq \varnothing$.

Suppose that $P_{\boldsymbol{y}} = N$. Since $\boldsymbol{I}_J \boldsymbol{A} \boldsymbol{I}_J$ is also a semi-PN matrix, the inverse $\boldsymbol{B} = (\boldsymbol{I}_J \boldsymbol{A} \boldsymbol{I}_J)^{-1}$ exists; moreover, any row of $\boldsymbol{B}$, say the first row $\boldsymbol{b}^1$, must contain some negative elements. Let t be the minimum of $(-y_i/b_{1i})$ over all negative elements of $\boldsymbol{b}^1$, and suppose that the minimum is obtained for $i = k$. Then $\boldsymbol{z}' = \boldsymbol{y}' + t \cdot \boldsymbol{b}^1$ is a semipositive but not positive solution to (23) since $t > 0$, $z_k = 0$, and $\boldsymbol{z}'(\boldsymbol{I}_J \boldsymbol{A} \boldsymbol{I}_J) = \boldsymbol{y}'(\boldsymbol{I}_J \boldsymbol{A} \boldsymbol{I}_J) + t \cdot \boldsymbol{e}^1$, where $\boldsymbol{e}^1 = (1, 0, 0, \ldots, 0)$. Thus without loss of generality we can assume that $\bar{P}_{\boldsymbol{y}} \neq \varnothing$. We begin by showing that both $\hat{J} \equiv P_{\boldsymbol{y}} \cap J$ and $\bar{J} \equiv P_{\boldsymbol{y}} \cap \bar{J}$ are nonempty, where $\bar{J}$ denotes the complement of J. Suppose on the contrary that $\hat{J} = \varnothing$. Then no $i \in J$ can be in $P_{\boldsymbol{y}}$. Hence

$$y_i = 0 \quad \text{for any } i \in J. \tag{24}$$

Thus

$$0 = \sum_{i \in J} a_{ij} y_i \geqq \sum_{i \notin J} a_{ij} y_i \geqq 0 \quad \text{for any } j \in J. \tag{25}$$

However, consistency requires that

$$\sum_{i \notin J} a_{ij} y_i = 0 \quad \text{for any } j \in J. \tag{26}$$

Since $a_{ij} > 0$ $(i \notin J, j \in J)$, (26) implies that

$$y_i = 0 \quad \text{for any } i \notin J. \tag{27}$$

From (24) and (27), $\boldsymbol{y}$ is the zero vector, contradicting the assumption that $P_{\boldsymbol{y}} \neq \varnothing$. Thus $P_y \cap J \neq \varnothing$. By a similar argument, $P_{\boldsymbol{y}} \cap \bar{J} \neq \varnothing$. Hence we may rewrite (23) as

$$\begin{aligned} \sum_{i \in \hat{J}} a_{ij} y_i &\geqq \sum_{i \in \hat{\hat{J}}} a_{ij} y_i \qquad (j \in J \supset \hat{J}), \\ \sum_{i \in \hat{\hat{J}}} a_{ij} y_i &\geqq \sum_{i \in \hat{J}} a_{ij} y_i \qquad (j \in \bar{J} \supset \hat{\hat{J}}). \end{aligned} \tag{28}$$

From (28), together with the facts that $\hat{J}$ and $\hat{\hat{J}}$ are complementary with respect to $P_{\boldsymbol{y}}$ and that $\hat{J} \cup \hat{\hat{J}} = P_{\boldsymbol{y}}$, it follows that there exists a positive vector $\boldsymbol{y}_{P_y}$ such that

$$\boldsymbol{y}_{P_y}(\boldsymbol{I}_{\hat{J}} \boldsymbol{A}_{P_y} \boldsymbol{I}_{\hat{J}}) \geqq \boldsymbol{0}, \tag{29}$$

where $\boldsymbol{y}_{P_y}$ is a subvector of $\boldsymbol{y}$ consisting of all y_i, $i \in P_{\boldsymbol{y}}$, and $\boldsymbol{A}_{P_y}$ is a principal submatrix of $\boldsymbol{A}$ corresponding to the set $P_{\boldsymbol{y}}$. From (29) and (i) of Corollary 1 in Chapter 1, $\boldsymbol{A}_{P_y}$ does not satisfy Condition [2]. But $P_{\boldsymbol{y}}$ is a proper subset of N, so that $\boldsymbol{A}_{P_y}$ is of order not greater than $n - 1$. Thus we have arrived at a contradiction. □

4 Matrices with the Minkowski or Metzler Property

Definition 11. A square matrix $\boldsymbol{A}$ is said to have the *Minkowski* (respectively, *Metzler*) *property* if the diagonal elements of $\boldsymbol{A}^{-1}$ are *nonnegative* (respectively, nonpositive) and if the off-diagonal elements of $\boldsymbol{A}^{-1}$ are nonpositive (respectively, nonnegative).

Lemma 10 (Chipman [1969]). *If a nonnegative square matrix* $\boldsymbol{A}$ *has the Minkowski property then both* $\boldsymbol{A}$ *and* $\boldsymbol{A}^{-1}$ *are P-matrices.*

PROOF. By assumption $\boldsymbol{A}^{-1}$ has nonpositive off-diagonal elements and a nonnegative inverse $\boldsymbol{A}$; hence $\boldsymbol{A}^{-1}$ must satisfy the Hawkins–Simon conditions (Theorem 7 in Chapter 1), that is, $\boldsymbol{A}^{-1}$ is a P-matrix. From Jacobi's Theorem on determinants it then follows that $\boldsymbol{A}$ also is a P-matrix. □

Remark. In the present context Jacobi's Theorem states that

$$\det A^{-1}\begin{pmatrix} i_1, \ldots, i_r \\ i_1, \ldots, i_r \end{pmatrix} = \frac{1}{\det A} \det A \begin{pmatrix} i_{r+1}, \ldots, i_n \\ i_{r+1}, \ldots, i_n \end{pmatrix}.$$

This is also known as Sylvester's Law.

Lemma 11. *If a nonnegative square matrix A has the Minkowski property then any principal submatrix of A has the same property.*

PROOF. From Lemma 10, A^{-1} is a P-matrix and therefore contains no zeros on the main diagonal. This enables us to base our proof on the familiar method of elimination. We begin by defining the matrix

$$\tilde{A} = \begin{pmatrix} 1 & 0 & & \cdots & 0 \\ -\dfrac{a^{21}}{a^{11}} & 1 & 0 & \cdots & 0 \\ \vdots & & & & \vdots \\ -\dfrac{a^{n1}}{a^{11}} & 0 & & \cdots & 1 \end{pmatrix} = \begin{pmatrix} & e^1 \\ -\alpha & I_{(n-1)} \end{pmatrix} \tag{30}$$

where a^{ij} denotes the ijth element of A^{-1}, $\alpha' = (a^{21}/a^{11}, \ldots, a^{n1}/a^{11})$ and $I_{(n-1)}$ is the identity matrix of order $n - 1$. Premultiplying both sides of

$$A^{-1}A = I \tag{31}$$

by $\tilde{A}$ we obtain

$$\begin{pmatrix} a^{11} & a^{12} & \cdots & a^{1n} \\ 0 & & & \\ \vdots & & \bar{a}^{ij} & \\ 0 & & & \end{pmatrix} \cdot A = \tilde{A} = \begin{pmatrix} & e^1 \\ -\alpha & I_{(n-1)} \end{pmatrix}, \tag{32}$$

where $\bar{a}^{ij} = a^{ij} - (a^{i1}/a^{11})a^{1j}$ $(i, j = 2, \ldots, n)$. Further manipulations of (32) yield

$$(\bar{a}^{ij}) \equiv \bar{A}_{22} = \left(A\begin{pmatrix}1\\1\end{pmatrix}\right)^{-1}, \tag{33a}$$

where, from the definition of $\bar{a}^{ij}$,

$$\bar{a}^{ij} \leqq 0 \quad \text{for } i \neq j;\, i, j \geqq 2. \tag{33b}$$

Moreover, it follows from (33a) that

$$\bar{a}^{ii} \cdot a_{ii} + \sum_{\substack{j=2 \\ j \neq i}}^{n} \bar{a}^{ij} \cdot a_{ji} = 1,$$

so that, since $\bar{a}^{ij} \cdot a_{ji} \leqq 0\,(i \neq j)$ and $a_{ii} > 0\,(i = 1, \ldots, n)$ by virtue of Lemma 10, we have

$$\bar{a}^{ii} = \frac{1}{a_{ii}}\left(1 - \sum_{\substack{j=2\\ j\neq i}}^{n} a^{ij} \cdot a_{ji}\right) > 0 \qquad (i = 2, \ldots, n). \tag{34}$$

That is,

$$\left(A\begin{pmatrix}1\\1\end{pmatrix}\right)^{-1}$$

has the Minkowski property. The proof is completed by noting that identical interchanges of rows and columns do not affect the properties under consideration and that any given principal submatrix of order $r\,(r = 1, \ldots, n-1)$ can be obtained from some principal submatrix of order $r + 1$ by deleting a suitable row and column. □

Remark. Lemma 11 is a generalization of part of Theorem 1 of Inada [1971], which is concerned with positive matrices. The method of proof is Inada's.

Before stating our next lemma we introduce some additional notation. Let $J = \{j_1, \ldots, j_s\}$ be any subset of N and $\boldsymbol{A}$ any $n \times n$ matrix. Then

$$\begin{pmatrix} a_{j_1 j_1} & \cdots & a_{j_1 j_s} \\ \vdots & & \vdots \\ a_{j_s j_1} & \cdots & a_{j_s j_s} \end{pmatrix} = \boldsymbol{A}\begin{pmatrix} j_1, \ldots, j_s \\ j_1, \ldots, j_s \end{pmatrix} = \boldsymbol{A}(J)$$

is the principal submatrix of $\boldsymbol{A}$ associated with J,

$$(a_{j_1 p}, \ldots, a_{j_s p})' = \boldsymbol{a}_p(J)$$

is the pth subcolumn of $\boldsymbol{A}$ associated with J, and

$$(a_{pj_1}, \ldots, a_{pj_s}) = \boldsymbol{a}^p(J)$$

is the pth subrow of $\boldsymbol{A}$ associated with J. Moreover, $a^{i_p i_s}$ is the $(i_p i_s)$th element of $(\boldsymbol{A}(J))^{-1}$ and $\boldsymbol{a}_{i_k}^{-1}(J(i_k))$ is the i_kth subcolumn of $(\boldsymbol{A}(J))^{-1}$ associated with $J(i_k)$.

Lemma 12. *Let $\boldsymbol{A}$ be an $n \times n$ nonnegative matrix with the Minkowski property. Then, for any $i_k \in J$, the equation*

$$\boldsymbol{A}(J(i_k)) \cdot \boldsymbol{x}(J(i_k)) = \boldsymbol{a}_{i_k}(J(i_k))$$

has a unique nonnegative solution $\boldsymbol{x}(J(i_k))$. Furthermore,

$$\boldsymbol{x}(J(i_k)) = -\frac{1}{a^{i_k i_k}}\boldsymbol{a}_{i_k}^{-1}(J(i_k))$$

and

$$\boldsymbol{a}^{i_k}(J(i_k)) \cdot \boldsymbol{x}(J(i_k)) < a_{i_k i_k}.$$

PROOF. From Lemma 11, $A(J)$ also has the Minkowski property. By considering the i_kth column of $A(J)(A(J))^{-1} = I$, we obtain

$$\sum_{i_j \in J} a_{i_p i_j} \cdot a^{i_j i_k} = a_{i_p i_k} \cdot a^{i_k i_k} + \sum_{i_j \neq i_k} a_{i_p i_j} \cdot a^{i_j i_k}$$

$$= \delta_{i_p i_k} \quad \text{for any } i_p \in J, \tag{35}$$

where $\delta_{i_p i_k}$ is Kronecker's delta. From Lemma 10, $a^{i_k i_k}$ is positive; hence

$$a_{i_p i_k} = \sum_{i_j \neq i_k} a_{i_p i_j} \left(- \frac{a^{i_j i_k}}{a^{i_k i_k}} \right)$$

$$= \sum_{i_j \neq i_k} a_{i_p i_j} x_{i_j}^{(i_k)} \qquad (i_p \neq i_k). \tag{36}$$

Since $a^{i_j i_k} \leqq 0 \; (i_j \neq i_k)$ and $a^{i_k i_k} > 0$, it is clear that $x_{i_j}^{(i_k)} \geqq 0$. From (35), with $i_p = i_k$,

$$a_{i_k i_k} = \frac{1}{a^{i_k i_k}} + \sum_{i_j \neq i_k} a_{i_k i_j} \cdot x_{i_j}^{(i_k)}, \tag{37}$$

which, since $a^{i_k i_k} > 0$, implies that

$$a_{i_k i_k} > \sum_{i_j \neq i_k} a_{i_k i_j} \cdot x_{i_j}^{(i_k)}$$

$$= \boldsymbol{a}^{i_k}(J(i_k)) \cdot \boldsymbol{x}(J(i_k)). \qquad \square$$

Remark. Lemma 12 is a variant of the necessity part of Lemma 4 of Uekawa [1971].

Theorem 6. *Let A be an $n \times n$ nonnegative matrix. Then the following assertions are equivalent to each other.*

i. *A has the Minkowski property;*
ii. *for any given k and J ($k \notin J$, $\emptyset \neq J \subset N$) there exist $x_j^{(k)} \geqq 0$ ($j \in J$) such that*

$$\sum_{j \in J} a_{ij} x_j^{(k)} \geqq a_{ik} \quad \textit{for any } i \in J, \tag{38a}$$

$$\sum_{j \in J} a_{ij} x_j^{(k)} \leqq a_{ik} \quad \textit{for any } i \notin J, \tag{38b}$$

with a strict inequality for $i = k$;

iii. *for any nonempty proper subset J of N and any given positive $\bar{\boldsymbol{x}}_{\bar{J}} > \boldsymbol{0}$,*

$$(\boldsymbol{I}_J \boldsymbol{A} \boldsymbol{I}_J) \begin{pmatrix} \boldsymbol{x}_J \\ \bar{\boldsymbol{x}}_{\bar{J}} \end{pmatrix} > \boldsymbol{0} \tag{39}$$

has a solution $\boldsymbol{x}_J > \boldsymbol{0}$.

Henceforth we shall refer to proposition (iii) as Condition [3].

Proof. We begin by showing that (i) $\Rightarrow$ (ii). For $J = N(k)$, the implication may be established by identifying N with J and k with i_k ($k = 1, \ldots, n$) in Lemma 12. It therefore suffices to establish the implication for $J = \{j_1, \ldots, j_r\}$, where $1 \leqq r \leqq n - 2$. For any given J and $i \notin J$ we define the principal submatrix of A

$$\hat{A} = \begin{pmatrix} a_{ii} & a_{ik} & a_{ij_1} & \cdots & a_{ij_r} \\ a_{ki} & a_{kk} & a_{kj_1} & \cdots & a_{kj_r} \\ a_{j_1 i} & a_{j_1 k} & a_{j_1 j_1} & \cdots & a_{j_1 j_r} \\ \vdots & \vdots & \vdots & & \vdots \\ a_{j_r i} & a_{j_r k} & a_{j_r j_1} & \cdots & a_{j_r j_r} \end{pmatrix}$$

$$= \begin{pmatrix} a_{ii} & a_{ik} \quad a_{ij_1} \quad \cdots \quad a_{ij_r} \\ a_{ki} & \\ a_{j_1 i} & A(\hat{J}(i)) \\ \vdots & \\ a_{j_r i} & \end{pmatrix}$$

where $\hat{J} = \{i, k, j_1, \ldots, j_r\}$. Applying Lemma 12 to $A(\hat{J}(i))$, there exist $x_s^{(k)} \geqq 0$, $s \in J$, such that

$$\sum_{s \in J} a_{js} x_s^{(k)} = a_{jk} \quad \text{for } j \in J, \tag{39a}$$

$$\sum_{s \in J} a_{ks} x_s^{(k)} < a_{kk}. \tag{39b}$$

Since $\hat{A}$ also has the Minkowski property,

$$\det \hat{A} > 0 \tag{40a}$$

and

$$0 \geqq \hat{a}^{12} = (-1)^3 \det \hat{A}\begin{pmatrix}2\\1\end{pmatrix} \Big/ \det \hat{A}, \tag{40b}$$

where $\hat{a}^{pq}$ is the (pq)th element of $\hat{A}^{-1}$. From (40a) and (40b),

$$0 \leqq \det \hat{A}\begin{pmatrix}2\\1\end{pmatrix}$$

and, expanding $\det \hat{A}\binom{2}{1}$ and taking into account (39a),

$$\det \hat{A}\begin{pmatrix}2\\1\end{pmatrix} = \left(a_{ik} - \sum_{s \in J} a_{is} x_s^{(k)}\right) \det A(J).$$

Hence

$$\sum_{s \in J} a_{is} x_s^{(k)} \leqq a_{ik} \quad \text{for all } i \notin J,$$

where, by virtue of (39b), the inequality is strict for $i = k$.

We turn now to the implication (ii) ⇒ (iii). Since, by assumption, there exist $x_j^{(k)} \geqq 0$ $(j \in J)$ satisfying inequalities (38a) and (38b), it is obvious that

$$\sum_{j \in J} a_{ij}(\bar{x}_k \cdot x_j^{(k)}) \geqq \bar{x}_k \cdot a_{ik} \qquad (i \in J),$$

$$\sum_{j \in J} a_{ij}(\bar{x}_k \cdot x_j^{(k)}) \leqq \bar{x}_k \cdot a_{ik} \qquad (i \notin J),$$

with strict inequality for $i = k$, where $\bar{x}_k$ is an element of $\bar{\boldsymbol{x}}_{\bar{J}} > \boldsymbol{0}$. By summing the ith inequality over all $k \notin J$ and defining $y_j = \sum_{k \notin J} (\bar{x}_k \cdot x_j^{(k)})$ $(j \in J)$, we easily obtain

$$\sum_{j \in J} a_{ij} \cdot y_j \geqq \sum_{k \notin J} a_{ik} \cdot \bar{x}_k \qquad (i \in J),$$

$$\sum_{j \in J} a_{ij} \cdot y_j < \sum_{k \notin J} a_{ik} \cdot \bar{x}_k \qquad (i \notin J).$$

Let us define $x_j = y_j + \varepsilon, j \in J$, with

$$0 < \varepsilon < \min_{i \in J_0} \frac{\sum_{j \notin J} a_{ij}\bar{x}_j - \sum_{j \in J} a_{ij} y_j}{\sum_{j \in J} a_{ij}},$$

where $J_0 = \{i \notin J : \sum_{j \in J} a_{ij} > 0\}$. Then, since $\sum_{j \in J} a_{ij} > 0$ for any $i \in J$ and $\sum_{j \notin J} a_{ij}\bar{x}_j > \sum_{j \in J} a_{ij} y_j$ for any $i \notin J$, we have

$$\sum_{j \in J} a_{ij} x_j > \sum_{j \notin J} a_{ij} \bar{x}_j \qquad (i \in J),$$

$$\sum_{j \in J} a_{ij} x_j < \sum_{j \notin J} a_{ij} \bar{x}_j \qquad (i \notin J).$$

Finally, we verify the implication (iii) ⇒ (i). Since $\boldsymbol{A}$ satisfies Condition [3] and therefore Condition [1], $\boldsymbol{A}$ is a P-matrix. It therefore suffices to show that $A_{ij} \leqq 0, i \neq j$, where A_{ij} is the cofactor of a_{ij}. Consider now a given nonempty proper subset J of N containing at most $n - 2$ indices. We shall show that $A_{ij} \leqq 0$ for any $i, j \notin J$. Let $\tilde{\boldsymbol{A}}_t$ be the matrix obtained by interchanging the ith and jth rows and by replacing a_{si} with $a_{si} + ta_{sj}$, $s = 1, \ldots, n$, where $t = \bar{x}_j/\bar{x}_i$. (Since $\bar{x}_j$ and $\bar{x}_i$ are chosen arbitrarily, t is an arbitrary positive number.) Since $\boldsymbol{A}$ satisfies Condition [3], $\tilde{\boldsymbol{A}}_t\binom{j}{j} \equiv \tilde{\boldsymbol{B}}(t)$ satisfies Condition [1]. To see this, delete the jth inequality in Condition [3] after interchanging indices i and j, then rewrite the term

$$\sum_{s \notin J} a_{ks}\bar{x}_s \quad \text{as} \quad \sum_{\substack{s \in J \\ s \neq i, j}} a_{ks}\bar{x}_s + \bar{x}_i(a_{ki} + ta_{kj}).$$

It is then easy to see that

$$\sum_{s \in J} a_{ks} x_s > \sum_{\substack{s \notin J \\ s \neq i, j}} a_{ks}\bar{x}_s + \bar{x}_i(a_{ki} + ta_{kj}) \qquad (k \in J),$$

$$\sum_{s \in J} a_{ks} x_s < \sum_{\substack{s \in J \\ s \neq i, j}} a_{ks}\bar{x}_s + \bar{x}_i(a_{ki} + ta_{kj}) \qquad (k \notin J, k \neq j).$$

Since $\tilde{\boldsymbol{B}}(t)$ satisfies Condition [1], $\tilde{\boldsymbol{B}}(t)$ is a P-matrix for any $t > 0$ and

$$\det \tilde{\boldsymbol{B}}(t) > 0 \quad \text{for any } t > 0.$$

Furthermore,

$$\det \tilde{\boldsymbol{B}}(0) = \begin{cases} (-1)^{2i-1} A_{ij} & (i > j), \\ (-1)^{2j-1} A_{ij} & (i < j). \end{cases} \tag{41}$$

On the other hand,

$$\lim_{t \to 0} \det \tilde{\boldsymbol{B}}(t) = \det \tilde{\boldsymbol{B}}(0) \geqq 0 \tag{42}$$

since $\tilde{\boldsymbol{B}}(t)$ is continuous in t. Combining (41) and (42), we obtain $A_{ij} \leqq 0$ for $i, j \notin J$. The above procedure can be repeated until we have $A_{ij} \leqq 0$ for $i \neq j$. Thus $\boldsymbol{A}$ has the Minkowski property. □

Lemma 13. *Let $\boldsymbol{A}$ be an $n \times n$ nonnegative matrix with positive off-diagonal elements and with the Metzler property. Then:*

i. *for $n = 2$, $a^{ii} \leqq 0$ and $a^{ij} > 0$ $(i \neq j)$,*
   for $n \geqq 3$, $a^{ii} < 0$ and $a^{ij} \geqq 0$ $(i \neq j)$; and

ii. *for any given $k \in N$ and $J = N(k)$ there exist $x_j^{(k)} \geqq 0$, $j \in J$, such that*

$$\sum_{j \in J} a_{ij} x_j^{(k)} \leqq a_{ik} \qquad (i \in J = N(k));$$

$$\sum_{j \in J} a_{kj} x_j^{(k)} > a_{kk}.$$

Proof. (i) For $n = 2$ the proposition may be verified by direct calculation of $\boldsymbol{A}^{-1}$. Let $n \geqq 3$ and suppose the contrary, that there exists an index, say i_0, for which $a^{i_0 i_0} = 0$. Let $J_{i_0} = \{j : a^{i_0 j} > 0\}$. Then $J_{i_0} \neq \varnothing$; for if $J_{i_0} = \varnothing$ then $\boldsymbol{A}$ is singular, a contradiction. It follows that

$$\sum_{j \in J_{i_0}} a^{i_0 j} a_{jk} = 0 \quad \text{for any } k \in N(i_0).$$

Hence

$$a_{jk} = 0 \quad \text{for any } j \in J_{i_0},\ k \in N(i_0).$$

Now J_{i_0} is a subset of $N(i_0)$. In fact, J_{i_0} is a proper subset of $N(i_0)$. To see this, suppose the contrary, that $J_{i_0} = N(i_0)$. Then, for any $k \neq i_0$,

$$\begin{aligned} 0 &= \sum_{j \in N(i_0)} a^{i_0 j} a_{jk}, \\ &= \sum_{\substack{j \in N(i_0) \\ j \neq k}} a^{i_0 j} a_{jk} + a^{i_0 k} a_{kk}. \end{aligned}$$

However, when $n \geqq 3$ the term

$$\sum_{\substack{j \in N(i_0) \\ j \neq k}} a^{i_0 j} a_{jk}$$

exists and is, of course, positive. Thus we have arrived at a contradiction. Hence $J_{i_0} \subset N(i_0)$. It follows that there exists an index $k_0 \in \bar{J}_{i_0}$ such that $a_{jk_0} = 0$ for any $j \in J_{i_0}$, which contradicts the hypothesis. (Here $\bar{J}_{i_0}$ is the complement of J_{i_0} relative to $N(i_0)$.)

(ii) Suppose that $n = 2$. Then, since $\boldsymbol{A}$ has the Metzler property, det $\boldsymbol{A} < 0$. If $a_{ii} > 0$, $i = 1, 2$, the solutions to the asserted inequalities are given by the intervals $[a_{12}/a_{11}, a_{22}/a_{21})$ and $[a_{21}/a_{22}, a_{11}/a_{12})$. If one diagonal element is zero then the solution to the inequality containing the other diagonal element satisfies the system. If both diagonal elements are zero then any positive number satisfies the system.

Suppose now that $n \geqq 3$. Then, from (i), $a^{ii} < 0$ for any $i \in N$. Consider the (ik)th element of $\boldsymbol{A}\boldsymbol{A}^{-1} = \boldsymbol{I}$. We have

$$\sum_{j=1}^{n} a_{ij}a^{jk} = 0 \qquad (i \neq k), \tag{43a}$$

$$\sum_{j=1}^{n} a_{ij}a^{jk} = 1 \qquad (i = k). \tag{43b}$$

From (43a),

$$\begin{aligned} a_{ik} &= -\frac{1}{a^{kk}} \sum_{j \neq k} a_{ij}a^{jk} \\ &= \sum_{j \neq k} a_{ij}\left(-\frac{a^{jk}}{a^{kk}}\right) \\ &= \sum_{j \neq k} a_{ij}x_j^{(k)} \end{aligned} \tag{44}$$

where $x_j^{(k)} \equiv -a^{jk}/a^{kk}$, $j \neq k$. Now $a^{kk} < 0$ and $a^{jk} \geqq 0$; hence $x_j^{(k)} \geqq 0$ for any $j \neq k$. Moreover, there exists an index $j_0 \in N(k)$ such that $x_{j_0}^{(k)} > 0$; for otherwise, from (44) and the asserted nonnegativity of $x_j^{(k)}$ and a_{ij}, we would have $a_{ik} = 0$, a contradiction. From (43b) and the definition of $x_j^{(k)}$ we easily obtain

$$a_{kk} - \sum_{j \neq k} a_{kj}x_j^{(k)} = \frac{1}{a^{kk}} < 0. \qquad \square$$

Lemma 14. *Let $\boldsymbol{A}$ be an $n \times n$ nonnegative matrix with positive off-diagonal elements and with the Metzler property. Then*:

i. *every principal submatrix of $\boldsymbol{A}$ of order $r \geqq 2$ has the Metzler property*; *and*
ii. *$\boldsymbol{A}$ is a semi-*PN *matrix.*

PROOF. (i) For $n = 2$ the assertion is trivial. Suppose then that $n \geqq 3$. Proceeding as in the proof of Lemma 11, we arrive again at (32), with

$$(\bar{a}^{ij}) = \boldsymbol{A}_{22} = \left(\boldsymbol{A}\begin{pmatrix}1\\1\end{pmatrix}\right)^{-1} \qquad (i, j = 2, \ldots, n),$$

and, since $a^{11} < 0$ and $a^{ij} \geqq 0$ $(i \neq j; i, j = 1, \ldots, n)$,

$$\bar{a}^{ij} \geqq 0 \qquad (i \neq j, i \leqq n, j \geqq 2).$$

Computing the off-diagonal elements of $\bar{A}_{22}\,(A(\begin{smallmatrix}1\\1\end{smallmatrix}))$ and bearing in mind these inequalities, we obtain

$$\bar{a}^{ii} = -\frac{1}{a_{ik}} \sum_{\substack{j \geqq 2 \\ j \neq i}} \bar{a}^{ij} a_{jk} \leqq 0 \qquad (k \neq i, i = 2, \ldots, n).$$

(ii) Let $J_r = \{j_1, \ldots, j_{r+1}\}$ be any subset of N. From (i) and Lemma 13(i),

$$\frac{\det A(J_r(i_k))}{\det A(J_r)} < 0 \quad \text{for } 2 \leq r \leq n-1, \tag{45}$$

where $i_k \in J_r$. Let $r = 2$. Then, by direct calculation of $(A(J_r(i_k)))^{-1}$ and from the fact that $A(J_r(i_k))$ has the Metzler property, $\det A(J_r(i_k)) < 0$. Thus every principal minor of order s has the sign of $(-1)^{s-1}$, $s = 2, \ldots, n$. □

Remark. Lemma 14, like Lemma 11, generalizes part of Theorem 1 of Inada [1971]. The method of proof is Inada's.

Theorem 7. *Let A be an $n \times n$ nonnegative matrix with positive off-diagonal elements. Then the following statements are equivalent.*

i. *A has the Metzler property;*
ii. *for any given k and J $(k \notin J, \varnothing \neq J \subset N)$ there exist $x_j^{(k)} \geqq 0, j \in J$, such that*

$$\sum_{j \in J} a_{ij} x_j^{(k)} \leqq a_{ik} \qquad (i \in J),$$

$$\sum_{j \in J} a_{ij} x_j^{(k)} \geqq a_{ik} \qquad (i \notin J),$$

with a strict inequality for $i = k$;
iii. *for any nonempty proper subset J of N and any given positive $\bar{x}_{\bar{J}} > \mathbf{0}$, the inequality*

$$(I_J A I_J)\begin{pmatrix} x_J \\ \bar{x}_{\bar{J}} \end{pmatrix} < \mathbf{0}$$

has a solution $x_J > \mathbf{0}$.

The condition (iii) will be referred to as Condition [4].

PROOF. (i) ⇒ (ii) For $n = 2$ the implication follows from Lemma 13(ii). We therefore may assume that $n \geqq 3$. Again from Lemma 13(ii), the implication is obvious for $J = N(k)$, $k = 1, \ldots, n$. Thus it remains to demonstrate the implication for $J = \{j_1, \ldots, j_r\}$, where $1 \leqq r \leqq n-2$. The demonstration is based on the fact that every principal submatrix of A has the Metzler property if A has it.

Suppose then that we are given a subset $J = \{j_1, \ldots, j_r\}$ of N, where $1 \leqq r \leqq n-2$. For any given $i \notin J$ and $k \notin J$, we define $\hat{J} = \{i, k, j_1, \ldots, j_r\}$. Then, by virtue of Lemma 14(i), $\hat{A} \equiv A(\hat{J})$ and $A(\hat{J}(i))$ have the Metzler property. For $r \geqq 2$, the application of Lemma 13(ii) to $A(\hat{J}(i))$ assures us of the existence of $x_s^{(k)} \geqq 0$, $s \in J$, such that

$$\sum_{s \subset J} a_{js} x_s^{(k)} = a_{jk} \qquad (j \in J), \tag{46a}$$

$$\sum_{s \in J} a_{ks} x_s^{(k)} > a_{kk}. \tag{46b}$$

(That (46a) consists of exact equalities is clear from the proof of Lemma 13(ii).) Since $\hat{A}$ has the Metzler property,

$$\operatorname{sign}(\det \hat{A}) = (-1)^{r+1}$$

and

$$0 \leqq \hat{a}^{12} = (-1)^3 \cdot \det \hat{A}\binom{2}{1} \Big/ \det \hat{A}.$$

It follows that

$$\det \hat{A}\binom{2}{1} = 0 \quad \text{or} \quad \operatorname{sign}\left(\det \hat{A}\binom{2}{1}\right) = (-1)^r. \tag{47}$$

In view of (46a), $\det \hat{A}\binom{2}{1}$ can be expressed as

$$\det \hat{A}\binom{2}{1} = \left(a_{ik} - \sum_{s \in J} a_{is} x_s^{(k)}\right) \cdot \det A(J). \tag{48}$$

Combining (47) and (48), and noting that sign $(\det A(J)) = (-1)^{r-1}$, we obtain

$$a_{ik} \leqq \sum_{s \in J} a_{is} x_s^{(k)} \quad \text{for } i \notin J. \tag{49}$$

Inequalities (46) and (49) yield the desired result.

For $r = 1$, we define $\hat{J} = \{i_0, k, j_1\}$, where

$$\max_{\substack{i \in N(j_1) \\ i \neq k}} \frac{a_{ik}}{a_{ij_1}} = \frac{a_{i_0 k}}{a_{i_0 j_1}}.$$

Since $A(\hat{J})$ has the Metzler property,

$$\det A(\hat{J}(i_0)) < 0$$

and

$$\hat{A}_{21} = (-1)\det \hat{A}\binom{2}{1} \geqq 0.$$

Let $a_{j_1j_1} > 0$. Then, in view of the above pair of inequalities, the proof may be completed by defining $x_j^{(k)} = a_{j_1k}/a_{j_1j_1}$. On the other hand, if $a_{j_1j_1} = 0$, it is easily shown that any positive number greater than $\max\{a_{kk}/a_{kj_1}, a_{i_0k}/a_{i_0j_1}\}$ satisfies the inequalities in question.

(ii) ⇒ (iii) Suppose that for any given J ($\emptyset \neq J \subset N$) and $k \notin J$ we are given some positive $\bar{x}_k > 0$. Then, from (ii), there exist $y_j^{(k)} = \bar{x}_k \cdot x_j^{(k)}$, $j \in J$, such that

$$\begin{aligned} &\sum_{j\in J} a_{ij}y_j^{(k)} \leqq \bar{x}_k a_{ik} \qquad (i \in J), \\ &\sum_{j\in J} a_{ij}y_j^{(k)} \geqq \bar{x}_k a_{ik} \qquad (i \notin J), \end{aligned} \tag{50}$$

with a strict inequality for $i = k$. Summing the ith inequality over all $k \notin J$, we obtain

$$\sum_{j\in J} a_{ij}y_j \leqq \sum_{k\notin J} a_{ik}\bar{x}_k \qquad (i \in J), \tag{51a}$$

$$\sum_{j\in J} a_{ij}y_j > \sum_{k\notin J} a_{ik}\bar{x}_k \qquad (i \notin J), \tag{51b}$$

where $y_j = \sum_{k\notin J} y_j^{(k)}$ for any $j \in J$.

Let $I_J = \{i \in J : \sum_{j\in J} a_{ij}y_j = \sum_{k\notin J} a_{ik}\bar{x}_k\}$. If $I_J = \emptyset$ then, by choosing sufficiently small $\varepsilon > 0$ and replacing y_j by $x_j = y_j + \varepsilon$ for any $j \in J$, the proof may be completed. Suppose that $I_J \neq \emptyset$. It then suffices to show that there exists $\mathbf{z}_J \geq \mathbf{0}$ which satisfies both (51a) and (51b) as strict inequalities, where $\mathbf{z}_J$ is a vector the components of which are z_j, $j \in J$. Let us define $J_i = \{j \in J : a_{ij}y_j > 0\}$ for any $i \in I_J$ and let $\tilde{J} = \bigcup_{i\in I_J} J_i$. Suppose that $J_{i_0} = \emptyset$ for some $i_0 \in I_J$. Then we arrive at a contradiction:

$$0 = \sum_{j\in J} a_{i_0j}y_j = \sum_{k\notin J} a_{i_0k}\bar{x}_k > 0.$$

Hence $\tilde{J} = \bigcup_{i\in I_J} J_i \neq \emptyset$. Let us choose a positive number ε such that

$$\min\left\{\min_{j\in\tilde{J}} y_j,\ \min_{i\notin J} \frac{\sum_{j\in J} a_{ij}y_j - \sum_{k\notin J} a_{ik}\bar{x}_k}{\sum_{j\in\tilde{J}} a_{ij}}\right\} > \varepsilon > 0$$

and define

$$z_j = \begin{cases} y_j - \varepsilon & \text{if } j \in \tilde{J} \\ y_j & \text{otherwise.} \end{cases}$$

Then z_j, $j \in J$, satisfy the inequalities

$$\sum_{j\in J} a_{ij}z_j < \sum_{k\notin J} a_{ik}\bar{x}_k \qquad (i \in J),$$

$$\sum_{i\in J} a_{ij}z_j > \sum_{k\notin J} a_{ik}\bar{x}_k \qquad (i \notin J).$$

(iii) ⇒ (i) Since A satisfies Condition [4], it satisfies Condition [2] and is therefore a semi-PN matrix. It therefore suffices to show that either $A_{ij} = 0$

or A_{ij} has the same sign as det $\boldsymbol{A}$. Let us define $\boldsymbol{A}_t$ as in the verification that (iii) ⇒ (i) in the proof of Theorem 6. Then $\boldsymbol{B}(t) \equiv \boldsymbol{A}_t(^i_j)$ satisfies Condition [2], and so is a semi-PN matrix for any $t > 0$. Hence

$$\text{sign}(\det \tilde{\boldsymbol{B}}(t)) = (-1)^{n-2} \quad \text{for any } t > 0.$$

Moreover, it is easy to see that

$$\det \tilde{\boldsymbol{B}}(0) = \begin{cases} (-1)^{2i-1}A_{ij} & \text{for } i > j \\ (-1)^{2j-1}A_{ij} & \text{for } j > i. \end{cases}$$

On the other hand, from the continuity of det $\tilde{\boldsymbol{B}}(t)$ in t, $\tilde{\boldsymbol{B}}(0)$ has the sign of $(-1)^{n-2}$ if it does not vanish. It follows that either $A_{ij} = 0$ or A_{ij} has the sign of $(-1)^{n-3} = (-1)^{n-1}$. □

Theorems 5–7 carry some useful implications. These are summarized in

Corollary 5. *Let $\boldsymbol{A}$ be an $n \times n$ nonnegative matrix and $\boldsymbol{C}$ an $n \times n$ matrix with the properties that $c_{ij} \leqq 0$ for $i \neq j$ and that there exists $\boldsymbol{\xi} > \boldsymbol{0}$ such that $\boldsymbol{C\xi} > \boldsymbol{0}$; let $\boldsymbol{D} \equiv \boldsymbol{C}^{-1}\boldsymbol{A}$; let $\boldsymbol{a}_j$ and $\boldsymbol{e}_j$ be the jth column of $\boldsymbol{A}$ and the jth unit vector of order n, respectively; let J be a nonempty proper subset of $N \equiv \{1, \ldots, n\}$ and $\bar{J}$ the set complementary to J relative to N; and let $\boldsymbol{A}_J$ be the submatrix obtained from $\boldsymbol{A}$ by deleting the columns of $\boldsymbol{A}$ with indices in $\bar{J}$ (and similarly for $\boldsymbol{A}_{\bar{J}}$, $\boldsymbol{x}_J$ and $\boldsymbol{x}_{\bar{J}}$, where $\boldsymbol{x}$ is an n-dimensional row vector). Then:*

i. *$\boldsymbol{D}$ is a semi PN matrix if and only if the equation*

$$\boldsymbol{A}_J\boldsymbol{x}_J - \boldsymbol{A}_{\bar{J}}\boldsymbol{x}_{\bar{J}} = \boldsymbol{C}_{\bar{J}}\boldsymbol{y}_{\bar{J}} - \boldsymbol{C}_J\boldsymbol{y}_J$$

has a positive solution $\boldsymbol{x} > \boldsymbol{0}$, $\boldsymbol{y} > \boldsymbol{0}$;

ii. *$\boldsymbol{D}$ has the Minkowski property if and only if, for any given $\bar{\boldsymbol{x}}_{\bar{J}} > \boldsymbol{0}$, the equation*

$$\boldsymbol{A}_J\boldsymbol{x}_J - \boldsymbol{A}_{\bar{J}}\bar{\boldsymbol{x}}_{\bar{J}} = \boldsymbol{C}_J\boldsymbol{y}_J - \boldsymbol{C}_{\bar{J}}\boldsymbol{y}_{\bar{J}}$$

has a positive solution $\boldsymbol{x}_J > \boldsymbol{0}$, $\boldsymbol{y} > \boldsymbol{0}$; and

iii. *if, in addition, $\boldsymbol{C}$ is indecomposable then $\boldsymbol{D}$ has the Metzler property if and only if for any given $\bar{\boldsymbol{x}}_{\bar{J}} > \boldsymbol{0}$ the equation*

$$\boldsymbol{A}_J\boldsymbol{x}_J - \boldsymbol{A}_{\bar{J}}\bar{\boldsymbol{x}}_{\bar{J}} = \boldsymbol{C}_{\bar{J}}\boldsymbol{y}_{\bar{J}} - \boldsymbol{C}_J\boldsymbol{y}_J$$

has a positive solution $\boldsymbol{x}_J > \boldsymbol{0}$, $\boldsymbol{y} > \boldsymbol{0}$.

PROOF. (i) From Theorem 7 of Chapter 1, $\boldsymbol{C}^{-1}$ is nonnegative; hence $\boldsymbol{D}$ also is nonnegative. Applying Theorem 5, $\boldsymbol{D}$ is a semi-PN matrix if and only if there exists an $\boldsymbol{x} > \boldsymbol{0}$ such that

$$[\boldsymbol{I}_J\boldsymbol{C}^{-1}\boldsymbol{A}\boldsymbol{I}_{\bar{J}}]\boldsymbol{x} < \boldsymbol{0}.$$

Defining $\boldsymbol{y} = -[\boldsymbol{I}_J\boldsymbol{C}^{-1}\boldsymbol{A}\boldsymbol{I}_{\bar{J}}]\boldsymbol{x}$, the above inequality can be rewritten as

$$[\boldsymbol{I}_J\boldsymbol{C}^{-1}\boldsymbol{A}\boldsymbol{I}_{\bar{J}}]\boldsymbol{x} = -\boldsymbol{y}.$$

Noticing that $\boldsymbol{I}_J \boldsymbol{I}_J = \boldsymbol{I}$, we obtain

$$\boldsymbol{A I}_J \boldsymbol{x} = -\boldsymbol{C I}_J \boldsymbol{y},$$

from which it follows that

$$\boldsymbol{A}_J \boldsymbol{x}_J - \boldsymbol{A}_{\bar{J}} \boldsymbol{x}_{\bar{J}} = \boldsymbol{C}_{\bar{J}} \boldsymbol{y}_{\bar{J}} - \boldsymbol{C}_J \boldsymbol{y}_J.$$

(ii) In view of (iii) of Theorem 6, by which $\boldsymbol{D}$ has the Minkowski property if and only if there exists an $\boldsymbol{x}_J > 0$ such that

$$\boldsymbol{y} \equiv [\boldsymbol{I}_J \boldsymbol{C}^{-1} \boldsymbol{A I}_J] \begin{pmatrix} \boldsymbol{x}_J \\ \bar{\boldsymbol{x}}_{\bar{J}} \end{pmatrix} > \boldsymbol{0},$$

we obtain

$$\boldsymbol{A}_J \boldsymbol{x}_J - \boldsymbol{A}_{\bar{J}} \bar{\boldsymbol{x}}_{\bar{J}} = \boldsymbol{C}_J \boldsymbol{y}_J - \boldsymbol{C}_{\bar{J}} \boldsymbol{y}_{\bar{J}}.$$

(iii) (*Necessity*). Since $\boldsymbol{D} \equiv \boldsymbol{C}^{-1}\boldsymbol{A}$ has the Metzler property, $\boldsymbol{A}$ must be nonsingular, which in turn implies that every column of $\boldsymbol{A}$ is semipositive. Moreover, the indecomposability of $\boldsymbol{C}$ ensures that $\boldsymbol{C}^{-1} > \boldsymbol{0}$. Hence every off-diagonal element of $\boldsymbol{D}$ is positive. It follows from (iii) of Theorem 7 that there exist $\boldsymbol{x}_J > \boldsymbol{0}$ and $\boldsymbol{y} > \boldsymbol{0}$ such that

$$[\boldsymbol{I}_J \boldsymbol{C}^{-1} \boldsymbol{A I}_J] \begin{pmatrix} \boldsymbol{x}_J \\ \bar{\boldsymbol{x}}_{\bar{J}} \end{pmatrix} \equiv -\boldsymbol{y} < \boldsymbol{0}$$

for any nonempty proper subset J of N and any given $\bar{\boldsymbol{x}}_{\bar{J}} > \boldsymbol{0}$. Then the same procedure as in the proof of (ii) yields

$$\boldsymbol{A}_J \boldsymbol{x}_J - \boldsymbol{A}_{\bar{J}} \bar{\boldsymbol{x}}_{\bar{J}} = \boldsymbol{C}_{\bar{J}} \boldsymbol{y}_{\bar{J}} - \boldsymbol{C}_J \boldsymbol{y}_J.$$

(iii) (*Sufficiency*). Suppose that $\boldsymbol{A}$ contains a column of zeros and let $J = \{j \in N : \boldsymbol{a}_j = \boldsymbol{0}\}$. Clearly J is a proper subset of N; for otherwise, from the assumed condition, the equation $\boldsymbol{C I}_J \boldsymbol{y} = \boldsymbol{0}$ has a positive solution $\boldsymbol{y}$, contradicting the fact that $\boldsymbol{C I}_J$ is nonsingular. Again from the assumed condition,

$$\boldsymbol{0} < \boldsymbol{C}^{-1}(\boldsymbol{A}_{\bar{J}} \bar{\boldsymbol{x}}_{\bar{J}}) = \begin{pmatrix} \boldsymbol{y}_J \\ -\boldsymbol{y}_{\bar{J}} \end{pmatrix},$$

a contradiction since $-\boldsymbol{y}_{\bar{J}} < \boldsymbol{0}$. Thus all columns of $\boldsymbol{A}$ are semipositive, whence every off-diagonal element of $\boldsymbol{D}$ is positive. The implication then follows from (iii) of Theorem 7. □

Notes to Chapter 3

1. As is well known, the roots of the polynomial equation $\sum_{j=0}^{n} a_j x^{n-j} = 0$ are continuous in $(a_0, a_1, \ldots, a_n)$. Moreover, the coefficients of the characteristic equation of $\boldsymbol{D}(t)\boldsymbol{A}$ are continuous in t. Hence the eigenvalues of $\boldsymbol{D}(t)\boldsymbol{A}$ are continuous in t.

Stability Analysis of Some Dynamic Systems 4

1 Systems of Linear Differential Equations: Solutions, Stability, and Saddle-point Property

In this section we are concerned mainly with systems of linear differential equations with coefficients independent of time t. However, it is convenient to introduce at the outset the notion of a system of differential equations in general. Let $f_j(x, t)$ be a real valued function defined on the $(n + 1)$-dimensional vector space $\mathbb{R}^{n+1}$ and let $\dot{x}$ be a vector consisting of the time derivatives of the x_j. Then the required system is

$$\dot{x} = f(x, t) = (f_1(x, t), \ldots, f_n(x, t))'. \tag{1}$$

Definition 1. By a *solution* or *state of* (1), defined on some given time interval, we simply mean a vector function $x(t)$ of t such that, for any t of the time interval,

$$\dot{x}(t) = f(x(t), t).$$

The point of time at which the motion begins is denoted by t_0 and, by appropriate choice of time origin, may be set equal to zero. A solution $x(t)$ which passes through a given state or point x_0 at the beginning of motion is denoted by $x(x_0, t)$; x_0 is called the *initial state*. By definition, $x_0 = x(x_0, t_0)$.

In the special linear case, $f_j(x) = \sum_{k=1}^{n} a_{jk} x_k + b_j$, $j = 1, \ldots, n$, and (1) reduces to

$$\dot{x} = Ax + b, \tag{2}$$

where A and b are, respectively, the matrix with elements a_{jk} and the vector with elements $b_j (j, k = 1, \ldots, n)$. In general, the elements of A and b depend on time t. In the present chapter, however, attention is confined to the case in which both $\boldsymbol{A}$ and $\boldsymbol{b}$ are independent of time. The system of linear differential equations (2) is said to be *homogeneous* if $\boldsymbol{b} = \boldsymbol{0}$, otherwise *nonhomogeneous*.

Our first task is to find the solutions of system (2). For the time being we confine our attention to the homogeneous system

$$\dot{\boldsymbol{x}} = \boldsymbol{A}\boldsymbol{x}. \tag{3}$$

Evidently $\boldsymbol{x} = \boldsymbol{0}$ is a solution of (3), so that the set S of all solutions of (3) is nonempty. Let $\boldsymbol{x}_i$, $i = 1, 2$, be any two solutions of (3) and let c_i, $i = 1, 2$, be any two scalars (or complex numbers). Then a direct calculation shows that $c_1\boldsymbol{x}_1 + c_2\boldsymbol{x}_2$ is also a solution of (3). Thus we have obtained

Lemma 1. *The set S of all solutions of* (3) *is a linear subspace in* $\mathbb{R}^n$.

To obtain the solutions of (3) explicitly, a new concept is needed.

Definition 2. Let $\boldsymbol{A}$ be an $n \times n$ matrix. Then an *exponential* $\boldsymbol{e}^{\boldsymbol{A}}$ of $\boldsymbol{A}$ is defined by a series $\sum_{j=0}^{\infty} (1/j!)\boldsymbol{A}^j$, where $j!$ denotes the factorial of j, that is, $j! = \prod_{i=1}^{j} i$, and $\boldsymbol{A}^0 = \boldsymbol{I}$, the identity matrix of order n. Since it is well known that the series $\sum_{j=0}^{\infty} (1/j!)\boldsymbol{A}^j$ is absolutely convergent, the definition of $\boldsymbol{e}^{\boldsymbol{A}}$ is meaningful.

Remark Concerning Definition 2. For the concept of absolute convergence, see Rudin [1964: p. 62–68]. Hartman [1964: p. 54–58], in fact, proves the absolute convergence of this series.

Lemma 2. *If two $n \times n$ matrices $\boldsymbol{A}$ and $\boldsymbol{B}$ are commutative, that is, if $\boldsymbol{AB} = \boldsymbol{BA}$, then $\boldsymbol{e}^{\boldsymbol{A}+\boldsymbol{B}} = \boldsymbol{e}^{\boldsymbol{A}}\boldsymbol{e}^{\boldsymbol{B}}$.*

PROOF. Since both $e^{\boldsymbol{A}}$ and $e^{\boldsymbol{B}}$ are absolutely convergent, we can rearrange terms in the expansion of $e^{\boldsymbol{A}}e^{\boldsymbol{B}}$ to obtain

$$\begin{aligned}
\boldsymbol{e}^{\boldsymbol{A}}\boldsymbol{e}^{\boldsymbol{B}} &= \left(\boldsymbol{I} + \frac{\boldsymbol{A}}{1!} + \frac{\boldsymbol{A}^2}{2!} + \frac{\boldsymbol{A}^3}{3!} + \cdots\right)\left(\boldsymbol{I} + \frac{\boldsymbol{B}}{1!} + \frac{\boldsymbol{B}^2}{2!} + \frac{\boldsymbol{B}^3}{3!} + \cdots\right) \\
&= \boldsymbol{I} + (\boldsymbol{A} + \boldsymbol{B}) + \frac{\boldsymbol{A}^2 + 2\boldsymbol{AB} + \boldsymbol{B}^2}{2!} + \cdots + \frac{\sum_{j=0}^{n} {}_nC_j \boldsymbol{A}^{n-j}\boldsymbol{B}^j}{n!} + \cdots,
\end{aligned}$$

where

$$ {}_nC_j = \frac{n!}{j!(n-j)!}.$$

By induction on n, and recalling that $\boldsymbol{AB} = \boldsymbol{BA}$, we find that $\sum_{j=0}^{n} {}_nC_j \boldsymbol{A}^{n-j}\boldsymbol{B}^j = (\boldsymbol{A} + \boldsymbol{B})^n$, which yields the desired result, that is, $\boldsymbol{e}^{\boldsymbol{A}}\boldsymbol{e}^{\boldsymbol{B}} = \sum_{n=0}^{\infty} (\boldsymbol{A} + \boldsymbol{B})^n/n! = \boldsymbol{e}^{\boldsymbol{A}} + \boldsymbol{e}^{\boldsymbol{B}}$. □

Theorem 1. *The general solution of the system of simultaneous differential equations* (3) *is given by*

$$x(t) = e^{At}c \tag{4}$$

where c *is an arbitrary* $n \times 1$ *vector independent of* t.

PROOF. Differentiating both sides of $e^{At} = \sum_{n=0}^{\infty} (At)^n/n!$ with respect to t, we have[1]

$$\frac{d(e^{At})}{dt} = A \sum_{j=0}^{\infty} \frac{(At)^j}{j!} = Ae^{At}.$$

Hence each column of e^{At} is a solution of (3). Since At and $-At$ are commutative, it follows from Lemma 1 that $e^{At}e^{-At} = e^{At-At} = I$ for any t. Therefore, e^{At} is surely nonsingular. S being a linear subspace by virtue of Lemma 1, any solution $\xi(t)$ of (3) can be expressed as a linear combination of all columns of e^{At}; that is, for some vector $c(t)$,

$$\xi(t) = e^{At}c(t). \tag{5}$$

It thus remains to verify that $c(t)$ is in fact independent of t. Differentiation of both sides of (5) yields

$$\begin{aligned}\dot{\xi}(t) &= Ae^{At}c(t) + e^{At}\dot{c}(t)\\ &= A\xi(t) + e^{At}\dot{c}(t).\end{aligned}$$

Noticing that $\xi(t)$ belongs to S and that e^{At} is nonsingular, $c(t)$ must be constant for every t. □

Given an initial state x_0 from which a solution $x(t)$ of (3) is to start, c in (4) can be uniquely determined by

$$c = e^{-At_0}x_0. \tag{6}$$

Adopting the convention that $t_0 = 0$, $x(t)$ in (4) can be expressed simply as

$$x(t) = e^{At}x_0. \tag{7}$$

We turn now to the nonhomogeneous system with b constant. The following corollary is immediate.

Corollary 1. *If the system* (2) *has a stationary solution* $\bar{x}$ *such that* $A\bar{x} + b = 0$ *then the set of all solutions of* (2) *is a linear variety passing through the point* $\bar{x}$. *Equivalently, every solution* $x(t)$ *of* (2) *is given by*

$$x(t) = e^{At}(x_0 - \bar{x}) + \bar{x},$$

where x_0 *is a given initial state at* $t = 0$.

PROOF. Define $y = x - \bar{x}$. Then $\dot{y} = \dot{x}$ and $x = y + \bar{x}$. In view of the assumption $A\bar{x} + b = 0$, the system (2) can be transformed into

$$\dot{y} = Ay. \tag{8}$$

Applying Theorem 1 to (8), we obtain

$$\boldsymbol{y}(t) = e^{\boldsymbol{A}t}\boldsymbol{c}.$$

The transformation of variables then yields

$$\boldsymbol{x}(t) = e^{\boldsymbol{A}t}\boldsymbol{c} + \bar{\boldsymbol{x}}.$$

Since $\boldsymbol{x}(t)$ must start from $\boldsymbol{x}_0$, $\boldsymbol{c} = (\boldsymbol{x}_0 - \bar{\boldsymbol{x}})$.

Noticing that the set of solutions $\boldsymbol{y}(t)$ of (8) is a linear subspace and that $\boldsymbol{x}(t) = \boldsymbol{y}(t) + \bar{\boldsymbol{x}}$, the first assertion of the corollary is trivial. □

Given the explicit form of the solutions of (2) and/or (3), we can analyze the behavior of those solutions when t becomes infinitely large. In particular, we can examine the stability of the motion generated by (2) and/or (3). First, however, we must define the notion of stability.

Definition 3. A stationary solution (or an equilibrium state) $\boldsymbol{x} = \boldsymbol{0}$ of (1), if it exists, is said to be *locally stable in the sense of Liapunov* if, for any $\varepsilon > 0$, there exists a $\delta > 0$ such that every solution $\boldsymbol{x}(t)$ of (1) starting at an initial state $\boldsymbol{x}_0$ in the δ-neighborhood $B(\boldsymbol{0}, \delta)$ of $\boldsymbol{0}$ remains within the ε-neighborhood $B(\boldsymbol{0}, \varepsilon)$ of $\boldsymbol{0}$ for all $t \leqq 0$. If, in addition, $\boldsymbol{x}(t) \to \boldsymbol{0}$ as $t \to \infty$ (or $\lim_{t\to\infty} \boldsymbol{x}(t) = \boldsymbol{0}$) then $\boldsymbol{x} = \boldsymbol{0}$ is said to be *locally asymptotically stable*. Frequently in the literature of economics, and henceforth in this chapter, "locally asymptotically stable" is abbreviated to "locally stable." Conformably with this usage, a stationary solution is hereafter said to be *locally* L-*stable* if it is locally stable in the sense of Liapunov. In general, the stationary solution of (1) need not be the origin but any constant vector $\bar{\boldsymbol{x}}$. However, without loss of generality $\bar{\boldsymbol{x}}$ can always be chosen as the origin.

Local stability implies local L-stability, but the converse is not true. In particular, local L-stability but not local stability is compatible with limit cycles. That is, L-stability does not necessarily exclude the possibility that for some $\boldsymbol{x}(t)$ starting from $\boldsymbol{x}_0$ sufficiently close to $\boldsymbol{0}$ there exist $r \in (0, \varepsilon)$ and $t_1 > 0$ such that $d(\boldsymbol{x}(t), \boldsymbol{0})^2 \geqq r$ for $t \geqq t_1$. Figure 4.1a depicts a stationary

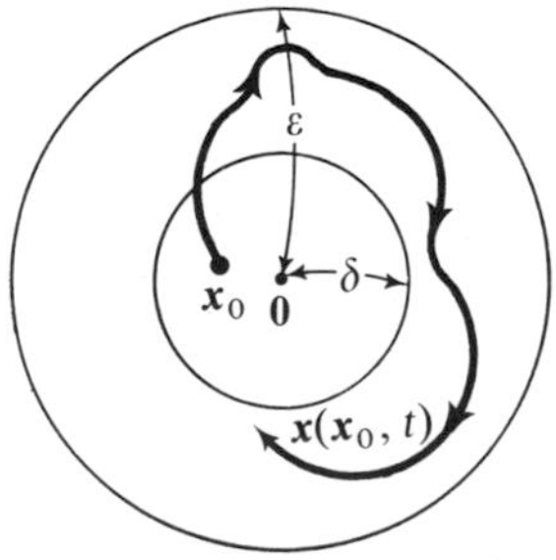

Figure 4.1a The origin is locally L-stable.

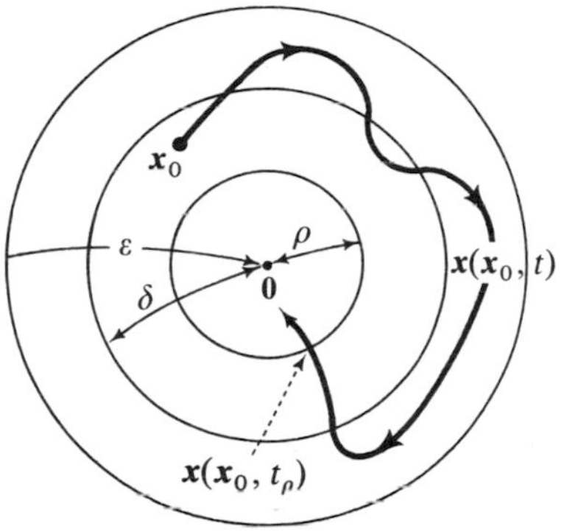

Figure 4.1b The origin is locally stable. Because of the additional condition that $\lim_{t\to\infty} x(x_0, t) = \mathbf{0}$ for any $\rho > 0$, there is $t_\rho > 0$ such that if $t > t_\rho$ then $d(x(x_0, t), \mathbf{0}) < \rho$.

solution which is locally L-stable but not locally stable, while Figure 4.1b displays a stationary solution which is locally stable.

The concepts of stability in Definition 3 refer to motions of system (1) which start sufficiently near the equilibrium state. Removing this restriction, we arrive at the concept of global stability.

Definition 4. If there exists an equilibrium state $\mathbf{0}$ of (1) such that $\lim_{t\to\infty} x(t) = \mathbf{0}$ for a solution $x(t)$ of (1) starting from an arbitrary initial state x_0 then the equilibrium state is said to be *globally stable* or to be *stable in the large.*

Remark Concerning Definition 4. It is often said that the system under investigation is globally (or locally) stable when what is meant is that the equilibrium state of the system is globally (or locally) stable. Moreover, by "stable" is always meant "asymptotically stable."

Having defined the concept of stability, we can return to the system of linear differential equations. As was shown earlier, a nonhomogeneous system can be reduced to a homogeneous one provided that the conditions enumerated in Corollary 1 are satisfied. We therefore content ourselves with a necessary and sufficient condition for an equilibrium state of (3) (typically the origin) to be globally stable. First, however, we introduce

Lemma 3. *Let* $\boldsymbol{K}$ *be an* $n \times n$ *matrix such that*

$$\boldsymbol{K} = (\mathbf{0}, \boldsymbol{e}_1, \ldots, \boldsymbol{e}_{n-1}),$$

where $\boldsymbol{e}_j$ *is the* jth *unit vector of order* n. *Then*

$$\boldsymbol{K}^i = (\overbrace{\mathbf{0}, \ldots, \mathbf{0}}^{i}, \boldsymbol{e}_1, \ldots, \boldsymbol{e}_{n-i}) \qquad (i = 2, \ldots, n-1) \tag{9}$$

and

$$\boldsymbol{K}^i = \mathbf{0} \qquad (i \geqq n). \tag{10}$$

PROOF. The proof of (9) is by induction on i, the power of $\boldsymbol{K}$. Suppose that $i = 2$. Then a direct calculation shows that

$$\boldsymbol{K}^2 = (\boldsymbol{K0}, \boldsymbol{Ke}_1, \ldots, \boldsymbol{Ke}_{n-1}) = (\boldsymbol{0}, \boldsymbol{0}, \boldsymbol{e}_1, \ldots, \boldsymbol{e}_{n-2}).$$

Suppose that (9) holds for some $r < n - 1$. Then, from the induction assumption,

$$\begin{aligned} \boldsymbol{K}^{r+1} &= \boldsymbol{K}^r \cdot \boldsymbol{K} = (\overbrace{\boldsymbol{0}, \ldots, \boldsymbol{0}}^{r}, \boldsymbol{e}_1, \ldots, \boldsymbol{e}_{n-r})\boldsymbol{K} \\ &= (\overbrace{\boldsymbol{0}, \ldots, \boldsymbol{0}}^{r+1}, \boldsymbol{e}_1, \ldots, \boldsymbol{e}_{n-(r+1)}), \end{aligned}$$

which establishes (9).

In view of (9), with $i = n - 1$, a direct calculation yields $\boldsymbol{K}^n = \boldsymbol{0}$, from which (10) is immediately derived. □

We now turn to the main proposition of this section.

Theorem 2. *An equilibrium state* $\boldsymbol{0}$ *of* (3) *is globally stable if and only if every eigenvalue of* $\boldsymbol{A}$ *has no nonnegative real part. (A square matrix with this property is said to be* stable.)

PROOF. It is well known that there is a nonsingular matrix $\boldsymbol{P}$ such that $\boldsymbol{P}^{-1}\boldsymbol{AP} = \boldsymbol{J}$ where $\boldsymbol{J}$ is the Jordan canonical form of $\boldsymbol{A}$. More specifically, let λ_j and μ_j, $j = 1, \ldots, m$, be eigenvalues of $\boldsymbol{A}$ and their multiplicity, respectively. Then

$$\boldsymbol{J} = \begin{bmatrix} \boldsymbol{J}_1 & & \boldsymbol{0} \\ & \ddots & \\ \boldsymbol{0} & & \boldsymbol{J}_m \end{bmatrix} \tag{11}$$

where each $\boldsymbol{J}_j$ is the Jordan diagonal block of order μ_j, that is,

$$\boldsymbol{J}_j = \begin{bmatrix} \lambda_j & 1 & & \boldsymbol{0} \\ & \ddots & \ddots & \\ & & \ddots & 1 \\ \boldsymbol{0} & & & \lambda_j \end{bmatrix} \qquad (j = 1, \ldots, m). \tag{12}$$

The change of variables

$$\boldsymbol{x} = \boldsymbol{Py} \tag{13}$$

transforms (3) into

$$\dot{\boldsymbol{y}} = \boldsymbol{Jy}. \tag{3'}$$

Let $\boldsymbol{x}_0$ be any initial state. Then, applying Theorem 1 to (3′), the solution $\boldsymbol{y}(t)$ of (3′), starting at the transformed initial state $\boldsymbol{y}_0 = \boldsymbol{P}^{-1}\boldsymbol{x}_0$, is

$$\boldsymbol{y}(t) = \boldsymbol{e}^{\boldsymbol{J}t}\boldsymbol{y}_0. \tag{14}$$

Taking into account (11) as well as the definition of $e^{\boldsymbol{J}t}$, it is clear that

$$e^{\boldsymbol{J}t} = \begin{bmatrix} e^{\boldsymbol{J}_1 t} & & \boldsymbol{0} \\ & \ddots & \\ \boldsymbol{0} & & e^{\boldsymbol{J}_m t} \end{bmatrix}. \tag{15}$$

Note that $\boldsymbol{J}_j$, $j = 1, \ldots, m$, can be rewritten as

$$\boldsymbol{J}_j = \lambda_j \boldsymbol{I}_{\mu_j} + \boldsymbol{K}_{\mu_j}$$

where

$$\boldsymbol{K}_{\mu_j} = \begin{bmatrix} 0 & 1 & & \boldsymbol{0} \\ & \ddots & \ddots & \\ & & \ddots & 1 \\ \boldsymbol{0} & & & 0 \end{bmatrix}$$

and $\boldsymbol{I}_{\mu_j}$ is an identity matrix of order μ_j $(j = 1, \ldots, m)$. Since $\lambda_j \boldsymbol{I}_{\mu_j}$ and $\boldsymbol{K}_{\mu_j}$ are commutative and since

$$e^{\lambda_j \boldsymbol{I}_{\mu_j} t} = e^{\lambda_j t} \boldsymbol{I}_{\mu_j},$$

it follows from Lemma 1 that

$$e^{\boldsymbol{J}_j t} = e^{\lambda_j t} e^{\boldsymbol{K}_{\mu_j} t}.$$

Expanding $e^{\boldsymbol{K}_{\mu_j} t}$ with the aid of Lemma 3, we obtain

$$e^{\boldsymbol{J}_j t} = e^{\operatorname{Re}(\lambda_j)t} e^{i \operatorname{Im}(\lambda_j)t} \begin{bmatrix} 1 & t & \dfrac{t^2}{2!} & \cdots & \dfrac{t^{\mu_j - 2}}{(\mu_j - 2)!} & \dfrac{t^{\mu_j - 1}}{(\mu_j - 1)!} \\ & \ddots & \ddots & \ddots & & \dfrac{t^{\mu_j - 2}}{(\mu_j - 2)!} \\ & & \ddots & \ddots & \ddots & \vdots \\ & & & \ddots & \ddots & \dfrac{t^2}{2!} \\ & \boldsymbol{0} & & & \ddots & t \\ & & & & & 1 \end{bmatrix} \quad j = 1, \ldots, m. \tag{16}$$

Let us define $\alpha_j = \sum_{i=1}^{j} \mu_i$ and $N_j = \{\alpha_{j-1} + 1, \ldots, \alpha_j\}$ with $\alpha_0 = 0$ $(j = 1, \ldots, m)$, and let us denote the (pg)th element of $e^{\boldsymbol{K}_{\mu_j} t}$ and the sth element of $\boldsymbol{y}_0$ by $K_{pg}(\mu_j, t)$ and y_{0s}, respectively. Then, from (15) and (16),

$$y_k(t) = e^{\operatorname{Re}(\lambda_j)t} e^{i \operatorname{Im}(\lambda_j)t} \sum_{q \in N_j} K_{p_k q}(\mu_j, t) y_{0q}$$

$$\text{for every } k \in N_j \qquad (j = 1, \ldots, m), \tag{17}$$

where $p_k = k - \alpha_{j-1}$ and $y_k(t)$ is of course any kth element of $\boldsymbol{y}(t)$.

(*Sufficiency*). Let μ be the largest of μ_j, $j = 1, \ldots, m$. Then we can rewrite (17) as

$$y_k(t) = e^{\text{Re}(\lambda_j)t} e^{i\,\text{Im}(\lambda_j)t} t^{\mu} \sum_{q \in N_j} K_{p_{kq}}(\mu_j, t) t^{-\mu} y_{0q}.$$

Since $|e^{i\,\text{Im}(\lambda_j)t}| = 1$, as is shown in Section 1 of Appendix B, we have

$$|y_k(t)| = e^{\text{Re}(\lambda_j)t} t^{\mu} \left| \sum_{q \in N_j} K_{p_{kq}}(\mu_j, t) t^{-\mu} y_{0q} \right|.$$

In view of (16), every $K_{p_{kq}}(\mu_j, t)t^{-\mu}$ converges to 0 as $t \to \infty$; hence with any $\varepsilon > 0$ there is associated a $t_\varepsilon > 0$ such that if $t > t_\varepsilon$ then

$$\left| \sum_{q \in N_j} K_{p_{kq}}(\mu_j, t) t^{-\mu} y_{0q} \right| < \varepsilon \quad \text{for any } k \in N_j \quad (j = 1, \ldots, m). \tag{18}$$

Since $\text{Re}(\lambda_j) < 0$ for all j, it follows from the inequalities (18), coupled with the well-known inequalities

$$e^{-\text{Re}(\lambda_j)t} > t^{\mu} \qquad (j = 1, \ldots, m), \qquad \text{for sufficiently large } t$$

that $y_k(t) \to 0$ as $t \to \infty$. Recalling that $\boldsymbol{x}(t) = \boldsymbol{P}\boldsymbol{y}(t)$, $\boldsymbol{x}(t)$ converges to $\boldsymbol{0}$ as $t \to \infty$.

(*Necessity*). Suppose the contrary. Then, for some λ_{j_0}, $\text{Re}(\lambda_{j_0}) \geqq 0$. Consider an initial state $\hat{\boldsymbol{y}}_0$ such that $\hat{y}_{0q} = \delta > 0$ for $q \in N_{j_0}$ and $\hat{y}_{0s} = 0$ for $s \neq q$. If $\mu_{j_0} > 1$ or $\text{Re}(\lambda_{j_0}) > 0$ then, for any $k \in N_{j_0}$,

$$|y_k(t)| = e^{\text{Re}(\lambda_{j_0})t} \left| \sum_{q \in N_{j_0}} K_{p_{kq}}(\mu_{j_0}, t) \hat{y}_{0q} \right|$$

becomes infinitely large as $t \to \infty$. Otherwise, λ_{j_0} is a purely imaginary characteristic root of $\boldsymbol{A}$ which is simple. Hence, in view of (16),

$$|y_k(t)| = \hat{y}_{0k} = \delta > 0, \qquad k = \alpha_{j_0 - 1} + 1.$$

Thus it has been shown that there exists an initial state $\boldsymbol{y}_0$ such that a solution $\boldsymbol{y}(t)$ of (3′) starting at $\boldsymbol{y}_0$ can never converge to $\boldsymbol{0}$, provided that $\boldsymbol{A}$ has an eigenvalue with nonnegative real part. This clearly violates the assumption. □

Remark Concerning Theorem 2. (i) Since the positive number δ in the proof of necessity can be chosen as small as need be, it is clear that not only global stability but also local stability of the equilibrium state must be ruled out if $\boldsymbol{A}$ has an eigenvalue with nonnegative real part. However, if $\text{Re}(\lambda_j) \leqq 0$, $j = 1, \ldots, m$, $\max_j \text{Re}(\lambda_j) \equiv \text{Re}(\lambda_{j_0}) = 0$ and $\mu_{j_0} = 1$ then, as was shown in the proof of necessity, the equilibrium state $\boldsymbol{0}$ is yet locally L-stable.

(ii) The change of variables $\boldsymbol{P}\boldsymbol{y} = \boldsymbol{x}$, together with equations (14), (15), and (16), yield a scalar formula for the solution $\boldsymbol{x}(t)$ of (3):

$$x_k(t) = \sum_{\tau=1}^{m} p_{k\tau} y_\tau(t) = \sum_{j=1}^{m} \left(\sum_{\tau_j \in N_j} p_{k\tau} f_\tau^j(t) \right) e^{\lambda_j t} \qquad (k = 1, \ldots, n),$$

where $f_\tau^j(t)$, $\tau \in N_j$, $j = 1, \ldots, m$, are polynomials in t of order at most $\mu_j - 1$. The coefficients of $f_\tau^j(t)$ are of course dependent upon the initial state and $p_{k\tau}$ is the $(k\tau)$th element of $\boldsymbol{P}$. Thus the problem of determining the solutions of (3) is reduced to the algebraic problem of determining the Jordan canonical form $\boldsymbol{J}$ of $\boldsymbol{A}$ and a matrix $\boldsymbol{P}$ for which $\boldsymbol{PAP}^{-1} = \boldsymbol{J}$.

(iii) Notice further that a suitable change of variables enables us to reduce a differential equation of higher order, say n, to a system analogous to (2). Consider the differential equation of order n

$$\sum_{j=0}^{n} a_j x^{(n-j)} = c,$$

where $x^{(n-j)} \equiv d^{n-j}x/dt^{n-j}$, $j = 1, \ldots, n$, $x^0 \equiv x$, a_j and c are given constants, and a_0 is assumed to be unity. Let $y_j = x^{(j)}$, $j = 0, 1, \ldots, n$, $\boldsymbol{y} = (y_0, y_1, \ldots, y_{n-1})'$, $\boldsymbol{b} = (\overbrace{0, \ldots, 0}^{n-1}, c)'$ and

$$\boldsymbol{A} = \begin{bmatrix} 0 & 1 & & & \\ \vdots & \ddots & \ddots & & \mathbf{0} \\ \vdots & 0 & & \ddots & \\ \vdots & \vdots & \mathbf{0} & \ddots & \\ 0 & 0 & \cdots & 0 & 1 \\ -a_n & -a_{n-1} & & \cdots & -a_1 \end{bmatrix}.$$

Then we can transform the differential equation to obtain

$$\dot{\boldsymbol{y}} = \boldsymbol{A}\boldsymbol{y} + \boldsymbol{b}.$$

Theorem 2 requires us to seek necessary and sufficient conditions for a square matrix to be stable. In this connection the theorems of Routh–Hurwitz and Liapunov are especially famous and important. However, we state and prove only the latter theorem, since it seems to be more relevant to our subsequent discussion. (A statement of the Routh–Hurwitz Theorem can be found in Section 1 of Appendix B.)

Before Liapunov's Theorem can be stated, some additional terminology must be introduced.

Definition 5. Let $\boldsymbol{Q}$ be an $n \times n$ matrix with complex elements q_{jk}, $j, k = 1, \ldots, n$. Then:

i. $\boldsymbol{Q}$ is said to be *hermitian* (abbreviated to h henceforth) or an *hermitian matrix* (hm) if $\boldsymbol{Q}^* \equiv (\bar{q}_{kj}) = \boldsymbol{Q} = (q_{jk})$, $j, k = 1, \ldots, n$ *or*, in matrix notation, if $\boldsymbol{Q}^* \equiv (\bar{\boldsymbol{Q}})' = \boldsymbol{Q}$;
ii. an hermitian form $\boldsymbol{x}^*\boldsymbol{Q}\boldsymbol{x}$ associated with an hm $\boldsymbol{Q}$ is said to be *hermitian positive definite* (hpd) if for every $\boldsymbol{x} \neq \boldsymbol{0}$, $\boldsymbol{x}^*\boldsymbol{Q}\boldsymbol{x} > 0$ and an hm $\boldsymbol{Q}$ is called *hermitian positive definite* if and only if the hermitian form associated with it is hpd;

iii. if $QQ^* = I$, Q is called *unitary*; and, finally,
iv. an hm Q is said to be *hermitian negative definite* (hnd) if and only if $-Q$ is hpd, where x is any complex vector of order n.

Theorem 3. *An* $n \times n$ *complex matrix* A *is stable if and only if for any* hnd W *an* hpd V *satisfying*

$$A^*V + VA = W \tag{19}$$

is determined uniquely.

Proof (*Necessity*). Since A is stable, Theorem 2 ensures that $\lim_{t\to\infty} e^{At}x_0 = \mathbf{0}$ for every x_0, whence $\lim_{t\to\infty} e^{At} = \mathbf{0}$. Let $X(t) = -e^{A^*t}We^{At}$. Then $X(0) = -W$ and $\lim_{t\to\infty} X(t) = \mathbf{0}$, since $\lim_{t\to\infty} e^{At} = \mathbf{0}$ as was seen above. Noticing that $Ae^{At} = e^{At}A$, differentiation of $X(t)$ with respect to t yields

$$\begin{aligned} \dot{X}(t) &= -(A^*e^{A^*t}We^{At} + e^{A^*t}WAe^{At}) \\ &= (A^*X(t) + X(t)A). \end{aligned} \tag{20}$$

Integrating both sides of (20) with respect to t[3] and recalling that $X(0) = -W$ and that $\lim_{t\to\infty} X(t) = \mathbf{0}$, we have

$$\begin{aligned} \left[A^*\left(\int_0^\infty X(t)dt\right) + \left(\int_0^\infty X(t)dt\right)A\right] &= \int_0^\infty \dot{X}(t)dt \\ &= \lim_{t\to\infty} X(t) - X(0) = W, \end{aligned}$$

which, setting $V = \int_0^\infty X(t)dt$, ensures that there exists V such that $A^*V + VA = W$. Since $e^{A^*t} = (e^{At})^*$ and since $\overline{\int_0^\infty r(t)dt} = \int_0^\infty \overline{r(t)}dt$ (by direct calculation), we see that at $V^* = V$, where $r(t)$ denotes a complex variable dependent on t. It therefore remains to verify that V is a uniquely determined hpd matrix. We first prove the uniqueness of V. Let $\tilde{V}$ be another matrix with the property that $W = A^*\tilde{V} + \tilde{V}A$. Then

$$\begin{aligned} V = \int_0^\infty X(t)dt &= -\int_0^\infty (e^{A^*t}(A^*\tilde{V} + \tilde{V}A)e^{At})dt \\ &= -\int_0^\infty (d(e^{A^*t}\tilde{V}e^{At})/dt)dt \\ &= \tilde{V} - \lim_{t\to\infty} (e^{A^*t}\tilde{V}e^{At}) \\ &= \tilde{V} \end{aligned}$$

since, by virtue of the assumed stability of A, $\lim_{t\to\infty} e^{A^*t}\tilde{V}e^{At} = \mathbf{0}$. Suppose now that V is not hpd. Then there exists an $x_0 \neq \mathbf{0}$ such that $x_0^*Vx_0 \leqq 0$. Note that W is hnd and that $x(t) = e^{At}x_0 \neq \mathbf{0}$. Then it is easy to see that $d(x^*(t)Vx(t))/dt = x^*(t)Wx(t) < 0$. It follows that $x^*(t)Vx(t)$ is monotonically decreasing along $x(t)$ with nonpositive initial value $x_0^*Vx_0$, contradicting the fact that $\lim_{t\to\infty} x^*(t)Vx(t) = 0$.

(*Sufficiency*). Let x_0 be any nonzero initial state. Then $x(t) = e^{At}x_0$ is a solution of (3) starting at x_0. By virtue of (19) and the assumption that W is hnd, $d(x^*(t)Vx(t))/dt = x^*(t)Wx(t) < 0$ for all $t \geqq 0$. Hence

$$x_0^* V x_0 \geqq x^*(t)Vx(t) > 0 \quad \text{for all } t \geqq 0. \tag{21}$$

Note that as a further consequence of (19)

$$x^*(t)Vx(t) = x_0^* V x_0 + \int_0^t x^*(\tau)Wx(\tau)d\tau. \tag{22}$$

Suppose that $x(t)$ does not converge to $\mathbf{0}$, so that there exists an $\varepsilon > 0$ such that

$$N(x(t)) \geqq \varepsilon \quad \text{for any } t > 0,$$

where $N(x(t))$ denotes an arbitrary norm of $x(t)$. Then there exists $\delta > 0$ such that

$$x^*(t)Wx(t) < -\delta \quad \text{for any } t > 0. \tag{23}$$

This can be shown in the following way. Since W is hnd, all eigenvalues w_j, $j = 1, \ldots, n$, are negative and there exists a unitary matrix Q such that $Q^*WQ = \hat{W}$ where $\hat{W}$ is a matrix of which the diagonal elements are w_j $(j = 1, \ldots, n)$. Transforming the variables by

$$x(t) = Qy(t) \tag{24}$$

we obtain

$$\begin{aligned} x^*(t)Wx(t) &= y^*(t)\hat{W}y(t) \\ &= \sum_{j=1}^{n} w_j |y_j(t)|^2 \\ &\leqq \max_j w_j \left(\sum_j |y_j(t)|^2 \right). \end{aligned}$$

Since $N(x(t)) \geq \varepsilon$, (24) implies that

$$\varepsilon \leqq N(x(t)) \leqq N(Q) \cdot N(y(t))$$

from which it follows that

$$\frac{\varepsilon}{N(Q)} \leqq N(y(t)).$$

On the other hand, it is well known that there exists an $m > 0$ such that[4]

$$mN(y(t)) \leq \left(\sum_{j=1}^{n} |y_j(t)|^2 \right)^{1/2}.$$

Choose a $\delta > 0$ such that $\delta < (m\varepsilon/N(Q))$. Then, recalling that w_j, $j = 1, \ldots, n$, is negative, (23) follows. Combining (23) with (22), it is clear that, for t

sufficiently large, $x^*(t)Vx(t) \leqq x_0^*Vx_0 - \delta t < 0$, which contradicts (21). Thus $\lim_{t\to\infty} e^{At}x_0 = 0$. Theorem 2 therefore ensures that every eigenvalue of A has a negative real part. □

Remark Concerning Theorem 3. As is easily seen, the proof of sufficiency in Theorem 3 holds intact in the case $W = -I$. Moreover, from the necessity part of Theorem 3 it is immediate that if A is stable then there exists a unique hpd V such that

$$A^*V + VA = -I.$$

Theorem 3 is therefore often stated as

Theorem 3′. *An $n \times n$ complex matrix A is stable if and only if there exists a unique* hpd V *such that*

$$A^*V + VA = -I. \tag{19′}$$

Our final task in this section is to introduce the notion of the saddle-point property of a stationary solution of (3) above. To this end we recall some facts of matrix algebra.

Lemma 4*. *The eigenvectors x_i, $i = 1, \ldots, k$, corresponding to the distinct eigenvalues λ_i, $i = 1, \ldots, k$, of an $n \times n$ matrix A are linearly independent, and if $n = k$ then*

$$P^{-1}AP = \Lambda,$$

where P is a matrix with x_j as its jth *column*, $j = 1, \ldots, n$, *and Λ is a diagonal matrix with diagonal elements λ_i, $i = 1, \ldots, n$.*

PROOF. To prove the first half of the assertion, we temporarily assume that $\lambda_i \neq 0$ $(i = 1, \ldots, k)$. Let c_i, $i = 1, \ldots, k$, be such that

$$\sum_{i=1}^{k} c_i x_i = 0. \tag{25}$$

Multiplying both sides of (25) by λ_1, we obtain

$$\sum_{i=1}^{k} c_i \lambda_1 x_i = 0 \tag{26}$$

Noticing that $Ax_i = \lambda_i x_i$ for $i = 1, \ldots, k$, it follows from (25) that

$$\sum_{i=1}^{k} c_i \lambda_i x_i = 0. \tag{27}$$

Subtracting (26) from (27),

$$\sum_{i=2}^{k} c_i(\lambda_i - \lambda_1)x_i = 0. \tag{28}$$

Premultiplying both sides of (28) by A and λ_2, we obtain

$$\sum_{i=2}^{k} c_i(\lambda_i - \lambda_1)\lambda_i \boldsymbol{x}_i = \boldsymbol{0} \tag{29}$$

and

$$\sum_{i=2}^{k} c_i(\lambda_i - \lambda_1)\lambda_2 \boldsymbol{x}_i = \boldsymbol{0}, \tag{30}$$

respectively. Subtracting (30) from (29),

$$\sum_{i=3}^{k} c_i(\lambda_i - \lambda_1)(\lambda_i - \lambda_2)\boldsymbol{x}_i = \boldsymbol{0}. \tag{31}$$

Proceeding in this way, we obtain the general equation

$$\sum_{i=j}^{k} c_i\left(\prod_{l=1}^{j-1} (\lambda_i - \lambda_l)\right)\boldsymbol{x}_i = \boldsymbol{0} \qquad (j = 1, \ldots, k). \tag{32}$$

In the case $j = k$, (32) reduces to

$$c_k\left(\prod_{l=1}^{k-1} (\lambda_k - \lambda_l)\right)\boldsymbol{x}_k = \boldsymbol{0}.$$

Since the λ_i, $i = 1, \ldots, k$, are distinct and since $\boldsymbol{x}_k \neq \boldsymbol{0}$, $c_k = 0$. It therefore follows from (32), with $j = k - 1$ and $c_k = 0$, that $c_{k-1} = 0$. Similar backward use of (32) enables us to conclude that $c_i = 0$ for $i = 1, \ldots, k$.

Next, let us suppose that one of the λ_i, say λ_1, is zero. Then the remaining eigenvalues $\lambda_2, \ldots, \lambda_k$ are all nonzero. Hence, by the foregoing argument, the vectors $\boldsymbol{x}_2, \ldots, \boldsymbol{x}_k$ are linearly independent. To demonstrate the linear independence of $\boldsymbol{x}_i$, $i = 1, \ldots, k$, suppose the contrary. Then there exists $\boldsymbol{\alpha} = (\alpha_1, \ldots, a_k)' \neq \boldsymbol{0}$ such that

$$\sum_{i=1}^{k} \alpha_i \boldsymbol{x}_i = \boldsymbol{0}.$$

If $\alpha_1 = 0$ then $\boldsymbol{\alpha}$ must be a zero vector since $\boldsymbol{x}_2, \ldots, \boldsymbol{x}_k$ are linearly independent. Hence $\boldsymbol{x}_1$ can be expressed as a linear combination of $\boldsymbol{x}_2, \ldots, \boldsymbol{x}_k$, that is,

$$\boldsymbol{x}_1 = \sum_{i=2}^{k} \beta_i \boldsymbol{x}_i. \tag{33}$$

Clearly, $\boldsymbol{\beta} = (\beta_2, \ldots, \beta_k)' \neq \boldsymbol{0}$, for otherwise $\boldsymbol{x}_1 = \boldsymbol{0}$, contradicting the definition of $\boldsymbol{x}_1$. Premultiplying both sides of (33) by A, we obtain

$$\boldsymbol{0} = \lambda_1 \boldsymbol{x}_1 = A\boldsymbol{x}_1 = \sum_{i=2}^{k} \beta_i A\boldsymbol{x}_i = \sum_{i=2}^{k} (\lambda_i \beta_i)\boldsymbol{x}_i,$$

which again contradicts the linear independence of $\boldsymbol{x}_i$ $(i = 2, \ldots, k)$.

The first assertion of the lemma, coupled with the assumption that $n = k$, guarantees that the matrix $\boldsymbol{P}$ is nonsingular, and by definition

$$\boldsymbol{AP} = \boldsymbol{P\Lambda}.$$

Hence

$$\boldsymbol{P}^{-1}\boldsymbol{AP} = \boldsymbol{\Lambda}. \qquad \square$$

For the time being, let us assume that the coefficient matrix $\boldsymbol{A}$ has n distinct real eigenvalues λ_i, $i = 1, \ldots, n$. By $\boldsymbol{x}_i$, we again denote the eigenvector corresponding to λ_i, $i = 1, \ldots, n$, and by $\boldsymbol{\Lambda}$ we again denote the diagonal matrix with diagonal elements λ_i, $i = 1, \ldots, n$. Since each λ_i is a simple root, $\boldsymbol{\Lambda}$ is the Jordan canonical form of $\boldsymbol{A}$ and, by Lemma 4*, $\boldsymbol{P} = (\boldsymbol{x}_1, \ldots, \boldsymbol{x}_n)$ is the matrix corresponding to the similarity transformation yielding the Jordan canonical form of $\boldsymbol{A}$. With the aid of (13), (14), (15), and (16), the solution of (3) can be expressed as

$$\boldsymbol{x}(t) = \boldsymbol{P}e^{\boldsymbol{\Lambda}t}\boldsymbol{P}^{-1}\boldsymbol{x}_0. \tag{34}$$

Now suppose that some λ_i, say $\lambda_1, \ldots, \lambda_k$, are negative and that the others are positive. If $\boldsymbol{x}_0$ is a scalar multiple of $\boldsymbol{x}_j$, a motion starting from such an initial state can be expressed as

$$\boldsymbol{x}_j(t) = c_j e^{\lambda_j t}\boldsymbol{x}_j \qquad (j = 1, \ldots, n).$$

Hence if $j \leqq k$ then $\boldsymbol{x}_j(t)$ tends to $\boldsymbol{0}$ as $t \to \infty$; otherwise, $\boldsymbol{x}_j(t)$ tends to $\boldsymbol{0}$ as $t \to -\infty$. In other words, the stationary point $\boldsymbol{0}$ of (3) has the property that only a finite number of orbits tends to $\boldsymbol{0}$ as $t \to \infty$ or as $t \to -\infty$, where an orbit is defined as a set of points $\boldsymbol{x}$ on a solution $\boldsymbol{x}(t)$ of (3) without reference to a parameterization.

Definition 6. A stationary point of (3) with the above property is called a *saddle point*.

In the two-dimensional case it is possible to provide a geometric illustration of the saddle-point property. For concreteness, suppose that $\lambda_1 < 0 < \lambda_2$. Since, by Lemma 4*, $\boldsymbol{x}_1$ and $\boldsymbol{x}_2$ are two linearly independent vectors of order 2, they form a basis for $\mathbb{R}^2$. Hence the motion expressed by (34) can be depicted on a plane with coordinates $\boldsymbol{x}_1$ and $\boldsymbol{x}_2$. If $c_1 \neq 0$ and $c_2 = 0$, that is, if $\boldsymbol{x}_0 = c_1\boldsymbol{x}_1$ for some $c_1 \neq 0$, then the motion starting from this initial state converges to $\boldsymbol{0}$ as $t \to \infty$, and diverges as $t \to -\infty$. On the other hand, if $\boldsymbol{x}_0 = c_2\boldsymbol{x}_2$ for some $c_2 \neq 0$, the motion $\boldsymbol{x}_2(t)$ diverges from $\boldsymbol{0}$ as $t \to \infty$, and converges as $t \to -\infty$. The remaining possible sign combinations of c_1 and c_2 are:

I. $c_1 > 0, c_2 > 0$,
II. $c_1 > 0, c_2 < 0$;
III. $c_1 < 0, c_2 > 0$; and
IV. $c_1 < 0, c_2 < 0$.

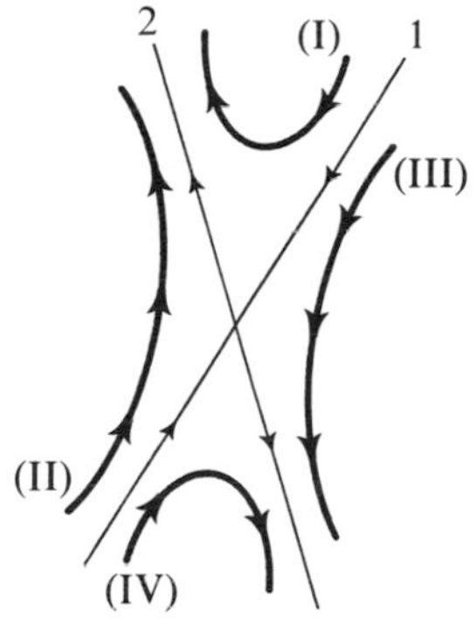

Figure 4.2a

Assuming $t \to \infty$, in Figure 4.2a there are depicted the motions corresponding to all possible sign combinations of c_1 and c_2. The basis $(\boldsymbol{x}_1, \boldsymbol{x}_2)$ is nonorthogonal in general. In the special case when $\boldsymbol{x}_1$ and $\boldsymbol{x}_2$ are perpendicular to each other, the diagram becomes a little easier to read.[5]

So far, we have assumed that all eigenvalues of $\boldsymbol{A}$ are distinct. However, the existence of multiple roots does not necessarily prevent the stationary solution from having the saddle-point property. To show this, suppose that $n = 3$ and that $\lambda_1 < 0 < \lambda_2 = \lambda_3 = \lambda$. Then, in view of (13), (14), (15), and (16), the solution may be expressed as:

$$\boldsymbol{x}(t) = \boldsymbol{P}\begin{bmatrix} c_1 e^{\lambda_1 t} \\ (c_2 + c_3 t)e^{\lambda t} \\ c_3 e^{\lambda t} \end{bmatrix}$$

where c_j denotes the jth component of $\boldsymbol{P}^{-1}\boldsymbol{x}_0$. Therefore, the motion starting from a scalar multiple of $\boldsymbol{x}_1$, the first column of $\boldsymbol{P}$, converges to $\boldsymbol{0}$ (diverges from $\boldsymbol{0}$) as $t \to \infty$ ($t \to -\infty$). On the other hand, if $\boldsymbol{x}_0$ is a scalar multiple of

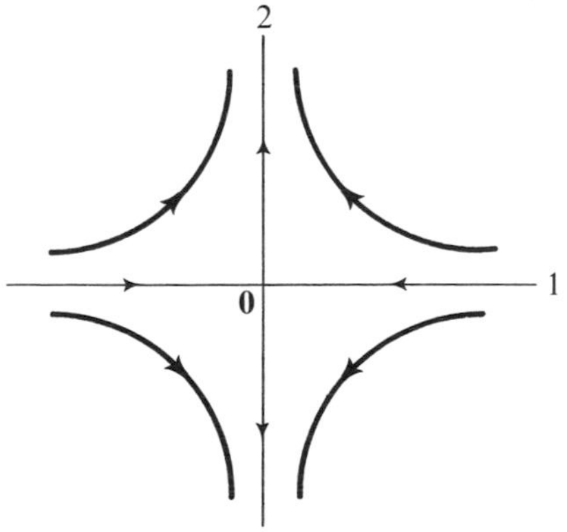

Figure 4.2b $t \to \infty$.

$\boldsymbol{x}_2$ then $\boldsymbol{x}_2(t) = (0, c_2, 0)'e^{\lambda t}$ diverges from (converges to) $\mathbf{0}$ as $t \to \infty$ ($t \to -\infty$). The motion $\boldsymbol{x}_3(t) = (0, c_3 t, c_3)'e^{\lambda t}$ clearly behaves in a different manner from $\boldsymbol{x}_j(t)$, $j = 1, 2$. However, in this case, the stationary solution $\mathbf{0}$ of (3) still possesses the saddle-point property. Figure 4.2b displays the motion of the solutions of (3) projected on the plane with axes $\boldsymbol{x}_1$ and $\boldsymbol{x}_2$, which we assume to be orthogonal to each other.

Remark Concerning Definition 6. The saddle point defined here must not be confused with the saddle point defined in Definition 6 of Chapter 1. The former concept is associated with a stationary solution of differential equations, the latter with optimizing problems. However, as is often seen in optimal control problems, the same point may have both properties.

2 Stability of Linear *Tâtonnement* Processes

We rest now from the purely mathematical development of the theory of linear differential equations to consider an important application of the theory, the stability analysis of linear *tâtonnement* processes.

We begin with some notation.

$\mathfrak{D}$: The set of all $n \times n$ diagonal matrices with real diagonal elements. Any element $\boldsymbol{D}$ of $\mathfrak{D}$ is called a *real diagonal matrix*, abbreviated to rdm. Similarly, a diagonal matrix, the diagonal elements of which are all positive, is said to be a *positive diagonal matrix*, abbreviated to pdm.

D_π: The set of all pdm.

$\mathfrak{E}$: The set of all nonsingular rdm.

$\boldsymbol{A}_{JK}$: A submatrix obtained from an $n \times n$ matrix $\boldsymbol{A}$ by extracting a_{jk} ($j \in J$, $k \in K$), where J and K are any two subsets of $N = \{1, \ldots, n\}$.

$\# S$: The number of elements belonging to the set S.

$\mu(\boldsymbol{A}, \lambda)$: The algebraic multiplicity of an eigenvalue λ of a square matrix $\boldsymbol{A}$ of order n.

$\mu(\boldsymbol{A}, S)$: The sum of the multiplicities of all the eigenvalues of $\boldsymbol{A}$ lying in a subset S of the set C of all complex numbers. Symbolically, $\mu(\boldsymbol{A}, S) = \sum_{\lambda \in S} \mu(\boldsymbol{A}, \lambda)$.

c^+: The set $\{\lambda \in C : \text{Re}(\lambda) > 0\}$.

c^0: The set $\{\lambda \in C : \text{Re}(\lambda) = 0\}$.

c^-: The set $\{\lambda \in C : \text{Re}(\lambda) < 0\}$.

$v^k(\boldsymbol{A})$: The set $\mu(\boldsymbol{A}, c^k)$, $k = +, 0, -$,

in $\boldsymbol{A}$: The inertia of $\boldsymbol{A}$, which is defined by in $\boldsymbol{A} \equiv (v^+(\boldsymbol{A}), v^0(\boldsymbol{A}), v^-(\boldsymbol{A}))$.

$\boldsymbol{\rho}$: A point-to-set mapping $\mathfrak{D}$ into N such that $\boldsymbol{\rho}(\boldsymbol{D}) = \{i \in N : d_{ii} \neq 0\}$. Hence, for any $S \subseteq N$, $\boldsymbol{\rho}^{-1}(S) = \{\boldsymbol{D} \in \mathfrak{D} : \boldsymbol{\rho}(\boldsymbol{D}) = S\}$.

$N(\boldsymbol{A})$: The norm of a matrix $\boldsymbol{A}$ which is defined by $\sup_{\boldsymbol{x} \neq \mathbf{0}} (N(\boldsymbol{A}\boldsymbol{x})/N(\boldsymbol{x}))$, the properties of which can be found in Section 2 of Appendix B.

We turn now to the properties of definite matrices.

Definition 7. An $n \times n$ complex matrix A is said to be quasi-hermitian positive (respectively, negative) definite, abbreviated as qhpd (respectively, qhnd), if $\frac{1}{2}x^*(A + A^*)x > 0$ (respectively, $\frac{1}{2}x^*(A + A^*)x < 0$) for every nonzero n-dimensional vector x.

Remark Concerning Definition 7. Since (x^*Ax) is a scalar, $(x^*A^*x) = (x^*Ax)^* = \overline{(x^*Ax)}$. It is therefore obvious that $\frac{1}{2}x^*(A + A^*)x = \frac{1}{2}\{x^*Ax + x^*A^*x\} = \frac{1}{2}\{(x^*Ax) + \overline{(x^*Ax)}\} = \text{Re}(x^*Ax)$. We hereafter denote the matrix $\frac{1}{2}(A + A^*)$ by $\grave{A}$, where $\overline{(x^*Ax)}$ denotes the conjugate of (x^*Ax).

Lemma 4. *If A is an hermitian matrix, then the following statements are equivalent.*

a. *A is* hnd;
b. *the eigenvalues λ_j ($j = 1, \ldots, n$) of A are all negative*;
c. *A is an NP-matrix or (Hicksian).*[6]

PROOF. (a) $\Rightarrow$ (b) Let x_j be an eigenvector of A corresponding to λ_j. Then we have

$$0 > x_j^*Ax_j = \lambda_j x_j^* x_j \qquad (j = 1, \ldots, n).$$

It is therefore clear that $\lambda_j < 0$ for $j = 1, \ldots, n$.

(b) $\Rightarrow$ (a) Suppose the contrary. Then there exists an $x \neq 0$ such that $x^*\grave{A}x = x^*Ax \geq 0$. Let Λ be the diagonal matrix with diagonal elements λ_j ($j = 1, \ldots, n$). Then, as is well known,[7] there exists a unitary matrix Q such that $Q^*AQ = \Lambda$. The change of variables $x = Qy$ therefore yields

$$x^*Ax = y^*Q^*AQy = y^*\Lambda y = \sum_{j-1}^{n} \lambda_j |y_j|^2.$$

Since a nonsingular matrix maps no nonzero vector to the origin, the hypothesis (b) implies that $\sum_{j=1}^{n} \lambda_j |y_j|^2 < 0$, yielding a contradiction:

$$0 \leqq x^*Ax = \sum_{j=1}^{n} \lambda_j |y_j|^2 < 0.$$

(a) $\Rightarrow$ (c) Since A is hnd, so is every principal submatrix of A. Suppose that A is not an NP-matrix. Then there exists a subset J of N such that $(-1)^{\#J} \det A_{JJ} \leqq 0$. Since A is supposed to be hnd, so is A_{JJ}; moreover, from the implication (a) $\Rightarrow$ (b), all eigenvalues μ_j ($j = 1, \ldots, \#J$) of A_{JJ} are negative. Together with the well-known identity $\det A_{JJ} = \prod_{j=1}^{\#J} \mu_j$, these facts yield the contradiction $(-1)^{\#J-1} = \text{sgn}(\det A_{JJ}) = (-1)^{\#J}$.

(c) $\Rightarrow$ (a) The proof is by induction on n. If $n = 1$, $A = a_{11} < 0$ since an hermitian matrix A is assumed to be Hicksian. Thus it is clear that $a_{11}|x|^2 < 0$ for any $x \neq 0$.

Let $J = \{1, \ldots, n-1\}$ and notice that $\boldsymbol{A}_{JJ}$ is also Hicksian under the hypothesis. Then by the induction assumption $\boldsymbol{A}_{JJ}$ is hnd. Partition $\boldsymbol{A}$ into

$$\boldsymbol{A} = \begin{pmatrix} \boldsymbol{A}_{JJ} & \boldsymbol{a} \\ \boldsymbol{a}^* & a_{nn} \end{pmatrix}$$

and define

$$\boldsymbol{E} = \begin{pmatrix} \boldsymbol{I}_{n-1} & -\boldsymbol{A}_{JJ}^{-1}\boldsymbol{a} \\ \boldsymbol{0} & 1 \end{pmatrix}.$$

A slight manipulation yields

$$\boldsymbol{AE} = \begin{pmatrix} \boldsymbol{A}_{JJ} & \boldsymbol{0} \\ \boldsymbol{a}^* & a_{nn} - \boldsymbol{a}^*\boldsymbol{A}_{JJ}^{-1}\boldsymbol{a} \end{pmatrix}.$$

Since $\det \boldsymbol{E} = 1$, it follows that

$$\det \boldsymbol{A} = (\det \boldsymbol{A}_{JJ})(a_{nn} - \boldsymbol{a}^*\boldsymbol{A}_{JJ}^{-1}\boldsymbol{a}).$$

Thus

$$a_{nn} - \boldsymbol{a}^*\boldsymbol{A}_{JJ}^{-1}\boldsymbol{a} < 0$$

since by hypothesis $\operatorname{sgn}(\det \boldsymbol{A}) = -1 \cdot (\operatorname{sgn}(\det \boldsymbol{A}_{JJ}))$. Corresponding to the partition of $\boldsymbol{A}$, partition an $n \times 1$ vector $\boldsymbol{x}$ into $\boldsymbol{x}' = (\boldsymbol{x}'_J, x_n)$. Then the change of variables $\boldsymbol{x}_J = \boldsymbol{y}_J - x_n\boldsymbol{A}_{JJ}^{-1}\boldsymbol{a}$ yields

$$\boldsymbol{x}^*\boldsymbol{A}\boldsymbol{x} = \boldsymbol{y}_J^*\boldsymbol{A}_{JJ}\boldsymbol{y}_J + (a_{nn} - \boldsymbol{a}^*\boldsymbol{A}_{JJ}^{-1}\boldsymbol{a})|x_n|^2.$$

Recalling that by the induction assumption $\boldsymbol{y}_J^*\boldsymbol{A}_{JJ}\boldsymbol{y}_J < 0$ and that $(a_{nn} - \boldsymbol{a}^*\boldsymbol{A}_{JJ}^{-1}\boldsymbol{a}) < 0$, $\boldsymbol{x}^*\boldsymbol{A}\boldsymbol{x} < 0$ for any nonzero $\boldsymbol{x}$. □

Remark Concerning Lemma 4. (i) Since the proof of the implication (c) ⇒ (a) only utilizes the partial assumption that every leading principal submatrix of $\boldsymbol{A}$ is Hicksian, it can be replaced by

c*. *Every leading principal submatrix of* $\boldsymbol{A}$ *is Hicksian.*

(ii) Noticing that $\boldsymbol{A}$ is hpd if and only if $-\boldsymbol{A}$ is hnd, the lemma can be restated as

Lemma 4′. *If* $\boldsymbol{A} = \boldsymbol{A}^*$, *then each one of the following statements implies the others:*

a′. $\boldsymbol{A}$ *is* hpd;
b′. *all eigenvalues of* $\boldsymbol{A}$ *are positive*;
c′. $\boldsymbol{A}$ *is a* P-*matrix.*

(iii) An hermitian matrix $\boldsymbol{A}$ is said to be *hermitian positive* (respectively, negative) semidefinite, abbreviated as hpsd (respectively, hnsd), if for any $\boldsymbol{x}$, $\boldsymbol{x}^*\boldsymbol{A}\boldsymbol{x} \geqq 0$ (respectively, $\boldsymbol{x}^*\boldsymbol{A}\boldsymbol{x} \leqq 0$).

Then Lemma 4 is relaxed to

Lemma 4″. If $\boldsymbol{A} = \boldsymbol{A}^*$, *then the following statements are mutually equivalent.*

a″. $\boldsymbol{A}$ *is* hnsd.
b″. *All eigenvalues of* $\boldsymbol{A}$ *are nonpositive.*
c″. $-\boldsymbol{A}$ *is a* P_0*-matrix.*

Proof. The equivalence of (a″) and (b″) is parallel to that of (a) and (b) *mutatis mutandis*. It therefore suffices to show the implications (b″) ⇒ (c″) ⇒ (a″).

(b″) ⇒ (c″) Notice that $\boldsymbol{A}$ is hnsd if and only if this is true of each principal submatrix of $\boldsymbol{A}$. From the equivalence of (a″) and (b″) it then follows that (b″) is equivalent to the statement that all eigenvalues of every principal submatrix of $\boldsymbol{A}$ are nonpositive. Then, all eigenvalues of each principal submatrix of $-\boldsymbol{A}$ are nonnegative. Recalling that the value of the determinant of a matrix equals the product of all eigenvalues of the matrix, the implication (b″) ⇒ (c″) is immediate.

(c″) ⇒ (a″) Careful scrutiny of the proof of the implication (i) ⇒ (ii) of Theorem 3 of Chapter 3 shows that the implication remains valid even when $\boldsymbol{A}$ is hermitian. Hence, by virtue of Lemma 4,

$$f(t, \boldsymbol{x}) = \boldsymbol{x}^*(-\boldsymbol{A} + t\boldsymbol{I})\boldsymbol{x} = \boldsymbol{x}^*(t\boldsymbol{I} - \boldsymbol{A})\boldsymbol{x} > 0 \quad \text{for all } t > 0 \text{ and all } \boldsymbol{x}.$$

Suppose that $\boldsymbol{A}$ is not hnsd. Then there exists an $\boldsymbol{x}_0 \neq \boldsymbol{0}$ such that $\boldsymbol{x}_0^*\boldsymbol{A}\boldsymbol{x} > 0$. Therefore, for any $t \in (0, (\boldsymbol{x}_0^*\boldsymbol{A}\boldsymbol{x}_0/\boldsymbol{x}_0^*\boldsymbol{x}_0))$, $f(t, \boldsymbol{x}_0) < 0$, a contradiction □

(iv) In this context, we must notice that the statement (c″) of Lemma 4″ cannot be replaced by:

c‴. *Every leading principal minor of* $\det(-\boldsymbol{A})$ *is nonnegative.*

In fact,

$$A = \begin{pmatrix} 0 & 0 \\ 0 & 1 \end{pmatrix}$$

surely satisfies (c‴). However, for any $\boldsymbol{x} = (x_1, x_2)'$ with $x_2 \neq 0$, $\boldsymbol{x}'\boldsymbol{A}\boldsymbol{x} = x_2^2 > 0$, that is, $\boldsymbol{A}$ is not hnsd.

Lemma 5. *Let* λ_j *and* μ_j *be eigenvalues of an* $n \times n$ *complex matrix* $\boldsymbol{A}$ *and of* $\grave{\boldsymbol{A}}$, *respectively. Then:*

i. $\mu_{\min} \leqq \operatorname{Re}(\lambda_j) \leqq \mu_{\max}$, $j = 1, \ldots, n$; *and*
ii. *if* $\boldsymbol{A}$ *is* qhpd *as well as real then it is a* P*-matrix, where* $\mu_{\min}$ *and* $\mu_{\max}$ *are respectively the smallest and the largest of* μ_j $(j = 1, \ldots, n)$.

Proof. (i) Let $\boldsymbol{x}$ be an eigenvector of $\boldsymbol{A}$ corresponding to λ_j. Then

$$\operatorname{Re}(\lambda_j \boldsymbol{x}^*\boldsymbol{x}) = \operatorname{Re}(\boldsymbol{x}^*\boldsymbol{A}\boldsymbol{x}) = \boldsymbol{x}^*\grave{\boldsymbol{A}}\boldsymbol{x} \qquad (j = 1, \ldots, n).$$

Since $\grave{A}$ is an hermitian matrix, there exists a unitary matrix Q such that $Q^*\grave{A}Q = \hat{M}$, where $\hat{M}$ is a diagonal matrix with diagonal elements μ_j $(j = 1, \ldots, n)$. Letting $x = Qy$, we obtain

$$\mathrm{Re}(\lambda_j x^* x) = \mathrm{Re}(\lambda_j) \sum_{i=1}^{n} |y_i|^2 = \sum_{i=1}^{n} \mu_i |y_i|^2 \qquad (j = 1, \ldots, n),$$

from which it follows that

$$\mu_{\min} \leqq \mathrm{Re}(\lambda_j) \leqq \mu_{\max}.$$

(ii) Suppose the contrary. Then there exists a subset J of N for which $\det A_{JJ} \leqq 0$. Assume first that $\det A_{JJ} = 0$. Then there is a nonzero x_J of order $\#J$ such that $A_{JJ} x_J = \mathbf{0}$, from which it follows that $x_J^* A_{JJ} x_J = 0$, contradicting the fact that

$$\mathrm{Re}(x_J^* A_{JJ} x_J) = \frac{1}{2} \begin{pmatrix} x_J \\ \mathbf{0} \end{pmatrix}^* \grave{A} \begin{pmatrix} x_J \\ \mathbf{0} \end{pmatrix} > 0.$$

Hence $\det A_{JJ} < 0$, which, in view of Lemma 6 in Chapter 3, implies that A_{JJ} has a negative eigenvalue λ_0. According to Definition 7, A_{JJ} is qhpd if and only if $\grave{A}_{JJ}$ is hpd. Applying Lemma 4 (equivalence of (a) and (b)), the smallest eigenvalue $v_{\min}$ of $\grave{A}_{JJ}$ is positive. This, in conjunction with (i), yields the selfcontradiction

$$0 < v_{\min} \leqq \lambda_0 < 0. \qquad \square$$

As a final preliminary to the principal business of this section we define several additional types of stability.

Definition 8. Let A be an $n \times n$ matrix defined on the complex field.

i. If DA is stable for every $D \in \mathbf{D}_\pi$ then A is said to be D-*stable*.
ii. If every principal submatrix A_{JJ} of A is D-stable then A is said to be *totally stable*.
iii. If MA is stable for any hpd M then A is said to be S-*stable*.

Remark Concerning Definition 8. (i) and (iii) of Definition 8 are due essentially to Arrow and McManus [1958; p. 449], while (ii) is found in Quirk and Ruppert [1965; p. 314].

The modern analysis of linear *tâtonnement* processes has its origin in the classical works of Hicks [1939] and Samuelson [1941; 1947]. In terms of the mathematics deployed, this analysis has two strands, one relying on Frobenius's theorem on nonnegative square matrices and the other (associated with Arrow [1974]) on certain set-theoretical results. We turn first to Arrow's work and begin with a theorem concerning a single-valued function of a connected set into an arbitrary range. To avoid offensive repetition, we note at the outset that matrices which appear subsequently are complex and of order n, unless otherwise specified.

The definition of a connected set and of a set which is open relative to a set including the set itself may be found in Section 2 of Appendix B.

Theorem 4. *Let f be a single-valued function of a connected set K into any range Y. If, for any $\mathbf{y} \in f(K)$, $f^{-1}(\mathbf{y})$ is open relative to K then f is a constant.*

PROOF. Suppose that the theorem is false. Then there exist $\mathbf{x}$ and $\mathbf{x}_1$ of K such that $\mathbf{x} \neq \mathbf{x}_1$ and $\mathbf{y}_{\mathbf{x}} \equiv f(\mathbf{x}) \neq f(\mathbf{x}_1) \equiv \mathbf{y}_{\mathbf{x}_1}$. Noticing that $K = \bigcup_{\mathbf{y} \in f(K)} f^{-1}(\mathbf{y})$ and that each $f^{-1}(\mathbf{y})$ is open relative to K, the assumed connectedness of K implies that

$$f^{-1}(\mathbf{y}_{\mathbf{x}}) \bigcap \left(\bigcup_{\mathbf{y} \neq \mathbf{y}_{\mathbf{x}}} f^{-1}(\mathbf{y}) \right) \neq \varnothing.$$

On the other hand, it is true that

$$f^{-1}(\mathbf{y}_{\mathbf{x}}) \bigcap f^{-1}(\mathbf{y}_{\mathbf{x}_1}) = \varnothing,$$

for otherwise there would be an $\mathbf{x}_2 \in f^{-1}(\mathbf{y}_{\mathbf{x}}) \cap f^{-1}(\mathbf{y}_{\mathbf{x}_1}) \subseteq K$. We therefore have

$$f(\mathbf{x}_2) = \mathbf{y}_{\mathbf{x}} \neq \mathbf{y}_{\mathbf{x}_1} = f(\mathbf{x}_2),$$

which is absurd because f is assumed to be single-valued. Since $f^{-1}(\mathbf{y}_{\mathbf{x}})$ still intersects $K_{\mathbf{x}} \equiv \bigcup_{\mathbf{y} \neq \mathbf{y}_{\mathbf{x}}} f^{-1}(\mathbf{y})$ there must be an $\mathbf{x}_3 \subset \bigcup_{\mathbf{y} \neq \mathbf{y}_{\mathbf{x}}} f^{-1}(\mathbf{y}) \subseteq K$ for which

$$f^{-1}(\mathbf{y}_{\mathbf{x}}) \cap f^{-1}(\mathbf{y}_{\mathbf{x}_3}) \neq \varnothing,$$

where $\mathbf{y}_{\mathbf{x}_3} \equiv f(\mathbf{x}_3)$. However this is impossible: $K_{\mathbf{x}} = \varnothing$. Hence, f is a constant on K. □

Before stating the next theorem, a lemma useful for the verification of the theorem is introduced.

Lemma 6. *Let F and A_0 be an open subset of C and a given square matrix of order n respectively. Then for a matrix A sufficiently close to A_0,*

$$\mu(A, F) \geqq \mu(A_0, F).$$

PROOF. By λ_i ($i = 1, \ldots, p$) and μ_i denote the eigenvalues of A_0 contained in F and their multiplicities. Then from the assumption it follows that with λ_i, $i = 1, \ldots, p$, there are associated p neighborhoods $B_i(\lambda_i, \varepsilon_i)$ of λ_i such that $B_i(\lambda_i, \varepsilon_i) \subseteq F$. Letting ε_i, $i = 1, \ldots, p$, be sufficiently small, we can further assume that if $i \neq j$ then $\lambda_j \notin B_i(\lambda_i, \varepsilon_i)$ and $B_i(\lambda_i, \varepsilon_i) \cap B_j(\lambda_j, \varepsilon_j) = \varnothing$. It is therefore obvious that

$$\mu(A, F) \geqq \sum_{i=1}^{p} \mu(A, B_i(\lambda_i, \varepsilon_i)) \quad \text{for any matrix } A.$$

Recall that an eigenvalue of A is continuous in A. Then corresponding to every λ_i, A has at least one eigenvalue λ_i' lying in $B(\lambda_i, \varepsilon_i)$ provided that

A is sufficiently close to A_0. Rewrite λ_i, so that $\lambda_i = \lambda_{i_1} = \lambda_{i2} = \cdots = \lambda_{i_{\mu_i}}$. Then the foregoing observation is trivially true for each λ_{i_τ}. Hence if A is sufficiently close to A_0 then

$$\mu(A, B_i(\lambda_i, \varepsilon_i)) \geqq \mu(A_0, B_i(\lambda_i, \varepsilon_i)) \qquad (i = 1, \ldots, p).$$

Combining these two sets of inequalities, we can assert that

$$\mu(A, F) \geqq \sum_{i=1}^{p} (A_0, B_i(\lambda_i, \varepsilon_i)) \quad \text{for } A \text{ sufficiently close to } A_0.$$

It remains only to notice that $\sum_{i=1}^{p} \mu(A_0, B_i(\lambda_i, \varepsilon_i)) = \mu(A_0, F)$. □

Theorem 5 (Arrow [1974; Theorem 4]). *Let C_j, $j = 0, \ldots, m$, be subsets of C such that*

i. $\bigcup_{j=0}^{m} C_j = C$,
ii. $C_i \cap C_j = \varnothing$ *for* $i \neq j$, *and*
iii. C_j, $j = 1, \ldots, m$, *are open relative to* C.

If $\mu(A, C_0)$ is a constant for every A of a connected set K of matrices, then $\mu(A, C_j)$ is invariant over K for each j.

Proof. Since by assumption $\mu(A, C_0)$ is constant over K, we can denote it by an integer m_0. A being an $n \times n$ matrix it is obvious that

$$n = \sum_{j=0}^{m} \mu(A, C_j) \quad \text{for any } A \in K.$$

Hence for every $A \in K$ we have

$$\sum_{j=1}^{m} \mu(A, C_j) = n - m_0.$$

Define a function $\boldsymbol{\mu}$ of K by $\boldsymbol{\mu}(A) = (\mu(A, C_1), \ldots, (\mu(A, C_m))'$. Then, by definition, $\boldsymbol{\mu}^{-1}(\boldsymbol{r}) \neq \varnothing$ for any $\boldsymbol{r} \in \boldsymbol{\mu}(K)$. Let A_0 be any matrix of $\boldsymbol{\mu}^{-1}(\boldsymbol{r})$. Then, by virtue of Lemma 6, there exists a $\delta > 0$ such that

$$\boldsymbol{\mu}(A) \geqq \boldsymbol{\mu}(A_0) \quad \text{for any } A \in (B(A_0, \delta) \cap K).$$

Recalling that $\sum_{j=1}^{m} \mu(A, C_j) = n - m_0$ for any $A \in K$, the last inequality shows that

$$\boldsymbol{\mu}(A) = \boldsymbol{\mu}(A_0) \quad \text{for any } A \in (B(A_0, \delta) \cap K)$$

since $n - m_0 = \sum_{i=1}^{m} \mu(A, C_j) \geqq \sum_{i=1}^{m} \mu(A_0, C_j) = n - m_0$, provided that $A \in (B(A_0, \delta) \cap K)$. Hence, for any $\boldsymbol{r} \in \boldsymbol{\mu}(K)$ and any $A_1 \in \boldsymbol{\mu}^{-1}(\boldsymbol{r})$,

$$(B(A_1, \delta) \cap K) \subseteq \boldsymbol{\mu}^{-1}(\boldsymbol{r}),$$

which, in turn, implies that $\boldsymbol{\mu}^{-1}(\boldsymbol{r})$ is open relative to K for any $\boldsymbol{r} \in \boldsymbol{\mu}(K)$. Thus, in view of Theorem 4, $\boldsymbol{\mu}(A)$ must be constant over K. □

An application of Theorem 5 yields

Theorem 6 (Arrow [1974; Theorem 5]). *Let F and K be sets of square matrices of order n. If K is a connected set containing $\boldsymbol{I}$ and if $v^0(\boldsymbol{DA}) = v^0(\boldsymbol{D})$ for all $\boldsymbol{D} \in F$ and all $\boldsymbol{A} \in K$ then* in $\boldsymbol{DA}$ = in $\boldsymbol{D}$ *for all $\boldsymbol{D} \in F$ and $\boldsymbol{A} \in K$.*

PROOF. Since the assertion is trivially true for $\boldsymbol{D} = \boldsymbol{0}$, we can assume that F contains no zero matrix. We show first of all that the set $K_{\boldsymbol{D}} = \{\boldsymbol{DA}: \boldsymbol{A} \in K\}$ is a connected set of matrices for every $\boldsymbol{D} \in F$. Suppose the contrary. Then there is a $\boldsymbol{D} \in F$ for which $K_{\boldsymbol{D}}$ is disconnected. Hence there exist nonempty subsets $K_{\boldsymbol{D}}^i$ $(i = 1, 2)$ of $K_{\boldsymbol{D}}$ with the properties (i) $K_{\boldsymbol{D}}^i$, $i = 1, 2$, is open relative to $K_{\boldsymbol{D}}$, (ii) $K_{\boldsymbol{D}}^1 \cap K_{\boldsymbol{D}}^2 = \emptyset$, and (iii) $K_{\boldsymbol{D}}^1 \cup K_{\boldsymbol{D}}^2 = K_{\boldsymbol{D}}$. Let $K^i = \{\boldsymbol{A} \in K: \boldsymbol{DA} \in K_{\boldsymbol{D}}^i\}$, $i = 1, 2$. Then, from (ii) and (iii), K^1 and K^2 are nonempty subsets of K such that $K^1 \cup K^2 = K$ and $K^1 \cap K^2 = \emptyset$. Suppose that either one of them, say K^1, is not open relative to K. Then there exists $\boldsymbol{A}_0 \in K^1$ such that for any $\delta > 0$, $(B(\boldsymbol{A}_0, \delta) \cap K) \nsubseteq K^1$. Since $\boldsymbol{DA}_0$ is a member of $K_{\boldsymbol{D}}^1$ which is open relative to $K_{\boldsymbol{D}}$, there exists a $\delta_0 > 0$ for which

$$(B(\boldsymbol{DA}_0, \delta_0) \cap K_{\boldsymbol{D}}) \subseteq K_{\boldsymbol{D}}^1.$$

Since K^1 is assumed not to be open relative to K, with $\delta \in (0, \delta_0/N(\boldsymbol{D}))$ there is associated a matrix $\boldsymbol{A}_1$ which lies in $(B(\boldsymbol{A}_0, \delta) \cap K)$ but not in K^1. However, it follows from the choice of δ that

$$\begin{aligned} d(\boldsymbol{DA}_0, \boldsymbol{DA}_1) &= N(\boldsymbol{D}(\boldsymbol{A}_0 - \boldsymbol{A}_1)) \\ &\leq N(\boldsymbol{D})N(\boldsymbol{A}_0 - \boldsymbol{A}_1) \\ &= N(\boldsymbol{D})d(\boldsymbol{A}_0, \boldsymbol{A}_1) < \delta_0. \end{aligned}$$

Hence, $\boldsymbol{DA}_1 \in K_{\boldsymbol{D}}^1$, which is a contradiction, for $\boldsymbol{A}_1 \notin K^1$. Thus both K^1 and K^2 are open relative to K, implying that K is disconnected, which again contradicts the assumed connectedness of K.

Let $\boldsymbol{D}$ be any matrix of F. Then, by hypothesis, $v^0(\boldsymbol{DA})$ is constant over $K_{\boldsymbol{D}}$. Hence in view of Theorem 5, in $\boldsymbol{DA}$ is constant over $K_{\boldsymbol{D}}$. Taking into account that $\boldsymbol{I} \in K$, $\boldsymbol{D}$ is clearly contained in $K_{\boldsymbol{D}}$. Therefore the constancy of in $\boldsymbol{DA}$ over $K_{\boldsymbol{D}}$ surely implies that in $\boldsymbol{DA}$ = in $\boldsymbol{D}$ for all $\boldsymbol{A} \in K$. □

The second condition of Theorem 6 is crucial in that it guarantees the constancy of in $\boldsymbol{DA}$ over $K_{\boldsymbol{D}}$. It is therefore of some value to establish a criterion for $v^0(\boldsymbol{DA}) = v^0(\boldsymbol{D})$ for all $\boldsymbol{D}$ of a subset $\mathfrak{F}$ of $\mathfrak{D}$.

Theorem 7. *Let $\mathfrak{F}$ and $\boldsymbol{A}$ be any subset of $\mathfrak{D}$ and a given square matrix of order n respectively. Then $v^0(\boldsymbol{DA}) = v^0(\boldsymbol{A})$ for all $\boldsymbol{D} \in \mathfrak{F}$ if and only if $\boldsymbol{A}_{JJ} + ia\boldsymbol{D}_{JJ}^{-1}$ is nonsingular for all $a \in \mathbb{R}$, all $J \subseteq N$, and all $\boldsymbol{D} \in (\boldsymbol{\rho}^{-1}(J) \cap \mathfrak{F})$ (where $i = \sqrt{-1}$).*

PROOF. We begin with a remark: if J is a subset of N and if $\boldsymbol{D} \in \boldsymbol{\rho}^{-1}(J)$ then $v^0(\boldsymbol{DA}) = v^0(\boldsymbol{D})$ if and only if $v^0(\boldsymbol{D}_{JJ}\boldsymbol{A}_{JJ}) = 0$, for the hypotheses imply that $v^0(\boldsymbol{D}) = n - \#J$ and that

$$\det(\lambda \boldsymbol{I} - \boldsymbol{DA}) = \det(\lambda \boldsymbol{I}_{JJ} - \boldsymbol{D}_{JJ}\boldsymbol{A}_{JJ})\lambda^{n - \#J}.$$

(*Necessity*). Let J be any subset of N. Then, by virtue of the above remark, $v^0(\boldsymbol{D}_{JJ}\boldsymbol{A}_{JJ}) = 0$ for any $\boldsymbol{D} \in (\rho^{-1}(J) \cap \mathfrak{F})$. Suppose that the implication is false. Then there exist a subset J of N and a real number a such that $\det(\boldsymbol{A}_{JJ} + ia\boldsymbol{D}_{JJ}^{-1}) = 0$ for some $\boldsymbol{D} \in (\rho^{-1}(J) \cap \mathfrak{F})$, which further implies that $\det((-ia)\boldsymbol{I}_{JJ} - \boldsymbol{D}_{JJ}\boldsymbol{A}_{JJ}) = 0$, contradicting the earlier observation.

(*Sufficiency*). Since, by assumption, $\det(\boldsymbol{A}_{JJ} + ia\boldsymbol{D}_{JJ}^{-1}) \neq 0$ for any $a \in \mathbb{R}$, any $J \subseteq N$ and any $\boldsymbol{D} \in (\rho^{-1}(J) \cap \mathfrak{F})$, we evidently have

$$v^0(\boldsymbol{D}_{JJ}\boldsymbol{A}_{JJ}) = 0$$

for any $J \subseteq N$ and $\boldsymbol{D} \in (\rho^{-1}(J) \cap \mathfrak{F})$. In the light of the remark made at the outset of the proof, the assertion is immediate. □

Theorem 8. *Let $\mathfrak{F}$ and i be the same as in Theorem 7. If K is a set of matrices such that $\boldsymbol{A}_{JJ} + ia\boldsymbol{D}_{JJ}^{-1} + b\boldsymbol{I}_{JJ}$ is nonsingular for every $J \subseteq N$, any $\boldsymbol{D} \in (\rho^{-1}(J) \cap \mathfrak{F})$, any $a \in \mathbb{R}$, and any $b \geqq 0$, then* $\operatorname{in} \boldsymbol{DA} = \operatorname{in} \boldsymbol{D}$ *for all $\boldsymbol{D} \in \mathfrak{F}$ and all $\boldsymbol{A} \in K$ where $\boldsymbol{I}$ is an identity matrix of order n.*

Proof. We first show that $\boldsymbol{I}$ belongs to K. For any $J \subseteq N$, any $\boldsymbol{D} \in (\rho^{-1}(J) \cap \mathfrak{F})$, any $a \in \mathbb{R}$, and any $b \geqq 0$ we easily see that $\det(\boldsymbol{I}_{JJ} + ia\boldsymbol{D}_{JJ}^{-1} + b\boldsymbol{I}_{JJ}) = \prod_{j \in J}(1 + b + iad_{jj}^{-1}) \neq 0$, for otherwise we would have the contradiction that $0 = 1 + b$. Thus $\boldsymbol{I}$ is surely contained in K; moreover, K is certainly nonempty.

Next we verify that K is a connected set. Let $[\boldsymbol{A}, \boldsymbol{I}]$ be the line segment joining any $\boldsymbol{A} \in K$ and $\boldsymbol{I}$. Then $K \subseteq \bigcup_{\boldsymbol{A} \in \boldsymbol{K}} [\boldsymbol{A}, \boldsymbol{I}]$. Conversely, to any $\boldsymbol{B} \in \bigcup_{\boldsymbol{A} \in \boldsymbol{K}} [\boldsymbol{A}, \boldsymbol{I}]$ there corresponds $\boldsymbol{A} \in K$ for which $\boldsymbol{B}$ can be expressed as $(1 - \lambda)\boldsymbol{A} + \lambda\boldsymbol{I}$, with $0 \leqq \lambda \leqq 1$. Since $\boldsymbol{B} = \boldsymbol{A}$ or $\boldsymbol{I}$ according as $\lambda = 0$ or 1 it suffices to verify that $\boldsymbol{B} \in K$ for $0 < \lambda < 1$. Noticing that $0 < \lambda < 1$, a slight manipulation yields

$$\begin{aligned} &\det((1 - \lambda)\boldsymbol{A}_{JJ} + \lambda\boldsymbol{I}_{JJ} + ia\boldsymbol{D}_{JJ}^{-1} + b\boldsymbol{I}_{JJ}) \\ &= (1 - \lambda)^{\#J} \det\left(\boldsymbol{A}_{JJ} + i\left(\frac{a}{1 - \lambda}\right)\boldsymbol{D}_{JJ}^{-1} + \left(\frac{b}{1 - \lambda}\right)\boldsymbol{I}_{JJ}\right) \neq 0 \end{aligned}$$

for any $J \subseteq N$, any $\boldsymbol{D} \in (\rho^{-1}(J) \cap \mathfrak{F})$, any $a \in \mathbb{R}$, and any $b \geqq 0$. Thus we have proved that $K = \bigcup_{\boldsymbol{A} \in \boldsymbol{K}} [\boldsymbol{A}, \boldsymbol{I}]$, which implies that K is a union of connected sets $[\boldsymbol{A}, \boldsymbol{I}]$, intersecting each other at a point $\boldsymbol{I}$ of K. In view of Theorem 9 in Subsection 2.2 of Appendix B, K is surely a connected set of matrices. Let $b = 0$. Then by Theorem 7 it is clear that $v^0(\boldsymbol{DA}) = v^0(\boldsymbol{D})$ for every $\boldsymbol{D} \in \mathfrak{F}$ and every $\boldsymbol{A} \in K$. Hence Theorem 8 is a straightforward consequence of Theorem 6. □

Theorems 7 and 8 are the fundamental theorems of this section. The following corollary lists several fruitful implications of those theorems.

Corollary 2. *Suppose that $\mathfrak{F}$ and i are the same as in Theorems 7 and 8. Then:*

i. *if* $\mathfrak{F} = \mathfrak{D}$ *then*

(i–1) $v^0(\boldsymbol{DA}) = v^0(\boldsymbol{D})$ *for every* $\boldsymbol{D} \in \mathfrak{D}$ *and a given square matrix* $\boldsymbol{A}$ *of order* n *if and only if* $(\boldsymbol{A} + i\boldsymbol{E})_{JJ}$ *is nonsingular for any* $J \subseteq N$ *and* $E \in (\mathfrak{E} \cup \{\boldsymbol{0}\})$, *and*

(i–2) *if every principal submatrix of* $\boldsymbol{A} + i\boldsymbol{E} + b\boldsymbol{I}$ *is nonsingular for every* $\boldsymbol{E} \in (\mathfrak{E} \cup \{\boldsymbol{0}\})$ *and every* $b \geqq 0$ *then* $\text{in}\, \boldsymbol{DA} = \text{in}\, \boldsymbol{D}$ *for all* $\boldsymbol{D} \in \mathfrak{D}$ *and all* $\boldsymbol{A}$ *satisfying the above condition*;

ii. *if* $\mathfrak{F} = \mathfrak{D}_\pi$ *then*

(ii–1) $v^0(\boldsymbol{DA}) = v^0(\boldsymbol{D})$ *for every* $\boldsymbol{D} \in \mathfrak{D}_\pi$ *if and only if* $\boldsymbol{A} + ia\boldsymbol{E}$ *is nonsingular for any* $a \in \mathbb{R}$ *and any* $E \in \mathfrak{D}_\pi$, *and*

(ii–2) $\text{in}\, \boldsymbol{DA} = \text{in}\, \boldsymbol{D}$ *for all* $\boldsymbol{D} \in \mathfrak{D}_\pi$ *and every* $\boldsymbol{A}$ *such that* $\boldsymbol{A} + ia\boldsymbol{E} + b\boldsymbol{I}$ *is nonsingular for every* $a \in \mathbb{R}$, *every* $b \geqq 0$, *and every* $\boldsymbol{E} \in \mathfrak{D}_\pi$.

PROOF. (i–1) Since $\mathfrak{F} = \mathfrak{D}$ implies that $\boldsymbol{\rho}^{-1}(J) = (\boldsymbol{\rho}^{-1}(J) \cap \mathfrak{F})$ for every $J \subseteq N$, Theorem 7 asserts that $v^0(\boldsymbol{DA}) = v^0(\boldsymbol{D})$ for all $\boldsymbol{D} \in \mathfrak{D}$ if and only if $\boldsymbol{A}_{JJ} + ia\boldsymbol{D}_{JJ}^{-1}$ is nonsingular for all $J \subseteq N$, all $a \in \mathbb{R}$, and all $\boldsymbol{D} \in \boldsymbol{\rho}^{-1}(J)$. Let $a = 0$. Then, for every J of N, $\det(\boldsymbol{A}_{JJ} + i\boldsymbol{0}_{JJ}) \neq 0$, which is the assertion in the case of $\boldsymbol{E} \in \{\boldsymbol{0}\}$. Suppose that $a \neq 0$ and that there exist an $\boldsymbol{E} \in \mathfrak{E}$ and a subset J of N such that $\boldsymbol{A}_{JJ} + i\boldsymbol{E}_{JJ}$ is singular. Then, for the matrix

$$\boldsymbol{D} = \begin{pmatrix} a^{-1}\boldsymbol{E}_{JJ}^{-1} & \boldsymbol{0} \\ \boldsymbol{0} & \boldsymbol{0} \end{pmatrix},$$

we clearly have

$$\boldsymbol{D} \in \boldsymbol{\rho}^{-1}(J)$$

and

$$0 = \det(\boldsymbol{A}_{JJ} + i\boldsymbol{E}_{JJ}) = \det(\boldsymbol{A}_{JJ} + ia\boldsymbol{D}_{JJ}^{-1}),$$

contradicting Theorem 7.

(i–2) Let K be the set of all matrices satisfying the assumed condition. Then, by the same reasoning as in the proof of Theorem 8, we see that K is a connected set of matrices containing $\boldsymbol{I}$. Since the condition of (i–1) is certainly assured by putting $b = 0$, the present assertion is an immediate consequence of Theorem 6.

(ii–1) From the hypothesis $\mathfrak{F} = \mathfrak{D}_\pi$, it follows that $(\boldsymbol{\rho}^{-1}(J) \cap \mathfrak{D}_\pi) = \varnothing$ or $\mathfrak{D}_\pi$ according as $J \subset N$ or $J = N$. The assertion then follows from Theorem 7.

(ii–2) The proof of (ii–2) is omitted since it is essentially the same as that of (i–2). □

Remark. The propositions (i–1) through (ii–2) in Corollary 2 respectively correspond to Theorem 7, Theorem 8, Corollary 7, and Corollary 9 in Arrow [1974]. On the other hand, our Theorems 7 and 8 correspond to Theorems 7 and 9 of Arrow [1974]. Needless to say, if a matrix $\boldsymbol{A}$ satisfies the condition of (ii–2) in Corollary 2 then $-\boldsymbol{A}$ is D-stable.

The next two corollaries are useful results derived from the assertion (i–2) of Corollary 2.

Corollary 3. *If* $\boldsymbol{A}$ *is* qhpd *then* in $\boldsymbol{HA}$ = in $\boldsymbol{H}$ *for every hermitian matrix* $\boldsymbol{H}$.

Proof. We first prove that if $\boldsymbol{A}$ is qhpd then in $\boldsymbol{DA}$ = in $\boldsymbol{D}$ for all $\boldsymbol{D} \in \mathfrak{D}$. Recalling the well-known fact that every principal submatrix of a qhpd matrix is nonsingular, it suffices to show that if $\boldsymbol{A}$ is qhpd so is $\boldsymbol{A} + i\boldsymbol{E} + b\boldsymbol{I}$ for any $\boldsymbol{E} \in (\mathfrak{E} \cup \{0\})$ and any $b \geqq 0$. A direct calculation shows that

$$\boldsymbol{x}^*(\boldsymbol{A} + i\grave{\boldsymbol{E}} + b\boldsymbol{I})\boldsymbol{x} = \boldsymbol{x}^*\grave{\boldsymbol{A}}\boldsymbol{x} + b\boldsymbol{x}^*\boldsymbol{x} > 0 \quad \text{for any } \boldsymbol{x} \neq \boldsymbol{0}$$

since the nonnegativity of b implies that $\boldsymbol{x}^*\grave{\boldsymbol{A}}\boldsymbol{x} + b\boldsymbol{x}^*\boldsymbol{x} \geq \boldsymbol{x}^*\grave{\boldsymbol{A}}\boldsymbol{x}$, which is positive by the hypothesis.

Since $\boldsymbol{H}$ is hermitian, there exists a unitary matrix $\boldsymbol{Q}$ such that $\boldsymbol{Q}^*\boldsymbol{HQ} = \boldsymbol{\Lambda}$ where $\boldsymbol{\Lambda}$ is a diagonal matrix, the diagonal elements of which are eigenvalues of $\boldsymbol{H}$. Since $\boldsymbol{QQ}^* = \boldsymbol{I}$, $\det(\lambda\boldsymbol{I} - \boldsymbol{HA}) = \det(\lambda\boldsymbol{I} - \boldsymbol{Q}^*\boldsymbol{HAQ})$. We thus obtain

$$\text{in } \boldsymbol{HA} = \text{in } \boldsymbol{Q}^*(\boldsymbol{HA})\boldsymbol{Q} = \text{in } \boldsymbol{Q}^*\boldsymbol{HQQ}^*\boldsymbol{AQ} = \text{in } \boldsymbol{\Lambda}(\boldsymbol{Q}^*\boldsymbol{AQ}).$$

Since $\boldsymbol{A}$ is qhpd, so is $\boldsymbol{Q}^*\boldsymbol{AQ}$. Moreover $\boldsymbol{\Lambda}$ belongs to $\mathfrak{D}$ for $\boldsymbol{H}$ is hermitian. Hence in $\boldsymbol{HA}$ = in $\boldsymbol{\Lambda}(\boldsymbol{Q}^*\boldsymbol{AQ})$ = in $\boldsymbol{\Lambda}$ = in $\boldsymbol{H}$. The final equality is, of course, by the definition of $\boldsymbol{\Lambda}$. □

Corollary 4. *Let* $\boldsymbol{A}^{**}$ *denote a matrix obtained from an* $n \times n$ *matrix* $\boldsymbol{A}$ *by replacing the diagonal element* a_{jj} *with its real part* $\text{Re}(a_{jj})$. *If* $\boldsymbol{A}^{**}$ *is indecomposable and has a positive semidominant diagonal* (psdd) *then*

$$\text{in } (\boldsymbol{DA})_{JJ} = \text{in } \boldsymbol{D}_{JJ} \quad \text{for any } J \subseteq N \text{ and any } \boldsymbol{D} \in \mathfrak{D}.$$

Proof. By hypothesis $\boldsymbol{A}^{**}$ is indecomposable and has a psdd. Hence from Corollary 2 in Chapter 1 there exist $d_j > 0$ $(j = 1, \ldots, n)$ such that

$$d_j|\text{Re}(a_{jj})| = d_j \text{Re}(a_{jj}) > \sum_{k \neq j} d_k|a_{kj}| \qquad (j = 1, \ldots, n).$$

Therefore, for any $\boldsymbol{E} \in (\mathfrak{E} \cup \{\boldsymbol{0}\})$ and any $b \geqq 0$, we have

$$d_j|(\boldsymbol{A} + i\boldsymbol{E} + b\boldsymbol{I})_{jj}| \geqq d_j|\text{Re}(a_{jj}) + b| = d_j(\text{Re}(a_{jj}) + b) > \sum_{k \neq j} d_k|a_{kj}| \qquad (j = 1, \ldots, n).$$

Hence $(\boldsymbol{A} + i\boldsymbol{E} + b\boldsymbol{I})_{JJ}$ is nonsingular for any $J \subseteq N$, any $b \geqq 0$, and any $\boldsymbol{E} \in (\mathfrak{E} \cup \{\boldsymbol{0}\})$. Noticing further that, under the present hypothesis every principal submatrix of $\boldsymbol{A}^{**}$ has a pdd, the corollary is easily derived from (i–2) of Corollary 2. □

Remarks Concerning Corollaries 3 and 4. (i) Corollaries 3 and 4 are, respectively, a tiny generalization and a slightly elaborated version of Corollaries 8.2 and 8.3 in Arrow [1974]. Moreover, the assertion that if $\boldsymbol{A}$ is qhpd then in $\boldsymbol{DA}$ = in $\boldsymbol{D}$ for all $\boldsymbol{D} \in \mathfrak{E}$ is in fact Corollary 8.1 of Arrow [1974; p. 199].

(ii) If an hermitian matrix $\boldsymbol{H}$ is replaced by $\boldsymbol{D} \in \mathfrak{D}$ then Corollary 3 can be strengthened to obtain the same result as in Corollary 4, since every principal submatrix of a qhpd matrix is qhpd.

(iii) A square matrix with a positive (respectively, negative) semidominant diagonal will be said to have a psdd (respectively, nsdd).

In view of the fact that the product of two diagonal matrices is also a diagonal matrix, one further generalization is possible.

Theorem 9. *If* in $\boldsymbol{DA}$ = in $\boldsymbol{D}$ *for all* $\boldsymbol{D} \in \mathfrak{D}$ *and if* $\boldsymbol{P} \in \mathfrak{D}_\pi$ *then* in $\boldsymbol{DPA}$ = in $\boldsymbol{D}$ *for any* $\boldsymbol{D} \in \mathfrak{D}$.

Proof. Since $\boldsymbol{DP} \in \mathfrak{D}$ for all $\boldsymbol{D} \in \mathfrak{D}$ and all $\boldsymbol{P} \in \mathfrak{D}_\pi$, the hypothesis ensures that in $\boldsymbol{DPA}$ = in $\boldsymbol{DP}$ = in $\boldsymbol{D}$, which completes the proof. □

Since the conditions stated in Corollaries 3 and 4 imply that in $\boldsymbol{DA}$ = in $\boldsymbol{D}$ for all $\boldsymbol{D} \in \mathfrak{D}$, Theorem 9 can be strengthened:

Corollary 5. *If* $\boldsymbol{P} \in \mathfrak{D}_\pi$ *and if either* $\boldsymbol{A}$ *is* qhpd *or* $\boldsymbol{A}^{**}$ *has a* pdd *then* in $\boldsymbol{DPA}$ = in $\boldsymbol{D}$ *for any* $\boldsymbol{D} \in \mathfrak{D}$.

Remark. In the latter case (that is, if $\boldsymbol{A}^{**}$ has a pdd), the conclusion takes a stronger form: in $(\boldsymbol{DPA})_{JJ}$ = in $\boldsymbol{D}_{JJ}$ for all $\boldsymbol{D}$ and all $J \subseteq N$.

The analytical tools provided so far enable us to synthesize the several propositions concerning the local stability of economic systems. We begin by listing those properties of matrices which appear in the propositions. The properties under consideration are:

P.1. $\boldsymbol{A}$ is stable.
P.2. $\boldsymbol{A}$ is D-stable.
P.3. $\boldsymbol{A}$ is totally stable.
P.4. $\boldsymbol{A}$ is totally S-stable, that is, every principal submatrix of $\boldsymbol{A}$ is S-stable.
P.5. $\boldsymbol{A}$ is Hicksian.
P.6. $\boldsymbol{A}$ is quasi-hermitian negative definite (qhnd).
P.7. $\boldsymbol{A}$ is hermitian.
P.8. $\boldsymbol{A}$ is Metzlerian, that is, $\boldsymbol{A}$ is real and every off-diagonal element of $\boldsymbol{A}$ is nonnegative.
P.9. $\boldsymbol{A}^{**}$ has a negative dominant diagonal (ndd).
P.10. $\boldsymbol{A}$ is a real matrix.

Theorem 10. P.6. *implies* P.4, *which, in turn, implies* P.3, P.2, *and* P.1.

PROOF. Since $\boldsymbol{A}$ is qhnd, every principal submatrix $\boldsymbol{A}_{JJ}$ of $\boldsymbol{A}$ is also qhnd, whence $-\boldsymbol{A}_{JJ}$ is qhpd. Applying Corollary 3 to $-\boldsymbol{A}_{JJ}$, it is easy to see that in $\boldsymbol{H} =$ in $\boldsymbol{H}(-\boldsymbol{A}_{JJ}) = -$in $\boldsymbol{H}\boldsymbol{A}_{JJ}$ where $\boldsymbol{H}$ is any hpd matrix of order $\#J$. Hence, in view of Lemma 4′, $(0, 0, \#J) = -$in $\boldsymbol{H} =$ in $\boldsymbol{H}\boldsymbol{A}_{JJ}$. Thus P.6 surely implies P.4.

Since every pdm $\boldsymbol{D}$ is hpd, P.3 follows from P.4. The remaining implications are trivial and hence their proofs are omitted. □

Remark. Using the notation a ⇒ b for the statement "condition a implies condition b," some of the history of Theorem 10 may be briefly sketched.

i. P.6 ⇒ P.1 (Samuelson [1947; p. 438])
ii. P.6 ⇒ P.4′ (Theorem 1 in Arrow and McManus [1958]), where P.4′ states that $\boldsymbol{A}$ is S-stable.
iii. P.6 ⇒ P.3 (Corollary to Theorem 1 in Arrow and McManus [1958; p. 450]).

Theorem 11. P.9 *implies* P.3, P.2, *and* P.1.

PROOF. Noticing that P.3 implies P.2, which, in turn, implies P.1, it suffices to verify the first implication alone. As is clear from the proof of Corollary 4, in $(\boldsymbol{D}\boldsymbol{X})_{JJ} =$ in $\boldsymbol{D}_{JJ}$ for a matrix $\boldsymbol{X}$ such that $\boldsymbol{X}^{**}$ has a pdd, any $J \subseteq N$ and any $\boldsymbol{D} \in \mathfrak{D}$. Let $\boldsymbol{X}$ be $-\boldsymbol{A}$. Then $\boldsymbol{X}^{**}$ has a pdd by hypothesis. Hence

$$(0, 0, \#J) = -(\text{in } \boldsymbol{D}_{JJ}) = -\text{in } (\boldsymbol{D}(-\boldsymbol{A}))_{JJ} = \text{in } (\boldsymbol{D}\boldsymbol{A})_{JJ} \quad \text{for any } \boldsymbol{D} \in \mathfrak{D}_\pi \text{ and any } J \subseteq N,$$

which is the first implication.

Remark. McKenzie [1960; Theorem 1] and Newman [1959–60; Theorem 10] prove that P.9 and P.10 imply P.1.

If, in addition, the matrix under consideration is assumed to be hermitian, we obtain:

Theorem 12. *Suppose that* P.7 *is given. Then* P.1, P.5, *and* P.6 *are equivalent to each other, and* P.9 *implies them all.*

PROOF. The first half of this theorem is simply Lemma 4. Combining Theorem 11 and Lemma 4, the second half follows at once. □

Theorem 13. *Given* P.10,

i. P.6 *implies* P.5 (Samuelson [1947; p. 14], Morishima [1952]),
ii. P.3 *implies* P.5 (Metzler [1945; p. 282]), *and*
iii. P.9 *implies* P.5 (Newman [1959–60; Theorem 9]).

PROOF. (i) According to assertion (ii) of Lemma 5, $-\boldsymbol{A}$ is a P-matrix, which implies P.5.

(ii) Since P.3 implies P.1, $-\boldsymbol{A}$ is a real nonsingular matrix, so that $\det(-\boldsymbol{A})$ is either positive or negative. Suppose that $\det(-\boldsymbol{A}) < 0$. Then there exists a positive eigenvalue ρ_0 of $\boldsymbol{A}$ since $\det(\rho \boldsymbol{I} - \boldsymbol{A})$ is continuous in ρ and since $\det(\rho \boldsymbol{I} - \boldsymbol{A}) > 0$ for some ρ sufficiently large. This, however, contradicts the stability of $\boldsymbol{A}$.

Applying the same argument to every principal submatrix of $\boldsymbol{A}$, the proof of (ii) is easily completed.

(iii) The assertion follows from Theorem 11 and assertion (ii) above. □

We turn now to the class of Metzlerian matrices.

Theorem 14. *If* P.8 *is postulated then* P.1, P.2, P.3, P.5, *and* P.9 *are equivalent to each other.*

Proof. Recalling that the implication P.9 ⇒ P.3 ⇒ P.2 ⇒ P.1 has been established in Theorem 11, it suffices to verify the implication P.1 ⇒ P.5 ⇒ P.9, under the assumption of P.8.

(i) P.1 ⇒ P.5. Since $\boldsymbol{A}$ is stable as well as Metzlerian, the Frobenius root λ^* of $\boldsymbol{A}$ must be negative. Let λ_J^* be the Frobenius root of any principal submatrix $\boldsymbol{A}_{JJ}$ of $\boldsymbol{A}$. Then, by Lemma 4 in Chapter 2 (see the proof of (i) of the Lemma), we observe that

$$a_{jj} \leqq \lambda_J^* \leqq \lambda^* < 0 \quad \text{for any } J \subseteq N.$$

Hence every first-order principal minor of $\det \boldsymbol{A}$ is surely negative. Let $\boldsymbol{A}_{JJ}$ be any principal submatrix of $\boldsymbol{A}$ of order not less than 2. Applying (vi) of Lemma 4 in Chapter 2, we easily see that $-\boldsymbol{A}_{JJ}^{-1} \geqq \boldsymbol{0}$, whence in view of the well-known Hawkins–Simon Theorem (Theorem 7 in Chapter 1), $\det(-\boldsymbol{A}_{JJ})$ must be positive. Hence $\operatorname{sgn}(\det \boldsymbol{A}_{JJ}) = (-1)^{\#J}$ or, equivalently, P.5 holds.

(ii) P.5. ⇒ P.9. By virtue of P.5, $-\boldsymbol{A}'$ is a P-matrix. Therefore

$$a_{jj} < 0 \quad (j = 1, \ldots, n).$$

By Theorem 3 in Chapter 3, there exists an $\boldsymbol{x} > \boldsymbol{0}$ such that

$$-\boldsymbol{A}'\boldsymbol{x} > \boldsymbol{0},$$

from which it follows that

$$-a_{jj}x_j - \sum_{i \neq j} a_{ij}x_j = |a_{jj}|x_j - \sum_{i \neq j} x_i|a_{ij}| > 0 \qquad (j = 1, \ldots, n).$$

Thus $\boldsymbol{A}$ certainly has an ndd. □

Remark. Postulating P.8, Metzler [1945] has shown the equivalence between P.1 and P.5, while McKenzie [1960; Theorem 2′] and Newman [1959–60; Theorem 16] have proved the equivalence of P.9 and P.1 under the same condition.

Corollary 6. *Let* $\boldsymbol{A}$ *be an* $n \times n$ *Metzlerian stable matrix. If* $\boldsymbol{D} \in \mathfrak{D}$ *then it is necessary and sufficient for* $\boldsymbol{DA}$ *to be stable that* $\boldsymbol{D} \in \mathfrak{D}_\pi$.

PROOF (*Necessity*). By Theorem 14, $\boldsymbol{A}$ satisfies P.9, which, in conjunction with Corollary 4 and the assumed stability of $\boldsymbol{DA}$, implies that

$$(n, 0, 0) = -\text{in } \boldsymbol{DA} = \text{in } \boldsymbol{D}(-\boldsymbol{A}) = \text{in } \boldsymbol{D}.$$

Since $\boldsymbol{D}$ is assumed to be rdm, $\boldsymbol{D}$ must be a pdm.

(*Sufficiency*). Since $\boldsymbol{A}$ satisfies P.9, Corollary 4 asserts that $-\text{in } \boldsymbol{DA} = \text{in } \boldsymbol{D}(-\boldsymbol{A}) = \text{in } \boldsymbol{D}$. However, by hypothesis, $\text{in } \boldsymbol{D} = (n, 0, 0)$. Hence we obtain

$$\text{in } \boldsymbol{DA} = (0, 0, n). \qquad \square$$

Remark Concerning Corollary 6. While the necessity part of Corollary 6 was stated by Arrow and Enthoven (Arrow and Enthoven [1956; p. 292]) the sufficiency part seems to be a new result.

As our final task in this section we investigate several conditions for the stability of multiple markets where commodities may be gross complements as well as gross substitutes. Mathematically, we shall be concerned with the family of generalized Metzlerian matrices defined as matrices reducible to the Metzlerian form under some suitable transformation. Of the generalized Metzlerian matrices, the so-called Morishima matrices are especially well known.

Definition 9. An $n \times n$ real matrix $\boldsymbol{A}$ is said to be a *Morishima matrix* if there exist subsets J and K of N such that

i. $J \cap K = \emptyset$,
ii. $J \cup K = N$, and
iii. if $i \neq j$ then

$$\begin{aligned} a_{ij} &\geqq 0 \qquad \text{for } i, j \in J \text{ or } i, j \in K \\ a_{ij} &\leqq 0 \qquad \text{otherwise.} \end{aligned}$$

Remark Concerning Definition 9. Since we can assume, without loss of generality, that the subset J in Definition 9 consists of the first $\#J$ indices of N, $\boldsymbol{A}$ is a Morishima matrix if and only if

$$\begin{pmatrix} \boldsymbol{I}_J & \boldsymbol{0} \\ \boldsymbol{0} & -\boldsymbol{I}_K \end{pmatrix} \boldsymbol{A} \begin{pmatrix} \boldsymbol{I}_J & \boldsymbol{0} \\ \boldsymbol{0} & -\boldsymbol{I}_K \end{pmatrix}$$

is Metzlerian, where $\boldsymbol{I}_J$ and $\boldsymbol{I}_K$ respectively denote identity matrices of order $\#J$ and $\#K$.

Corollary 7. *Theorem* 14 *also holds for Morishima matrices.*

PROOF. Let $\boldsymbol{A}$ be an $n \times n$ Morishima matrix. Then in view of the proof of Theorem 14, it suffices to verify the equivalence of the following three statements:

i. $\boldsymbol{A}$ is stable;
ii. $\boldsymbol{A}$ is Hicksian; and
iii. $\boldsymbol{A}$ has an ndd.

In view of the above remark concerning Definition 9, the matrix

$$F \equiv \begin{pmatrix} I_J & 0 \\ 0 & -I_K \end{pmatrix} A \begin{pmatrix} I_J & 0 \\ 0 & -I_K \end{pmatrix} = \begin{pmatrix} A_{JJ} & -A_{JK} \\ -A_{KJ} & A_{KK} \end{pmatrix}$$

is a Metzlerian matrix.

(i) ⇒ (ii) Noticing that

$$\begin{pmatrix} I_J & 0 \\ 0 & -I_K \end{pmatrix} = \begin{pmatrix} I_J & 0 \\ 0 & -I_K \end{pmatrix}^{-1},$$

it follows from the definition of F that all eigenvalues of A coincide with those of F. Since A is assumed to be stable, the Frobenius root of F must be negative. Therefore by Theorem 14 F is Hicksian. A slight manipulation then shows that A is Hicksian too.

(ii) ⇒ (iii) Since $\det F_{LL} = \det A_{LL}$ for any subset L of N, F becomes Hicksian under the present hypothesis. The matrix F, however, being Metzlerian, Theorem 14 ensures that F has an ndd. Since we have shown that

$$F = \begin{pmatrix} A_{JJ} & -A_{JK} \\ -A_{KJ} & A_{KK} \end{pmatrix},$$

A surely has an ndd.

In view of Theorem 11, no further proof is needed. □

Remarks Concerning Theorem 14 and Corollary 7. While the equivalence of (i) and (ii) was established by Morishima [1952; Theorem 1, p. 104.], Mukherji [1972; Lemma 10, p. 456.] proved the equivalence of (i) and (iii). Moreover, it is noteworthy that Theorem 14 and Corollary 7 show that Hicksian perfect stability is equivalent to total stability if the matrix is Metzlerian or if it is a Morishima matrix. In other words, our Theorem 14 and Corollary 7 can be viewed as sharpened versions of Metzler's Theorem (Metzler [1945; p. 282]).

Recently Ohyama [1972] and Mukherji [1972] have introduced more sophisticated transformations and demonstrated the stability of some additional non-Metzlerian matrices.

Theorem 15 (Ohyama [1972]). *Let* X *be an* $n \times n$ *real matrix such that*

a. *there exists a nonnegative positive definite matrix* G *such that* GX *is Metzlerian,*
b. $Xp < 0$ *and* $X'p < 0$ *for some* $p > 0$, *and*
c. *the vector* p *is an eigenvector of* G *corresponding to the Frobenius root* $\lambda^* > 0$ *of* G.

Then

i. X *is stable,*
ii. *every diagonal element of* X^{-1} *is negative, and*
iii. *if, in addition,* X *is symmetric then* X *is negative definite and hence totally S-stable.*

Proof. (i) Premultiplying both sides of $\boldsymbol{Xp} < \mathbf{0}$ by $\boldsymbol{G}$, we see that

$$\boldsymbol{GXp} < \mathbf{0}.$$

Since $\boldsymbol{Gp} = \lambda^* \boldsymbol{p}$ and $\boldsymbol{X}'\boldsymbol{p} < \mathbf{0}$, we obtain

$$\boldsymbol{X}'\boldsymbol{Gp} = \lambda^* \boldsymbol{X}'\boldsymbol{p} < \mathbf{0}.$$

Hence $(\boldsymbol{GX} + \boldsymbol{X}'\boldsymbol{G})\boldsymbol{p} < \mathbf{0}$ which, together with the fact that $\boldsymbol{GX}$ as well as $\boldsymbol{X}'\boldsymbol{G} \equiv (\boldsymbol{GX})'$ is Metzlerian, implies that $\boldsymbol{GX} + \boldsymbol{X}'\boldsymbol{G}$ has an ndd. In view of Theorem 11 and the assumed symmetry of $\boldsymbol{GX} + \boldsymbol{X}'\boldsymbol{G}$, the matrix $\boldsymbol{GX}$ is qnd. Therefore, by virtue of Theorem 10, $\boldsymbol{X}$ is stable, as desired.

(ii) As we have just seen, the Metzlerian matrix $\boldsymbol{GX}$ has an ndd, whence, by Theorem 7 in Chapter 1, $(\boldsymbol{GX})^{-1} = (\boldsymbol{X}^{-1}\boldsymbol{G}^{-1}) \leq \mathbf{0}$. Therefore the assumed nonnegativity of $\boldsymbol{G}$ ensures that $\boldsymbol{X}^{-1} \leq \mathbf{0}$. Suppose that there exists an index $i \in N$ such that $X_{ii}^{-1} = 0$. Then a slight manipulation yields the contradiction that

$$0 < G_{ii} = \sum_{j \neq i} (\boldsymbol{GX})_{ij} X_{ji}^{-1} \leqq 0,$$

where G_{rs} denotes the (rs)th element of $\boldsymbol{G}$ and so on.

(iii) Since $\boldsymbol{X}$ is stable as well as symmetric, the assertion follows directly from Lemma 4 and Theorem 10. □

We next turn to a weakened version of Mukherji's results (Mukherji [1972; Lemmas 1, 2, and 5]).

Theorem 16. *If an $m \times m$ real matrix $\boldsymbol{X}$ is such that*

a. $N_{\boldsymbol{X}} \equiv \{\boldsymbol{y} \neq \mathbf{0} : \boldsymbol{Xy} = \mathbf{0}\} \neq \varnothing$,
b. *there exists a nonsingular matrix $\boldsymbol{S}$ such that $\boldsymbol{SXS}^{-1}$ is Metzlerian, and*
c. *either $\boldsymbol{Sq}$ or $\boldsymbol{q}'\boldsymbol{S}^{-1}$ is positive for some $\boldsymbol{q} \in N_{\boldsymbol{X}}$,*

then

i. *any nonzero eigenvalue λ of $\boldsymbol{X}$ has a negative real part.*

If, in addition, $\boldsymbol{X} = \boldsymbol{X}'$ and the zero root of $\boldsymbol{X}$ is a simple root, then

ii. *$\boldsymbol{X}$ contains a principal submatrix $\tilde{\boldsymbol{X}}$ of order $m - 1$ that is* nd.

Proof. (i) Since $\boldsymbol{SXS}^{-1}$ is similar to $\boldsymbol{X}$ and is assumed to be Metzlerian, it suffices to show that the Frobenius root λ^* of $\boldsymbol{SXS}^{-1} \equiv \boldsymbol{F}$ is equal to zero. Let $\boldsymbol{y}^* \geq \mathbf{0}$ be a row eigenvector of $\boldsymbol{F}$ associated with λ^*. Then from the definition of $\boldsymbol{y}^*$ and $\boldsymbol{F}$ we obtain

$$\boldsymbol{y}^*\boldsymbol{SX} = \lambda^*(\boldsymbol{y}^*\boldsymbol{S}).$$

Postmultiplying both sides of the above equality by $\boldsymbol{q} \in N_{\boldsymbol{X}}$, we obtain

$$0 = \boldsymbol{y}^*\boldsymbol{S}(\boldsymbol{Xq}) = \lambda^*(\boldsymbol{y}^*\boldsymbol{Sq}).$$

Recalling that $\boldsymbol{Sq} > \mathbf{0}$ and that $\boldsymbol{y}^* \geq \mathbf{0}$, $\boldsymbol{y}^*\boldsymbol{Sq}$ is clearly positive. Hence $\lambda^* = 0$ as desired.

(ii) In view of Lemma 4″, the assumed symmetry of $\boldsymbol{X}$ together with (i), implies that $\boldsymbol{X}$ is negative semidefinite (nsd). Since $\lambda^* = 0$ is assumed to be a simple root, there exists a principal submatrix $\tilde{\boldsymbol{X}}$ which is nonsingular. $\tilde{\boldsymbol{X}}$ being nsd, it suffices to show that it is negative definite (nd). Suppose the contrary. Then there exists a nonzero vector $\boldsymbol{u}$ of order $m - 1$ satisfying $\boldsymbol{u}'\tilde{\boldsymbol{X}}\boldsymbol{u} = 0$. Let $\tilde{\boldsymbol{X}}\boldsymbol{u} \neq \mathbf{0}$. Then, for $\boldsymbol{p} = \tilde{\boldsymbol{X}}\boldsymbol{u} + t\boldsymbol{u}$, we have

$$0 \geqq \boldsymbol{p}'\tilde{\boldsymbol{X}}\boldsymbol{p} = (\tilde{\boldsymbol{X}}\boldsymbol{u})'\tilde{\boldsymbol{X}}(\tilde{\boldsymbol{X}}\boldsymbol{u}) + 2t(\tilde{\boldsymbol{X}}\boldsymbol{u})'(\tilde{\boldsymbol{X}}\boldsymbol{u}).$$

Since $(\tilde{\boldsymbol{X}}\boldsymbol{u})'(\tilde{\boldsymbol{X}}\boldsymbol{u}) > 0$, $\boldsymbol{p}'\tilde{\boldsymbol{X}}\boldsymbol{p} > 0$ for t sufficiently large, which is a contradiction. In view of the asserted nonsingularity of $\tilde{\boldsymbol{X}}$, $\boldsymbol{u}'\tilde{\boldsymbol{X}}\boldsymbol{u} = 0$ thus implies that $\boldsymbol{u} = \mathbf{0}$, whence $\tilde{\boldsymbol{X}}$ is surely nd. The remaining assertion is rather obvious. □

Whereas the transformations considered so far are linear, R. Sato [1972; 1973] introduced a nonlinear (power positive) transformation and was able to demonstrate the stability of yet other non-Metzlerian matrices.

Theorem 17 (R. Sato [1973]). *Corresponding to an $n \times n$ real matrix $\boldsymbol{A}$ and a positive number ρ, define $\boldsymbol{B} = \boldsymbol{A} + \rho\boldsymbol{I}$. If $\boldsymbol{B}$ is a power positive matrix of the kth degree then, by Lemma 6 of chapter 2, it has a real eigenvalue β_0 of greatest magnitude. Suppose that β_0 is positive. Then*

i. *the following statements are mutually equivalent:*
   (a) $\beta_0 < \rho$;
   (b) $\boldsymbol{C} \equiv \boldsymbol{B}^k - \rho^k\boldsymbol{I}$ *is stable; and*
   (c) $\boldsymbol{A}$ *is stable.*

Moreover,

ii. *if either $\boldsymbol{A}$ is Hicksian or the greatest column-sum (or row-sum) of $\boldsymbol{A}$ is negative then $\boldsymbol{A}$ is stable.*

Proof. (i) (a) ⇒ (b) Suppose the contrary. Then there exists an eigenvalue β of $\boldsymbol{B}$ such that $\mathrm{Re}(\beta^k - \rho^k) \equiv |\beta|^k \cdot \cos(k\theta) - \rho^k \geqq 0$ where

$$\beta = |\beta|(\cos\theta + i \cdot \sin\theta).$$

Since ρ is assumed to be positive, in order for $\mathrm{Re}(\beta^k - \rho^k)$ to be nonnegative it is necessary that

$$1 \geqq \cos(k\theta) > 0.$$

Hence

$$|\beta|^k - \rho^k \geqq |\beta|^k \cdot \cos(k\theta) - \rho^k \geqq 0,$$

from which it follows that $\beta_0 \geqq |\beta| \geqq \rho$, contradicting the hypothesis.

(i) (b) ⇒ (c) Noticing that $\beta_0^k - \rho^k$ is an eigenvalue of $\boldsymbol{C}$, the stability of $\boldsymbol{C}$, together with the assumed positivity of β_0 and ρ, implies the assertion (a).

Thus it suffices to establish the implication (a) $\Rightarrow$ (c). Let α be any eigenvalue of $\boldsymbol{A}$ and define $\alpha_0 = \beta_0 - \rho$. Since $\alpha + \rho$ and $\beta_0(=\alpha_0 + \rho)$ are eigenvalues of $\boldsymbol{B}$, we obtain

$$\operatorname{Re}(\alpha) \equiv \operatorname{Re}(\alpha + \rho) - \rho \leqq \beta_0 - \rho < 0.$$

(i) (c) $\Rightarrow$ (a) Since $\beta_0 - \rho$ is a real eigenvalue of $\boldsymbol{A}$, the assertion (a) follows directly from (c).

(ii) By virtue of the Hicksian conditions imposed on $\boldsymbol{A}$, every coefficient of the characteristic function $\det(\lambda\boldsymbol{I} - \boldsymbol{A})$ of $\boldsymbol{A}$ is positive. It is therefore clear that $\boldsymbol{A}$ can have no nonnegative eigenvalue. Hence $\beta_0 - \rho < 0$ which, by (i) above, implies the stability of $\boldsymbol{A}$.

Let $\boldsymbol{l}$ be an n-vector consisting of ones. Then from the hypothesis it follows that $\boldsymbol{l}'\boldsymbol{A} < \boldsymbol{0}'$ or, equivalently, $\boldsymbol{l}'\boldsymbol{B} < \rho\boldsymbol{l}'$. Applying Lemma 6 in Chapter 2, there exists an n-vector $\boldsymbol{x} > \boldsymbol{0}$ such that $\boldsymbol{B}\boldsymbol{x} = \boldsymbol{\beta}_0\boldsymbol{x}$. Hence

$$\rho\boldsymbol{l}'\boldsymbol{x} > \boldsymbol{l}'\boldsymbol{B}\boldsymbol{x} = \boldsymbol{\beta}_0(\boldsymbol{l}'\boldsymbol{x}).$$

Thus, by (i), $\boldsymbol{A}$ is stable □

Remarks Concerning Theorem 17. (i) Theorem 17 is a unified version of Theorems 1, 2, and 3 of R. Sato [1973]. It is noteworthy that the Hicksian conditions on $\boldsymbol{A}$ are not necessary for the stability of $\boldsymbol{A}$. To see this, it suffices to consider an example provided by R. Sato [1972; p. 498]. Let

$$\boldsymbol{B} = \begin{pmatrix} 10 & 0 & -5 & 5 \\ 1 & 3 & 3 & 2 \\ 6 & 1 & 0 & 3 \\ -4 & 2 & 11 & 1 \end{pmatrix}.$$

As shown by Sato, $\boldsymbol{B}^2 > \boldsymbol{0}$ and $9 < \beta_0 < \sqrt{99.6}$. For $\rho = \sqrt{99.6}$, $\boldsymbol{A} = \boldsymbol{B} - \rho\boldsymbol{I}$ is stable but not Hicksian, since $10 > \rho = \sqrt{99.6} > \beta_0$.

(ii) Applying Corollary 2 of Chapter 2, β_0 in Theorem 17 is positive if k is odd or if $\boldsymbol{B}$ contains at least one row (or column) that is nonnegative.

(iii) A direct expansion of $\boldsymbol{B}^k = (\boldsymbol{A} + \rho\boldsymbol{I})^k$ yields

$$\boldsymbol{C} = \boldsymbol{B}^k - \rho^k\boldsymbol{I} = \sum_{j=0}^{k-1} {}_kC_j\rho^j\boldsymbol{A}^{k-j}.$$

This enables us to interpret the matrix $\boldsymbol{C}$ as the weighted sum of $\boldsymbol{A}$ over certain periods k (see R. Sato [1973; p. 758]).

Since the so-called Hicksian conditions on $\boldsymbol{A}$ also ensure the stability of $\boldsymbol{A}$ in this case, a sufficient condition for $\boldsymbol{A}$ to be Hicksian is of interest. First we prove

Lemma 7 (Ostrowski [1955; Theorem 3]). *If an $n \times n$ real matrix $\boldsymbol{M} = (m_{ij})$, $i, j = 1, \ldots, n$, satisfies*

$$m_{ii} > nM_i^+ - \sum_{j \neq i} m_{ij} \qquad (i = 1, \ldots, n),$$

then $\boldsymbol{M}$ *is a* P-*matrix, where*

$$M_i^+ = \max_{j \neq i}\{\max(m_{ij}, 0)\} = \max_{j \neq i}\{\tfrac{1}{2}(m_{ij} + |m_{ij}|)\}.$$

PROOF. First, we verify that $\det \boldsymbol{M} > 0$. Define $\boldsymbol{B} \equiv (b_{ij}) \equiv (m_{ij} - M_i^+)$, $i, j = 1, \ldots, n$. Then $b_{ij} \leqq 0$ for $i \neq j$ and we obtain

$$b_{ii} - \sum_{j \neq i} |b_{ij}| = m_{ii} - M_i^+ - \sum_{j \neq i} (M_i^+ - m_{ij}) = m_{ii} - \left(nM_i^+ - \sum_{j \neq i} m_{ij}\right) > 0.$$

Since $m_{ii} - M_i^+ > \sum_{j \neq i} (M_i^+ - m_{ij}) \geqq 0$, the matrix $\boldsymbol{B}$ has a pdd. Hence by (ii) of Theorem 6 in Chapter 1, $\boldsymbol{B}$ is a P-matrix. Theorem 7 of Chapter 1 therefore asserts that $\boldsymbol{B}^{-1} \geq \boldsymbol{0}$, which in turn implies that all cofactors Δ_{ij} of b_{ij} are nonnegative. Recalling that $m_{ij} = b_{ij} + M_i^+$ for $i, j = 1, \ldots, n$, a direct calculation yields

$$\det \boldsymbol{M} = \det \boldsymbol{B} + \sum_{i=1}^{n} M_i^+ \left(\sum_{j=1}^{n} \Delta_{ij}\right).$$

Noticing that $\det \boldsymbol{B} > 0$ and that $\sum_{i=1}^{n} M_i^+ (\sum_{j=1}^{n} \Delta_{ij}) \geqq 0$, $\det \boldsymbol{M}$ is certainly positive.

Second, let J be any subset of N. For any $i \in J$, define $M_{Ji}^+ = \max_{j \in J(i)} \{\max(m_{ij}, 0)\}$ where $J(i)$ is a set obtained from J by removing the index i. Moreover, let $P_i = \{j \in N : m_{ij} \geqq 0\}$ and let $\bar{J} = \{j \in N : j \notin J\}$. Then, for any $i \in J$, we have

$$\begin{aligned} &\left(nM_i^+ - \sum_{j \neq i} m_{ij}\right) - \left((\# J)M_{Ji}^+ - \sum_{j \in J(i)} m_{ij}\right) \\ &= (\# J)(M_i^+ - M_{Ji}^+) + (\# \bar{J})M_i^+ + \sum_{\substack{i \in \bar{J} \\ j \notin P_i}} (-m_{ij}) - \sum_{\substack{j \in \bar{J} \\ j \in P_i}} m_{ij} \\ &\geqq (\# J)(M_i^+ - M_{Ji}^+) + \sum_{\substack{j \in \bar{J} \\ j \notin P_i}} (-m_{ij}) + ((\# \bar{J}) - \#(P_i \cap \bar{J}))M_i^+ \geq 0. \end{aligned}$$

Thus for any $J \subseteq N$ and every $i \in J$, we obtain

$$m_{ii} > nM_i^+ - \sum_{j \neq i} m_{ij} \geqq (\# J)M_{Ji}^+ - \sum_{j \in J(i)} m_{ij}.$$

Applying the foregoing argument to the principal submatrix $\boldsymbol{M}_{JJ}$, it is clear that

$$\det \boldsymbol{M}_{JJ} > 0 \quad \text{for any } J \subseteq N. \qquad \square$$

Remark Concerning Lemma 7. The conditions of Lemma 7 have been called the "nearly dominant diagonal" condition by R. Sato [1973; Theorem 4, p. 759].

Combining Theorem 17 with Lemma 7, we can assert:

Corollary 8. *Let $\boldsymbol{A}$ be the matrix satisfying the conditions in Theorem 17. If $\boldsymbol{M} \equiv -\boldsymbol{A}$ satisfies the conditions in Lemma 7 then $\boldsymbol{A}$ is stable, and if the matrix $\boldsymbol{C}$ in Theorem 17 is Metzlerian then conditions P.5 and P.9 on $\boldsymbol{C}$ are equivalent to the stability of $\boldsymbol{A}$.*

PROOF. Lemma 7, together with the definition of $\boldsymbol{M}$, asserts that $\boldsymbol{A}$ is Hicksian. Hence, by (ii) of Theorem 17, $\boldsymbol{A}$ is stable.

In view of Theorem 14, both P.5 and P.9 on $\boldsymbol{C}$ are necessary and sufficient for a Metzlerian matrix $\boldsymbol{C}$ to be stable. Therefore the stability of $\boldsymbol{A}$ immediately follows from the second half of (ii) of Theorem 17. □

3 An Existence Theorem for Systems of Ordinary Differential Equations

Our main purpose in this section is to state and verify a theorem ensuring the existence of solutions of (1). To this end, we need a new concept and a few lemmas.

Definition 10. Let E be a subset of $\mathbb{R}^k$ and let $\boldsymbol{f}_n$, $n = 1, 2, \ldots$, be functions of E into $\mathbb{R}^k$. Then the sequence $\{\boldsymbol{f}_n\}$ is said to *converge uniformly on E to $\boldsymbol{f}$*, if there exists a function $\boldsymbol{f}$ of E into $\mathbb{R}^k$ such that for every $\varepsilon > 0$ there exists an integer $N = N(\varepsilon)$ such that $N \leqq n$ implies

$$\|\boldsymbol{f}_n(\boldsymbol{x}) - \boldsymbol{f}(\boldsymbol{x})\| \leqq \varepsilon \quad \text{for all } \boldsymbol{x} \in E.$$

On the other hand, we say that $\{\boldsymbol{f}_n\}$ *converges pointwise, or simply converges, on E to $\boldsymbol{f}$*, if there exists $\boldsymbol{f}$ of E into $\mathbb{R}^k$ such that for every $\varepsilon > 0$, and for every $\boldsymbol{x} \in E$, there exists an integer $N' = N'(\varepsilon, \boldsymbol{x})$ for which if $n \geq N'$ then $\|\boldsymbol{f}_n(\boldsymbol{x}) - \boldsymbol{f}(\boldsymbol{x})\| \leqq \varepsilon$.

We now turn to the lemmas.

Lemma 8. *Let $f_i(t)$, $i = 1, \ldots, n$, be a real-valued continuous function defined on a closed interval $[r_1, r_2]$. For any given t_0, and $t \in [r_1, r_2]$, define the function $F_i(t)$ by*

$$F_i(t) = \int_{t_0}^{t} f_i(\tau) d\tau.$$

Corresponding to $\boldsymbol{f}(t) \equiv (f_1(t), \ldots, f_n(t))'$ we can define a vector function $\boldsymbol{F}(t) \equiv (F_1(t), \ldots, F_n(t))'$, which we express as

$$\boldsymbol{F}(t) = \int_{t_0}^{t} \boldsymbol{f}(\tau) d\tau.$$

Then

$$\left\| \int_{t_0}^{t} f(\tau)d\tau \right\| \leqq \left| \int_{t_0}^{t} \| f(\tau) \| d\tau \right|.$$

PROOF. Let m be an integer and define $\Delta = (t - t_0)/m$ and $t_k = t_0 + \Delta \cdot k$, $k = 1, \ldots, m$, Then, from the definition of integration as well as the well-known triangle inequality of norms,

$$\left\| \int_{t_0}^{t} f(\tau)d\tau \right\| = \left\| \lim_{m\to\infty} \sum_{k=1}^{m} f(t_k)\Delta \right\| \leqq \lim_{m\to\infty} \sum_{k=1}^{m} \| f(t_k) \| \cdot |\Delta| = \left| \int_{t_0}^{t} \| f(\tau) \| d\tau \right|.$$

□

Lemma 9. *Let $g_i(x)$, $i = 1, \ldots, n$, be a real-valued function defined on a convex set S of $\mathbb{R}^n$, and let $g(x) \equiv (g_1(x), \ldots, g_n(x_n))'$. If $g_i(x)$, $i = 1, \ldots, n$, is differentiable on S and if in addition there exists a $K > 0$ such that*

$$\left| \frac{\partial g_i(x)}{\partial x_j} \right| \leq K \quad \text{for } i, j = 1, \ldots, n \text{ and all } x \in S \tag{35}$$

then

$$\| g(x) - g(y) \|_e \leqq n^2 K \| x - y \|_e \quad \text{for all } x \text{ and } y \text{ of } S \tag{36}$$

where $\|\xi\|_e$ signifies the Euclidean norm of ξ, that is,

$$\|\xi\|_e = \left(\sum_{j=1}^{n} \xi_j^2 \right)^{1/2}.$$

PROOF. Let s be a real number in the closed interval $[0, 1]$. Then $z(s) = s \cdot x + (1 - s)y$ represents a point on the line segment joining x and y. Since S is a convex set, $z(s) \in S$ for all $s \in [0, 1]$, Applying the Mean Value Theorem[8] to each $g_i(x)$, there exists $\theta_i \in (0, 1)$ such that

$$\begin{aligned} g_i(x) - g_i(y) &= g_i(z(1)) - g_i(z(0)) \\ &= \left.\frac{dg_i}{ds}\right|_{\theta_i} = \sum_{j=1}^{n} (x_j - y_j) \left.\frac{\partial g_i}{\partial x_j}\right|_{z(\theta_i)} \qquad (i = 1, \ldots, n).^{9} \end{aligned}$$

Since $z(\theta_i) \in S$, the hypothesis (35), together with the triangle inequality of modulus, guarantees that

$$\begin{aligned} |g_i(x) - g_i(y)| &= \left| \sum_{j=1}^{n} (x_j - y_j) \cdot \left.\frac{\partial g_i}{\partial x_j}\right|_{z(\theta_i)} \right| \\ &\leqq \sum_{j=1}^{n} |x_j - y_j| \cdot K \leqq nK \| x - y \|_e \qquad (i = 1, \ldots, n). \end{aligned}$$

Hence

$$\| g(x) - g(y) \|_e^2 = \sum_{i=1}^{n} |g_i(x) - g_i(y)|^2 \leqq n^3 K^2 \| x - y \|_e^2,$$

from which it follows that

$$\|\boldsymbol{g}(\boldsymbol{x}) - \boldsymbol{g}(\boldsymbol{y})\|_e \leqq n^{3/2}K\|\boldsymbol{x} - \boldsymbol{y}\|_e \leqq n^2K\|\boldsymbol{x} - \boldsymbol{y}\|_e \quad \text{for all } \boldsymbol{x}, \boldsymbol{y} \in S. \quad \square$$

Remark Concerning Lemma 9. Let $\|\boldsymbol{x}\|$ be an arbitrary norm of a vector $\boldsymbol{x}$. Then, as was shown in Appendix A, there exist $\alpha, \beta > 0$ such that

$$\alpha \cdot \|\boldsymbol{x}\| \leqq \|\boldsymbol{x}\|_e \leqq \beta\|\boldsymbol{x}\| \quad \text{for all relevant } \boldsymbol{x}.$$

Defining $K' = (K\beta)/\alpha$, we can generalize Lemma 9 to obtain

Lemma 9′. *Under the hypothesis of Lemma 9 there exists $K' > 0$ such that*

$$\|\boldsymbol{g}(\boldsymbol{x}) - \boldsymbol{g}(\boldsymbol{y})\| \leqq n^2K'\|\boldsymbol{x} - \boldsymbol{y}\| \quad \textit{for all } \boldsymbol{x} \textit{ and } \boldsymbol{y} \textit{ of } S. \tag{36′}$$

Condition (36) or (36′) is the so-called *Lipschitz condition.*

The next lemma provides a sufficient condition for a sequence of functions to be uniformly convergent.

Lemma 10. *Let E be a compact subset of $\mathbb{R}^k$ and by C_E denote a family (or class) of continuous functions of E into $\mathbb{R}^k$. Define a special norm by*

$$\|\boldsymbol{f}\|\dagger = \max_{\boldsymbol{x} \in E}\|\boldsymbol{f}(\boldsymbol{x})\|_e \quad \textit{for any } \boldsymbol{f} \in C_E.$$

Then a sequence of functions $\{\boldsymbol{f}_m \in C_E\}$ is convergent uniformly on E to $\boldsymbol{f} \in C_E$ if there exist nonnegative numbers a_i, $i = 1, 2, \ldots$, such that $\sum_{i=1}^{\infty} a_i$ converges and

$$\|\boldsymbol{f}_{i+1} - \boldsymbol{f}_i\|\dagger \leqq a_i \qquad (i = 1, 2, \ldots). \tag{37}$$

Proof. Let $\alpha > 0$ be such that

$$\alpha\|\xi\| \leqq \|\xi\|_e \quad \text{for all } \xi, \tag{38}$$

and let $\varepsilon > 0$ be any positive number. Thus, in view of Theorem 14 in Section 3 of Appendix B and the assumed nonnegativity of a_i, with $\alpha\varepsilon > 0$ there is associated an integer N such that if $n \geq m \geq N$ then

$$\sum_{j=m}^{n} a_j < \alpha\varepsilon.$$

Hence, with the aid of (37) and (38)

$$\begin{aligned}
\alpha\|\boldsymbol{f}_n(\boldsymbol{x}) - \boldsymbol{f}_m(\boldsymbol{x})\| &\leqq \|\boldsymbol{f}_n(\boldsymbol{x}) - \boldsymbol{f}_m(\boldsymbol{x})\|_e \quad \text{for all } \boldsymbol{x} \in E \\
&\leqq \|\boldsymbol{f}_n - \boldsymbol{f}_m\|\dagger \\
&= \|\boldsymbol{f}_n - \boldsymbol{f}_{n-1} + \boldsymbol{f}_{n-1} - \boldsymbol{f}_{n-2} + \cdots + \boldsymbol{f}_{m+1} - \boldsymbol{f}_m\|\dagger \\
&\leqq \sum_{j=m}^{n} a_j < \alpha \cdot \varepsilon.
\end{aligned}$$

Thus for any $\varepsilon > 0$ there exists an integer N such that if $n \geq m \geq N$ then

$$\|f_n(x) - f_m(x)\| \leqq \varepsilon \quad \text{for all } x \in E.$$

Applying Theorem 15 in Section 3 of Appendix B, the assertion is immediate. □

We can now turn to our main theorem.

Theorem 18. *Let Γ be an open set of $\mathbb{R}^{n+1}$. If $f_j(x, t)$, $j = 1, \ldots, n$, in (1) is continuous on Γ and if its partial derivatives with respect to x_k, $k = 1, \ldots, n$, are also continuous on Γ, then*

i. *for any $(x_0, t_0) \in \Gamma$ there exists a solution $x(t)$ of (1) defined on some time interval (t_1, t_2) containing t_0, and*
ii. *if $x_1(t)$ and $x_2(t)$ are solutions satisfying $x_i(t_0) = x_0$, $i = 1, 2$, then $x_1(t) = x_2(t)$ for all $t \in (t_1^1, t_2^1) \cap (t_1^2, t_2^2)$, where (t_1^j, t_2^j), $j = 1, 2$, denotes the time interval on which $x_j(t)$ is defined.*

Proof. For the time being, we commit ourselves to a preliminary consideration. To any vector function $x(t) \equiv (x_1(t), \ldots, x_n(t))'$ of t, there corresponds a vector function $x^*(t) = x_0 + \int_{t_0}^{t} f(x(\tau), \tau)d\tau$ such that $(x^*(t), t)$ lies in Γ[10] too. For simplicity, we denote this correspondence by an operator $A: x^* = Ax$. Then $x(t)$ is a solution of (1) starting from x_0 if and only if $x = Ax$. Suppose $x(t)$ is a solution of (1) with the property that $x(t_0) = x_0$. Then from $\dot{x}(t) = f(x(t), t)$ and from $x(t_0) = x_0$ it follows that

$$x(t) = x_0 + \int_{t_0}^{t} f(x(\tau), \tau)d\tau \quad \text{for any } t.$$

Conversely, assume that the above equality holds. Then, setting $t = t_0$, we easily see that $x(t_0) = x_0$ and that the differentiation of both sides with respect to t yields $\dot{x}(t) = f(x(t), t)$.

We can now turn to our main argument.

(i) Since (x_0, t_0) is a point of an open set Γ, there exist $p, q > 0$ such that

$$\Pi \equiv \{(x, t): \|x - x_0\|_e \leqq p \quad \text{and} \quad |t - t_0| \leqq q\} \subseteq \Gamma.$$

Noticing that Π is a compact set[11] in $\mathbb{R}^{n+1}$, Theorem 9 in Appendix A ensures that continuous functions[12] $\|f(x, t)\|_e$ and $|\partial f_i/\partial x_j|_{(x,t)}|$, $i, j = 1, \ldots, n$, attain their maximum on the set Π. Hence, there exist $K, M > 0$ such that

$$\|f(x, t)\|_e \leqq M \quad \text{for all } (x, t) \in \Pi, \tag{39.a}$$

and

$$\left| \frac{\partial f_i}{\partial x_j}\bigg|_{(x,t)} \right| \leqq K \quad \text{for all } (x, t) \in \Pi \text{ and } i, j = 1, \ldots, n. \tag{39.b}$$

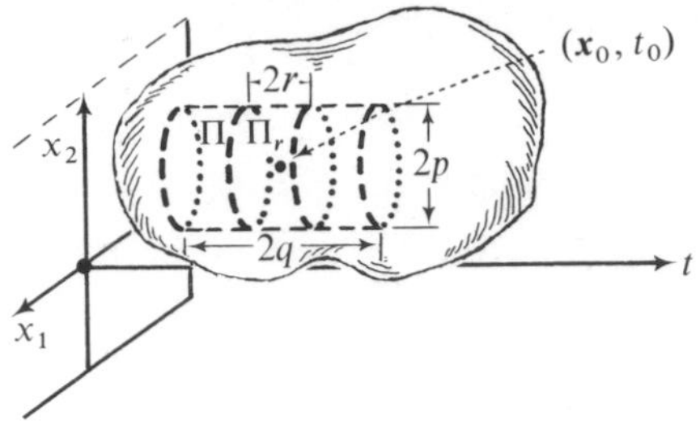

Figure 4.3

Define a set $\Pi_r = \{(x, t): \|x - x_0\|_e \leq p$ and $|t - t_0| \leq r\}$. Then Π_r is a subset of Π if and only if $r \leqq q$. (See Figure 4.3.) Let $\phi(t) = (\phi_1(t), \ldots, \phi_n(t))$ be any continuous vector function of which the components are defined on a time interval $[t_0 - r, t + r]$, and by F_r denote a family of $\phi(t)$ such that their graphs are in Π_r, that is, $F_r = \{\phi(t): (\phi(t), t) \in \Pi_r\}$. Then there exists an $r > 0$ such that

a. if $\phi \in F_r$ then $\phi^* = A\phi \in F_r$, and
b. if $k \in (0, 1)$ then

$$\|A\phi - A\chi\|^\dagger \leqq k\|\phi - \chi\|^\dagger \quad \text{for all } \phi \text{ and } \chi \text{ of } F_r.$$

In fact, if we choose an $r > 0$ such that

$$r \leqq \min\left\{q, \frac{p}{M}, \frac{k}{n^2 K}\right\} \tag{40}$$

then the conditions (a) and (b) are surely met. This can be shown as follows. From the definition of F_r it follows that $\phi \in F_r$ if and only if

$$\|\phi(t) - x_0\|_e \leqq p. \tag{41}$$

On the other hand, we have, by definition,

$$\phi^*(t) = x_0 + \int_{t_0}^{t} f(\phi(\tau), \tau) d\tau. \tag{42}$$

Hence, applying Lemma 8,

$$\|\phi^*(t) - x_0\|_e = \left\| \int_{t_0}^{t} f(\phi(\tau), \tau) d\tau \right\|_e = \left| \int_{t_0}^{t} \|f(\phi(\tau), \tau)\|_e d\tau \right|. \tag{43}$$

In view of (40), (41), and the definition of F_r, it is clear that $(\phi(t), t) \in \Pi_r \subseteq \Pi$, which, together with (43), implies that

$$\|\phi^*(t) - x_0\|_e \leqq \left| \int_{t_0}^{t} M \, d\tau \right| \leqq Mr \leqq p.$$

Therefore, from (41) again, $\phi^* \in F_r$.

Let $\boldsymbol{\phi}$ and $\boldsymbol{\chi}$ be any two members of F_r. Making use of Lemmas 8 and 9, as well as (39.b), we obtain

$$\begin{aligned}\|\boldsymbol{\phi}^*(t) - \boldsymbol{\chi}^*(t)\|_e &= \left\| \int_{t_0}^{t} \{\boldsymbol{f}(\boldsymbol{\phi}(\tau), \tau) - \boldsymbol{f}(\boldsymbol{\chi}(\tau), \tau)\} d\tau \right\|_e \\ &\leqq \left| \int_{t_0}^{t} \|\boldsymbol{f}(\boldsymbol{\phi}(\tau), \tau) - \boldsymbol{f}(\boldsymbol{\chi}(\tau), \tau)\|_e d\tau \right| \\ &\overset{13}{\leqq} \left| n^2 K \int_{t_0}^{t} \|\boldsymbol{\phi}(\tau) - \boldsymbol{\chi}(\tau)\|_e d\tau \right|.\end{aligned}$$

The definition of the norm $\| \quad \|\dagger$ therefore yields

$$\|\boldsymbol{\phi}^* - \boldsymbol{\chi}^*\|\dagger \leqq n^2 K \cdot |t - t_0| \cdot \|\boldsymbol{\phi} - \boldsymbol{\chi}\|\dagger \leqq n^2 K r \|\boldsymbol{\phi} - \boldsymbol{\chi}\|\dagger.$$

Hence, condition (b) is satisfied for r satisfying (40). Henceforth, we choose an r that satisfies (40).

Let $\boldsymbol{\phi}_0(t)$ be a vector function defined on the time interval $[t_0 - r, t_0 + r]$ such that

$$\boldsymbol{\phi}_0(t) = \boldsymbol{x}_0 \quad \text{for all } t \in [t_0 - r, t_0 + r]. \tag{44}$$

Thus it is clear that $\boldsymbol{\phi}_0 \in F_r$. Suppose that

$$\boldsymbol{\phi}_{i+1} = \boldsymbol{A\phi}_i \qquad (i = 0, 1, 2, \ldots). \tag{45}$$

Then every $\boldsymbol{\phi}_i$ is a member of F_r, since $\boldsymbol{\phi}_0 \in F_r$ and since r is chosen so that condition (a) is satisfied. With the aid of (45) and of condition (b), we obtain

$$\|\boldsymbol{\phi}_{i+1} - \boldsymbol{\phi}_i\| = \|\boldsymbol{A\phi}_i - \boldsymbol{A\phi}_{i-1}\|\dagger \leqq k\|\boldsymbol{\phi}_i - \boldsymbol{\phi}_{i-1}\|\dagger \qquad (i = 1, 2, \ldots). \tag{46}$$

From this, bearing (41) and (44) in mind, it follows that

$$\|\boldsymbol{\phi}_{i+1} - \boldsymbol{\phi}_i\|\dagger \leqq pk^i \quad (i = 0, 1, 2, \ldots). \tag{47}$$

Hence, by virtue of Lemma 10, there exists a vector function $\boldsymbol{\phi} \in F_r$ to which the sequence $\{\boldsymbol{\phi}_i\}$ converges uniformly on the interval $[t_0 - r, t_0 + r]$. Noticing that (46), together with (47), implies that the sequence $\{\boldsymbol{A\phi}_i\}$ also converges to $\boldsymbol{A\phi}$ uniformly on the same interval, it now remains to prove that $\boldsymbol{A\phi} = \boldsymbol{\phi}$. To verify this, suppose the contrary. Then $\|\boldsymbol{A\phi} - \boldsymbol{\phi}\|\dagger > 0$. Set $\varepsilon = \|\boldsymbol{A\phi} - \boldsymbol{\phi}\|\dagger$. Since $\{\boldsymbol{\phi}_i\}$ and $\{\boldsymbol{A\phi}_i\}$ are uniformly convergent, with $\delta = \frac{1}{3}\varepsilon$, there is associated an integer N_δ such that if $i \geq N_\delta$ then

$$\|\boldsymbol{A\phi}_i - \boldsymbol{A\phi}\|\dagger < \tfrac{1}{3}\varepsilon \tag{48.a}$$

and

$$\|\boldsymbol{\phi}_{i+1} - \boldsymbol{\phi}\|\dagger < \tfrac{1}{3}\varepsilon \tag{48.b}$$

Taking into account (45), (48.a), and (48.b), we obtain

$$\begin{aligned}\varepsilon &= \|A\phi - \phi\|^\dagger \\ &= \|A\phi - A\phi_i + \phi_{i+1} - \phi\|^\dagger \\ &\leqq \|A\phi_{i+1} - A\phi\|^\dagger + \|\phi_{i+1} - \phi\|^\dagger < \tfrac{2}{3}\varepsilon \quad \text{for } i \geq N_\delta,\end{aligned}$$

which is a self-contradiction. This completes the proof of (i).

(ii) By (r_1, r_2) denote the intersection of (t_1^1, t_2^1) and (t_1^2, t_2^2). Since $t_0 \in (t_1^j, t_2^j)$, $j = 1, 2$, (r_1, r_2) is nonempty. Suppose that there exists $t^* \in (r_1, r_2)$ such that $\boldsymbol{x}_1(t^*) \neq \boldsymbol{x}_2(t^*)$. Then clearly $t_0 \neq t^*$. $\boldsymbol{x}_j(t_0) = \boldsymbol{x}_0$, $j = 1, 2$. For concreteness, suppose that $t_0 < t^*$. Let $\mathfrak{M} = \{t \in [t_0, t^*] : \boldsymbol{x}_1(t) = \boldsymbol{x}_2(t)\}$. Since $\boldsymbol{x}_j(t)$, $j = 1, 2$, passes through $\boldsymbol{x}_0$ at t_0, the set $\mathfrak{M}$ is nonempty. Let $\{t_i\}$ be a sequence of $\mathfrak{M}$ convergent to τ. Then, from the continuity of $\boldsymbol{x}_j(t)$,[14] $j = 1, 2$,

$$\boldsymbol{x}_1(\tau) = \lim_{i \to \infty} \boldsymbol{x}_1(t_i) = \lim_{i \to \infty} \boldsymbol{x}_2(t_i) = \boldsymbol{x}_2(\tau).$$

Hence, τ belongs to $\mathfrak{M}$, which, in turn, implies that the set $\mathfrak{M}$ is closed.[15] By $\hat{t}$ denote the supremum of $\mathfrak{M}$.[16] Then, clearly, $\hat{t} \in \mathfrak{M}$, whence $\boldsymbol{x}_1(\hat{t}) = \boldsymbol{x}_2(\hat{t})$, which henceforth we denote by $\hat{\boldsymbol{x}}$.

Since by definition the point $(\hat{\boldsymbol{x}}, \hat{t})$ belongs to an open set Γ, there exist $p', q' > 0$ such that if $\|\boldsymbol{x} - \hat{\boldsymbol{x}}\|_e \leq p'$ and if $|t - \hat{t}| \leq q'$ then $(\boldsymbol{x}, t)$ also belongs to Γ. Noting the continuity of $\boldsymbol{x}_j(t)$ at t, there exists $s > 0$ such that

$$\|\boldsymbol{x}_j(t) - \hat{\boldsymbol{x}}\|_e \leq p' \quad \text{for all } t \in [\hat{t} - s, \hat{t} + s].$$

Thus, choosing an $r' > 0$ such that

$$r' \leqq \min\left\{s, q', \frac{p'}{M}, \frac{k}{n^2 K}\right\},$$

the vector functions $\boldsymbol{x}_j(t), j = 1, 2$, defined on the time interval $[\hat{t} - r', \hat{t} + r']$ belong to the family $F_{r'}$. Therefore, from condition (b), it follows that

$$\|A\boldsymbol{x}_2 - A\boldsymbol{x}_1\|^\dagger = \|\boldsymbol{x}_2 - \boldsymbol{x}_1\|^\dagger \leqq k\|\boldsymbol{x}_2 - \boldsymbol{x}_1\|^\dagger. \tag{49}$$

Hence

$$\|\boldsymbol{x}_2 - \boldsymbol{x}_1\|^\dagger = 0,$$

or, equivalently,

$$\boldsymbol{x}_1(t) = \boldsymbol{x}_2(t) \quad \text{for all } t \in [\hat{t} - r', \hat{t} + r']. \tag{50}$$

However, (50) contradicts the definition of $\hat{t}$, which completes the proof. □

Remark Concerning Theorem 18. Theorem 18 seems to depend on the special norm $\|\quad\|^\dagger$. In fact, however, the use of this norm detracts in no way from the generality of the theorem. The reason for this may be given briefly. First, as was shown in Lemma 10, the uniform convergence guaranteed by (47) holds irrespective of the norm considered. Second, in view of Lemma 2 in Appendix A, inequality (49), which is the key to the uniqueness of the solution of (1), also holds for any norm.

4 Liapunov's Stability Theorems for Systems of Ordinary Differential Equations

Sections 1 and 2 of this chapter were devoted to the stability theory appropriate to systems of linear differential equations. In the present section we return to more general systems of nonlinear differential equations, already encountered in (1). The task of deriving stability conditions is approached by means of what is usually called *Liapunov's second* (or *indirect*) *method.*

It was shown in Section 1 that the concept of the stability of a system of differential equations is closely related to the concept of the boundedness of the solutions of the system. We therefore begin with a definition of boundedness.

Throughout the present and subsequent sections the symbol $\boldsymbol{x}_e$ will denote a stationary solution of the dynamic system under consideration.

Definition 11. A solution $\boldsymbol{x}(t)$ of (1) is said to be

i. *bounded*, if there exists a $b(\boldsymbol{x}_0, t_0) > 0$ such that

$$\|\boldsymbol{x}(t) - \boldsymbol{x}_e\| \leqq b(\boldsymbol{x}_0, t_0) \quad \text{for all } t \geqq t_0,$$

ii. *equibounded*, if, for any $\alpha > 0$, there exists a $b(\alpha, t_0) > 0$ such that

$\|\boldsymbol{x}_0 - \boldsymbol{x}_e\| \leqq \alpha$ implies that

$$\|\boldsymbol{x}(t) - \boldsymbol{x}_e\| \leqq b(\alpha, t_0) \quad \text{for all } t \geqq t_0,$$

and

iii. *uniformly bounded*, if $b(\alpha, t_0)$ in (ii) is independent of t_0

To make clear the difference between boundedness and stability, we introduce the following

Remark Concerning Definition 11. For boundedness it is required that the solution $\boldsymbol{x}(t)$ of (1) remains in some neighborhood of $\boldsymbol{x}_e$ throughout the time interval on which it is defined. On the other hand, for stability (especially local L-stability in Definition 3) it is required that if the motion starts at $\boldsymbol{x}_0$ sufficiently near to $\boldsymbol{x}_e$, the motion after t_0 remains in a neighborhood of $\boldsymbol{x}_e$, however small it is.

Next we define a Liapunov function and various concepts of stability associated with nonautonomous systems of differential equations.

Definition 12. A real valued function $V(\boldsymbol{x}, t)$ defined on $\mathbb{R}^{n+1}$ is said to be a *Liapunov function*, if

i. $V(\boldsymbol{x}, t)$ has continuous partial derivatives with respect to $\boldsymbol{x}$ and t,[17]
ii. $V(\boldsymbol{0}, t) = 0$ for all t,[18] and
iii. $V(\boldsymbol{x}, t) > 0$ for all t and all $\boldsymbol{x} \neq \boldsymbol{0}$.

Definition 13. A stationary solution $\boldsymbol{x}_e$ of (1), if it exists, is said to be:

i. L-*stable*, if for any $\varepsilon > 0$ and for any t_0 there exists a $\delta(t_0, \varepsilon) > 0$ such that $\|\boldsymbol{x}_0 - \boldsymbol{x}_e\| \leq \delta(t_0, \varepsilon)$ implies that $\|\boldsymbol{x}(t) - \boldsymbol{x}_e\| \leqq \varepsilon$ for all $t \geqq t_0$;
ii. *uniformly* L-*stable*, if $\delta(t_0, \varepsilon)$ in (i) is independent of t_0;
iii. *stable*, if it is L-stable and if for any t_0 there exists an $r(t_0) > 0$ such that if $\|\boldsymbol{x}_0 - \boldsymbol{x}_e\| \leq r(t_0)$ then for any $\mu > 0$ there exists a $T(\mu, \boldsymbol{x}_0, t_0) > 0$ with the property that

$$\|\boldsymbol{x}(t) - \boldsymbol{x}_e\| \leq \mu \quad \text{for all } t \geqq t_0 + T(\mu, \boldsymbol{x}_0, t_0),$$

iv. *equistable*, if it is L-stable and if $T(\mu, \boldsymbol{x}_0, t_0)$ in (iii) is of the form $T(\mu, r(t_0), t_0)$; and, finally,
v. *uniformly stable*, if it is uniformly L-stable and if in addition $r(t_0)$ in (iii) is independent of t_0 and consequently $T(\mu, r(t_0), t_0)$ in (iv) is of the form $T(\mu, r)$.

Definition 13 is concerned with concepts of local stability. It remains to define stability in the large, or global stability.

Definition 14. An L-stable stationary solution $\boldsymbol{x}_e$ of (1), if it exists, is said to be:

i. *stable in the large*, or *globally stable*, if, for any $\mu > 0$, any $\boldsymbol{x}_0$ and any t_0, there exists a $T(\mu, \boldsymbol{x}_0, t_0) > 0$ such that

$$\|\boldsymbol{x}(t) - \boldsymbol{x}_e\| \leqq \mu \quad \text{for all } t \geqq t_0 + T(\mu, \boldsymbol{x}_0, t_0);$$

ii. *equistable in the large*, or *globally equistable*, if, for any $r > 0$, any $\mu > 0$ and any t_0, there exists a $T(\mu, r, t_0) > 0$ such that if $\|\boldsymbol{x}_0 - \boldsymbol{x}_e\| \leqq r$ then

$$\|\boldsymbol{x}(t) - \boldsymbol{x}_e\| \leqq \mu \quad \text{for all } t \geqq t_0 + T(\mu, r, t_0); \text{ and}$$

iii. *uniformly stable in the large*, or *globally uniformly stable*, if $\boldsymbol{x}_e$ is uniformly L-stable, if $T(\mu, r, t_0)$ in (ii) is independent of t_0, and if every solution of (1) is uniformly bounded.

Remark Concerning Definition 14. The difference between global stability and global equistability must be emphasized. For simplicity, let t_0 be fixed. Suppose that $\boldsymbol{x}_e$ is globally stable. Then, by virtue of its dependence on $\boldsymbol{x}_0$, the time interval during which the solution of (1) remains outside $B(\boldsymbol{x}_e, \mu)$ may differ from one solution to another. On the other hand, if $\boldsymbol{x}_e$ is globally equistable then the same value of $T(\mu, r, t_0)$ is applicable to all solutions starting at $\boldsymbol{x}_0 \in B(\boldsymbol{x}_e, r)$. Furthermore, it is clear from the definition that (iii) implies (ii), and that (ii) implies (i).

Before turning to our main theorems, an additional notation is needed. Let $\boldsymbol{x}(t)$ be any solution of (1) and let $V(\boldsymbol{x}, t)$ be a Liapunov function. Then, applying the chain rule of differentiation, we obtain

$$\frac{d(V(\boldsymbol{x}(t), t))}{dt} = \sum_{j=1}^{n} \frac{\partial V}{\partial x_j} \frac{dx_j(t)}{dt} + \frac{\partial V}{\partial t} = \sum_{j=1}^{n} \frac{\partial V}{\partial x_j} f_j(\boldsymbol{x}(t), t) + \frac{\partial V}{\partial t}, \tag{51}$$

which we call the *trajectory derivative of a Liapunov function* and denote by $\dot{V}(\boldsymbol{x}, t)$.

We can now state and prove our main theorems. Since stability theory without a solution would be meaningless, we henceforth postulate the conditions of Theorem 8 as well as the existence of a stationary solution. Moreover, in the subsequent theorems we again follow the convention that $\boldsymbol{x}_e = \boldsymbol{0}$.

Theorem 19. *Let R be an arbitrarily chosen positive number and let $V(\boldsymbol{x}, t)$ be a Liapunov function defined on $\|\boldsymbol{x}\| < R$,*[19] *$0 \leqq t < \infty$. Then the stationary solution of* (1) *is:*

i. L-*stable under conditions* (a) *and* (b);
ii. *uniformly* L-*stable under conditions* (a), (b) *and* (c); *and*
iii. *uniformly stable under conditions* (a), (c) *and* (d), *where conditions* (a) *through* (d) *are as follows—*

(a) *there exists a continuous, nondecreasing scalar function $\alpha(\xi)$ such that*

$$\alpha(0) = 0$$

and

$$0 < \alpha(\|\boldsymbol{x}\|) \leqq V(\boldsymbol{x}, t) \quad \textit{for all } t \textit{ and all } \boldsymbol{x} \neq \boldsymbol{0}, \tag{52}$$

(b) $\dot{V}(\boldsymbol{x}, t) \leqq 0$,
(c) *there exists a continuous, nondecreasing scalar function $\beta(\xi)$ such that*

$$\beta(0) = 0$$

and

$$V(\boldsymbol{x}, t) \leqq \beta(\|\boldsymbol{x}\|) \quad \textit{for all } t, \tag{53}$$

and, finally,
(d) *there exists a continuous scalar function $\gamma(\xi)$ such that*

$$\gamma(0) = 0,$$
$$\gamma(\xi) > 0 \quad \textit{for all } \xi \neq 0,$$

and

$$\dot{V}(\boldsymbol{x}, t) \leqq -\gamma(\|\boldsymbol{x}\|) \quad \textit{for all } t \textit{ and } \boldsymbol{x} \neq \boldsymbol{0}. \tag{54}$$

PROOF. (i) Suppose the contrary. Then there exists an $\varepsilon_0 > 0$ such that for any $\delta > 0$, $\|\boldsymbol{x}_0\| < \delta$ and $\|\boldsymbol{x}(t_1)\| \geqq \varepsilon_0$ at some $t = t_1 \geqq t_0$. In view of (52), we obtain

$$0 < \alpha(\varepsilon_0) \leqq V(\boldsymbol{x}, t) \quad \text{for all } t \text{ and } \boldsymbol{x} \in C_{\varepsilon_0} = \{\boldsymbol{x}: \|\boldsymbol{x}\| = \varepsilon_0\}. \tag{55}$$

Let t_0 be fixed. Then, by virtue of the fact that $V(\boldsymbol{0}, t_0) = 0$ and from the continuity of the Liapunov function, corresponding to $\alpha(\varepsilon_0) > 0$ there exists a $\delta(\varepsilon_0, t_0) > 0$ such that if $\|\boldsymbol{x}_0\| < \delta(\varepsilon_0, t_0)$ then $V(\boldsymbol{x}_0, t_0) < \alpha(\varepsilon_0)$. Hence, by virtue of condition (b), we obtain

$$0 < \alpha(\varepsilon_0) \leqq V(\boldsymbol{x}(t_1), t_1) \leqq V(\boldsymbol{x}_0, t_0) < \alpha(\varepsilon_0),$$ [20]

which is a contradiction. Thus $\boldsymbol{x}_e = \boldsymbol{0}$ is L-stable.

(ii) From (i), $\boldsymbol{x} = \boldsymbol{0}$ is L-stable. Hence it suffices to show that the $\delta(t_0, \varepsilon_0)$ in the proof of (i) can be replaced by a δ independent of t_0. From the continuity of $\beta(\xi)$ and the fact that $\beta(0) = 0$ it follows that for $\varepsilon_0 > 0$ there exists a $\delta(\varepsilon_0)$ such that $\beta(\delta(\varepsilon_0)) < \alpha(\varepsilon_0)$. Hence, from (52) and condition (c), $\delta(\varepsilon_0) < \varepsilon_0$, whence if $\|\boldsymbol{x}_0\| < \delta(\varepsilon_0)$ then $V(\boldsymbol{x}_0, t_0) \leqq \beta(\delta(\varepsilon_0)) < \alpha(\varepsilon_0)$. Thus, applying the same argument as in the proof of (i), we verify the assertion.

(iii) Noticing that condition (d) implies condition (b), uniform L-stability follows from (ii). Hence with $R > 0$ there is associated an $r(R) > 0$ such that if $\|\boldsymbol{x}_0\| < r(R)$ then

$$\|\boldsymbol{x}(t)\| < R \quad \text{for all } t \geqq t_0. \tag{56}$$

Let $\boldsymbol{x}(t)$ be any solution of (1) with initial state $\boldsymbol{x}_0 \in B(\boldsymbol{0}, r(R))$. Then, from uniform L-stability, corresponding to any given $\mu > 0$ there exists a $\delta(\mu) > 0$ such that $\|\boldsymbol{x}_0\| < \delta(\mu)$ implies that

$$\|x(t)\| < \mu \quad \text{for all } t \geqq t_0. \tag{57}$$

We wish to show that the solution under consideration has the property that $\|x(t)\| < \delta(\mu)$ at some $t \geqq t_0$. Suppose the contrary. Then, since $\|\boldsymbol{x}_0\| < r(R)$,

$$\delta(\mu) \leqq \|\boldsymbol{x}(t)\| < R \quad \text{for all } t \geqq t_0. \tag{58}$$

Define c to be the minimum of the continuous function $\gamma(\xi)$ on the compact set $[\delta(\mu), R]$. Then, from condition (d), c is positive and (54) enables us to assert that

$$V(\boldsymbol{x}(t), t) \leqq V(\boldsymbol{x}_0, t_0) - c(t - t_0) \quad \text{for all } t \geqq t_0. \tag{59}$$

Defining

$$T = \frac{\beta(r(R)) - \alpha(\delta(\mu))}{c}$$ [21]

and noticing that $V(\boldsymbol{x}_0, t_0) \leqq \beta(r(R))$, we obtain

$$V(\boldsymbol{x}_0, t_0) - c(t - t_0) < \alpha(\delta(\mu)) \quad \text{for all } t > t_0 + T. \tag{60}$$

Hence, if $t > t_0 + T$ then (combining (58), (59), and (60))

$$\alpha(\delta(\mu)) \leqq V(\mathbf{x}(t), t) < \alpha(\delta(\mu)),$$

which is a contradiction. Therefore, there exists a $t_1 \in [t_0, t_0 + T]$ such that $\|\mathbf{x}(t_1)\| < \delta(\mu)$. Considering $\mathbf{x}(t_1)$ as the initial state of $\mathbf{x}(t)$ for $t \geqq t_1$, the uniform L-stability of $\mathbf{x}_e = \mathbf{0}$ again assures us that if $\|\mathbf{x}_0\| < r(R)$ then $\|x(t)\| < \mu$ for all $t \geqq t_0 + T \geqq t_1$. Moreover, by definition $r(R)$ and T are independent of t_0. Thus the uniform stability of $\mathbf{x}_e = \mathbf{0}$ has been established.□

Theorem 19 is concerned with local stability, but it also provides information that is useful in the analysis of global stability. In the definition of global stability, no substantial restriction is imposed on the region in which the initial state lies. In establishing conditions for global stability, therefore, the Liapunov function must be defined on $\mathbb{R}^{n+1}$. Moreover, reference will be made to two new conditions:

d′. there exists a positive constant a such that

$$\dot{V}(\mathbf{x}, t) \leqq -a \cdot V(\mathbf{x}, t) \quad \text{for all } t \geqq t_0,$$

and

e. $\lim_{\xi \to \infty} \alpha(\xi) = \infty$.

Theorem 20. *Let $V(\mathbf{x}, t)$ be a Liapunov function defined on $\mathbb{R}^{n+1}$. Then*

i. *the stationary solution $\mathbf{x}_e = \mathbf{0}$ of* (1) *is uniformly stable in the large if conditions* (a), (c), (d), *and* (e) *are satisfied, and*

ii. $\mathbf{x}_e = \mathbf{0}$ *is equistable in the large under conditions* (a) *and* (d′).

Proof. (i) In view of (iii) in Theorem 19, uniform stability is obvious. Thus, it remains to show (i–1) that $r(R)$ in the proof of (iii) in Theorem 19 can be chosen arbitrarily large and to verify (i–2) uniform boundedness.

(i–1) Let r be any positive number. Then there exists an $R_r > 0$ such that $\beta(r) < \alpha(R_r)$. Suppose the contrary. Then there exists an $r_0 > 0$ such that $\beta(r_0) \geqq \alpha(R)$ for any $R > 0$. This, however, contradicts condition (e). Identifying r and R_r respectively with $r(R)$ and R in the proof of (iii) in Theorem 19, and regarding $V(\mathbf{x}, t)$ as a Liapunov function defined on $\|\mathbf{x}\| < R_r, 0 \leq t \leq \infty$, it follows, from the same reasoning as in the proof of (iii) in Theorem 19, that $T(r, \mu, t_0)$ in (ii) of Definition 14 is independent of t_0 for any $r > 0$.

(i–2) Suppose the contrary. Then there exist an $r_0 > 0$ and a solution $\mathbf{x}(t)$ of (1) such that, for any $R > 0$, $\|\mathbf{x}_0\| < r_0$ and $\|\mathbf{x}(t_1)\| \geqq R$ for some $t_1 \geqq t_0$. However, by the foregoing argument, there exists $R_{r_0} > 0$ such that $\beta(r_0) < \alpha(R_{r_0})$. Hence, with the aid of conditions (a) and (b) implied by condition (d), we obtain

$$\alpha(R_{r_0}) \leqq V(\mathbf{x}(t_1), t_1) \leqq V(\mathbf{x}_0, t_0) \leqq \beta(r_0),$$

which contradicts the earlier observation.

(ii) In view of (d′), a direct calculation shows that

$$V(\boldsymbol{x}(t), t) \leqq V(\boldsymbol{x}_0, t_0)e^{-a(t-t_0)}.$$

For any $r > 0$, define $M(r, t_0) = \max_{\|\boldsymbol{x}_0\| \leqq r} V(\boldsymbol{x}_0, t_0)$. Moreover, let μ be any positive number and let $T(\mu, r, t_0) = (1/a) \cdot \log(M(r, t_0)/\alpha(\mu))$. If $t > t_0 + T(\mu, r, t_0)$ then

$$V(\boldsymbol{x}(t), t) \leqq M(r, t_0)e^{-aT(\mu, r, t_0)} \leqq \alpha(\mu) \quad \text{for } t > t_0 + T(\mu, r, t_0).$$

Again, noting that condition (d′) implies condition (b), L-stability of $\boldsymbol{x}_e = \mathbf{0}$ follows from the assertion (i) of Theorem 19. □

Remark Concerning Theorem 20. As was shown in the proof of assertion (i–2), uniform boundedness follows from conditions (a), (b), (c), and (d). Moreover, Kalman and Bertram [1960; Corollary 1.1, p. 378] prove that conditions (a) and (d) ensure equistability and that equistability in the large follows from conditions (a), (d) and (e). Since condition (d′) be regarded as a special case of condition (d) and since local equistability is implied by equistability in the large, Theorem 20 has a wider applicability than their Corollary 1.1 (especially, assertions (b) and (c)).

In the special case in which the time variable t disappears in the right-hand side of (1), the system is called *autonomous*. Thus an autonomous system of differential equations is expressed as

$$\dot{\boldsymbol{x}} = \boldsymbol{f}(\boldsymbol{x}) = (f_1(\boldsymbol{x}), \ldots, f_n(\boldsymbol{x}))'. \tag{1'}$$

Correspondingly, the Liapunov function depends solely on $\boldsymbol{x}$ and is written $V(\boldsymbol{x})$. As a consequence of the above modification, Definition 12 becomes

Definition 12′. A real valued function $V(\boldsymbol{x})$ defined on $\mathbb{R}^n$ is said to be a *Liapunov function* if

i. $\dot{V}(\boldsymbol{x})$ has continuous partial derivatives,
ii. $V(\mathbf{0}) = \mathbf{0}$, and
iii. $V(\boldsymbol{x}) > 0$ for all $\boldsymbol{x} \neq \mathbf{0}$.

Henceforth, we shall be concerned with the stability conditions for autonomous systems of differential equations. Before proceeding to the theorem, we list the several conditions to be considered:

a′. $\dot{V}(\boldsymbol{x}) \leqq 0$ for all $\boldsymbol{x}$;
a″. in addition to (a′), $\dot{V}(\boldsymbol{x}) < 0$ for all $\boldsymbol{x} \neq \mathbf{0}$;
b′. $V(\boldsymbol{x}) \to \infty$ as $\|\boldsymbol{x}\| \to \infty$.

Theorem 21. *Suppose that a Liapunov function* $V(\boldsymbol{x})$ *in Definition* 12′ *exists. Then a stationary solution* $\boldsymbol{x}_e = \mathbf{0}$ *of* (1′) *is*

i. L-*stable under condition* (a′),
ii. *stable under condition* (a″) *and finally,*
iii. *stable in the large under conditions* (a″) *and* (b′).

PROOF. (i) Let ε be any positive number. Then there exists an $\eta_\varepsilon > 0$ such that $V(\boldsymbol{x}) > \eta_\varepsilon$ for any $\boldsymbol{x}$ satisfying $\|\boldsymbol{x}\| = \varepsilon$, for, by virtue of (i) through (iii) of Definition 12′, $V(\boldsymbol{x})$ attains its positive minimum $2\eta_\varepsilon$ on the set $\{\boldsymbol{x}: \|\boldsymbol{x}\| = \varepsilon\}$. Again resorting to the continuity of $V(\boldsymbol{x})$ and condition (i) of Definition 12′, corresponding to η_ε there exists a $\delta_\varepsilon \in (0, \varepsilon)$[22] such that if $\|\boldsymbol{x}\| < \delta_\varepsilon$ then $V(\boldsymbol{x}) < \eta_\varepsilon$. Let Ω_0 be the open set $\{\boldsymbol{x}: \|\boldsymbol{x}\| < \varepsilon \text{ and } V(\boldsymbol{x}) < \eta_\varepsilon\}$. Then the boundary set $\partial\Omega_0$ of Ω_0 is contained in the set $\{\boldsymbol{x}: \|\boldsymbol{x}\| < \varepsilon \text{ and } V(\boldsymbol{x}) = \eta_\varepsilon\}$. This can be shown as follows. Let $\boldsymbol{y}$ be any point of $\partial\Omega_0$. Suppose that $\|\boldsymbol{y}\| = \varepsilon$. Then, by virtue of the definition of η_ε, $V(\boldsymbol{y}) > \eta_\varepsilon$. Since $V(\boldsymbol{x})$ is continuous at $\boldsymbol{x} = \boldsymbol{y}$, there exists a $\rho > 0$ such that $V(\boldsymbol{z}) > \eta_\varepsilon$ for any $\boldsymbol{z} \in B(\boldsymbol{y}, \rho)$.[23] Hence $\boldsymbol{y}$ is an exterior point of Ω_0, contradicting the definition of a boundary. Moreover, if $\|\boldsymbol{y}\| > \varepsilon$ then from the continuity of the norm it follows that there exists a $\rho' > 0$ such that $\|\boldsymbol{z}\| > \varepsilon$ for all $\boldsymbol{z} \in B(\boldsymbol{y}, \rho')$, which again contradicts the definition of a boundary. Thus $\|\boldsymbol{y}\| < \varepsilon$. Similarly, $V(\boldsymbol{y}) \neq \eta_\varepsilon$ can never occur.

Now consider any solution $\boldsymbol{x}(t)$ of (1) starting at $\boldsymbol{x}_0$ such that $\|\boldsymbol{x}_0\| < \delta_\varepsilon$. Then, clearly, $V(\boldsymbol{x}_0) < \eta_\varepsilon$. In view of condition (a′), $\eta_\varepsilon > V(\boldsymbol{x}_0) \geqq V(\boldsymbol{x}(t))$ for all $t \geqq t_0$. Hence the motion $\boldsymbol{x}(t)$ can never reach a point on $\partial\Omega_0$, for there $V(\boldsymbol{x}) = \eta_\varepsilon$. Recalling that $\partial\Omega_0 \subseteq B(\boldsymbol{0}, \varepsilon)$, this establishes the L-stability of $\boldsymbol{x}_e = \boldsymbol{0}$.

(ii) Let $\boldsymbol{x}(t)$ be a solution of (1′) with $\|\boldsymbol{x}_0\| < \delta_\varepsilon$. Since $\dot{V}(\boldsymbol{x}) \leqq 0$ and since $V(\boldsymbol{x}) \geqq 0$ for all $\boldsymbol{x}$, $V(\boldsymbol{x}(t))$ tends to a nonnegative limit c as $t \to \infty$.

First, assume that $c = 0$. Then $\lim_{t\to\infty} \boldsymbol{x}(t) = \boldsymbol{0}$, for otherwise, there would exist an $\varepsilon_0 > 0$ such that

$$\varepsilon_0 \leqq \|\boldsymbol{x}(t)\| \leqq \varepsilon\text{[24]} \quad \text{for } t \geqq \tau, \tag{61}$$

where τ is arbitrarily chosen. Choose a $\mu_0 > 0$, so that

$$\mu_0 < \min_{\varepsilon_0 \leqq \|\boldsymbol{x}\| \leqq \varepsilon} V(\boldsymbol{x}).$$

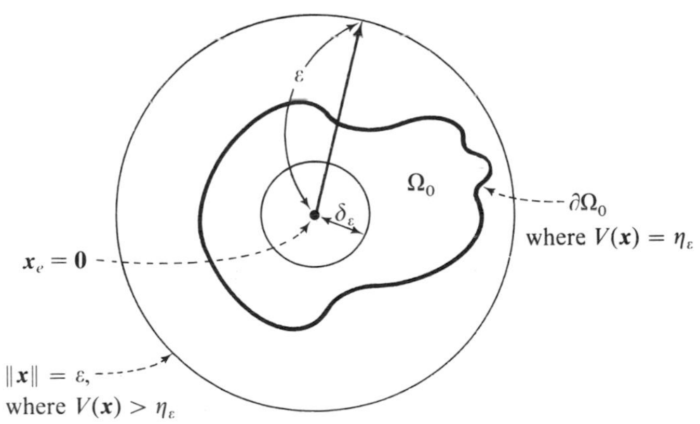

Figure 4.4

Then, from (61),

$$V(\boldsymbol{x}(t)) > \mu_0 \quad \text{for } t \geqq \tau.$$

This clearly contradicts the observed convergence, $\lim_{t\to\infty} V(\boldsymbol{x}(t)) = 0$.

Second, suppose that $c > 0$. Then the continuity of $V(\boldsymbol{x})$ again implies that, for $c/2 > 0$, there exists a $\delta(c/2) \in (0, \varepsilon)$ such that if $\|\boldsymbol{x}\| < \delta(c/2)$ then $V(\boldsymbol{x}) < c/2$. Noticing that $V(\boldsymbol{x}(t))$ is nonincreasing with respect to t, we see that

$$V(\boldsymbol{x}_0) \geqq V(\boldsymbol{x}(t)) \geqq c \quad \text{for all } t \geqq t_0. \tag{62}$$

Hence, if $\|\boldsymbol{x}_0\| < \min(\delta(c/2), \delta_\varepsilon)$ then

$$\varepsilon \geqq \|\boldsymbol{x}(t)\| \geqq \delta\left(\frac{c}{2}\right) \quad \text{for } t \text{ sufficiently large.}$$

On the other hand, recalling that

$$\dot{V}(\boldsymbol{x}) = \sum_{j=1}^{n} \frac{\partial V}{\partial x_j} \cdot f_j(\boldsymbol{x}),$$

we can choose a $v > 0$ such that if $\delta(c/2) \leqq \|\boldsymbol{x}\| \leqq \varepsilon$ then $\dot{V}(\boldsymbol{x}) \leqq -v < 0$.[25] Therefore, there exist a $t_1 \geqq t_0$ and a solution $\boldsymbol{x}(t)$ with $\|\boldsymbol{x}_0\| < \min(\delta(c/2), \delta_\varepsilon)$ such that

$$\dot{V}(\boldsymbol{x}(t)) \leq -v < 0 \quad \text{for } t \geqq t_1.$$

A direct calculation therefore shows that for $t > (V(\boldsymbol{x}_0) - c + vt_1)/v$,

$$\begin{aligned} V(\boldsymbol{x}(t)) &= V(\boldsymbol{x}_0) + \int_{t_0}^{t} \dot{V}(\boldsymbol{x}(\tau))d\tau \\ &= V(\boldsymbol{x}_0) + \int_{t_0}^{t_1} \dot{V}(\boldsymbol{x}(\tau))d\tau + \int_{t_1}^{t} \dot{V}(\boldsymbol{x}(\tau))d\tau \\ &\leqq V(\boldsymbol{x}_0) - v(t - t_1) < c, \end{aligned}$$

which contradicts (62).

(iii) Since condition (a″) includes condition (a′), L-stability is immediate. Hence it remains to show that δ_ε in the proof of assertion (ii) can be taken arbitrarily large.

Let δ be any positive number, and let $\eta_\delta > 0$ be such that $\eta_\delta > \max_{\|\boldsymbol{x}\| \leqq \delta} V(\boldsymbol{x})$. In view of condition (b′), we can find an ε_δ such that

$$\eta_\delta < \min_{\|\boldsymbol{x}\| = \varepsilon_\delta} V(\boldsymbol{x}).$$

Thus, to complete the proof it suffices to replace ε, η_ε and δ_ε in the proof of assertion (ii) with ε_δ, η_δ and δ respectively. □

Finally we turn to some useful corollaries of Theorem 21.

Corollary 9. *By* $J(x)$ *denote the Jacobian matrix of* $f(x)$, *that is,*

$$J(x) = \left(\frac{\partial f_i}{\partial x_j}\bigg|_x\right).$$

If there exists a real pd *matrix* G *such that* $G \cdot J(x)$ *is* qnd *for all* $x \neq x_e(=0)$ *then the stationary solution of* (1′) *is stable in the large.*

PROOF. Set $V(x) = \frac{1}{2}x'Gx$. Then clearly $V(0) = 0$, $V(x) > 0$ for all $x \neq 0$ and $\lim_{\|x\| \to \infty} V(x) = \infty$. By direct differentiation, we obtain

$$\dot{V}(x) = x'Gf(x).$$

Choose a $\tau \in [0, 1]$. Direct differentiation then yields

$$\frac{df(\tau x)}{d\tau}^{26} = J(\tau x) \cdot x.$$

Hence, recalling that $f(0) = 0$, we obtain

$$f(x) = \int_0^1 J(\tau x)x \, d\tau.$$

Therefore, for any $x \neq 0$,

$$\dot{V}(x) = x'G \int_0^1 J(\tau x)x \, d\tau$$

$$= \int_0^1 x'GJ(\tau x)x \, d\tau < 0. \qquad \square$$

Remark Concerning Corollary 9. Once the matrix G in the corollary is known, a Liapunov function can be written down immediately.

Our next corollary requires a lemma from matrix algebra.

Lemma 11. *Corresponding to an* $n \times n$ *real matrix* A, *there exists a nonsingular matrix* T *such that*

$$T^{-1}AT = B,$$

where B *is a triangular matrix of which the diagonal elements are the eigenvalues* λ_i, $i = 1, \ldots, n$, *of* A *and the absolute value of its nonzero elements can be made as small as needed.*

PROOF. We proceed by induction on n, the order of A. Suppose that $n = 2$. Let x_1 be an eigenvector associated with λ_1 and let x_2 be a vector such that $T_2 = (x_1, x_2)$ is nonsingular.[27] Then a direct calculation yields

$$T_2^{-1}AT_2 = \begin{pmatrix} \lambda_1 & b_{12} \\ 0 & b_{22} \end{pmatrix}.$$

Again, direct calculation of the characteristic equation of A shows that

$$\det(\lambda I - A) = \det(\lambda I - T_2^{-1}AT_2) = \det\begin{pmatrix} \lambda - \lambda_1 & -b_{12} \\ 0 & \lambda - b_{22} \end{pmatrix} = 0,$$

from which it follows at once that $\lambda_2 = b_{22}$.

Next we show that if the lemma holds for a real square matrix of order $n - 1$ then it holds for an $n \times n$ real matrix. Let ξ_1 be an eigenvector corresponding to an eigenvalue λ_1 of A. Since $\xi_1 \neq 0$ is linearly independent, there exist $n - 1$ vectors $\xi_2, \ldots, \xi_n$ such that $(\xi_1, \ldots, \xi_n) = T_1$ is nonsingular. Thus we obtain

$$T_1^{-1}AT_1 = \begin{pmatrix} \lambda_1 & b'_{12} & \cdots & b'_{1n} \\ 0 & b'_{22} & \cdots & b'_{2n} \\ \vdots & \vdots & & \vdots \\ 0 & b'_{n2} & \cdots & b'_{nn} \end{pmatrix} \equiv \begin{pmatrix} \lambda_1 & b'_{12} \quad \cdots \quad b'_{1n} \\ \mathbf{0} & B_{n-1} \end{pmatrix}.$$

Moreover, from the identity

$$\det(\lambda I - A) = \det(\lambda I - T_1^{-1}AT_1) = (\lambda - \lambda_1)\det(\lambda I - B_{n-1})$$

it follows that the eigenvalues of B_{n-1} are also eigenvalues of A, which we denote by $\lambda_2, \ldots, \lambda_n$. Applying the induction assumption to the matrix B_{n-1} of order $n - 1$, there exists a nonsingular matrix T_{n-1} such that

$$T_{n-1}^{-1}B_{n-1}T_{n-1} = \begin{pmatrix} \lambda_2 & b''_{23} & & \cdots & b''_{2n} \\ & \lambda_3 & b''_{34} & \cdots & b''_{3n} \\ & & \ddots & \ddots & \vdots \\ & \mathbf{0} & & \ddots & b''_{n-1n} \\ & & & & \lambda_n \end{pmatrix}.$$

Defining

$$T_n = \begin{pmatrix} 1 & \mathbf{0} \\ \mathbf{0} & T_{n-1} \end{pmatrix},$$

it is easy to see that

$$T_n^{-1}(T_1^{-1}AT_1)T_n = \begin{pmatrix} \lambda_1 & b_{12} & b_{13} & \cdots & b_{1n} \\ & \lambda_2 & b_{23} & \cdots & b_{2n} \\ & & \ddots & & \vdots \\ & \mathbf{0} & & \ddots & b_{n-1n} \\ & & & & \lambda_n \end{pmatrix}$$

Setting $T = T_1 T_n \hat{S}$, the assertion is immediate, where $\hat{S}$ is a diagonal matrix with s^k as its kth diagonal element and s is a positive number small enough.

Corollary 10. *The stationary solution* $\mathbf{0}$ *of* (1′) *is stable, if the Jacobian matrix* $A = J(\mathbf{0})$ *of* $f(x)$ *evaluated at* $x = \mathbf{0}$ *is a stable matrix.*

Proof. In view of Lemma 11, there exists a nonsingular matrix $\boldsymbol{T}$ such that

$$\boldsymbol{T}^{-1}\boldsymbol{A}\boldsymbol{T} = \begin{pmatrix} \lambda_1 & b_{12} & & \cdots & b_{1n} \\ & \lambda_2 & b_{23} & \cdots & b_{2n} \\ & & \ddots & & \vdots \\ & \mathbf{0} & & \ddots & b_{n-1n} \\ & & & & \lambda_n \end{pmatrix}. \tag{63}$$

The Taylor expansion of $\boldsymbol{f}(\boldsymbol{x})$ at $\boldsymbol{x} = \mathbf{0}$ yields

$$\dot{\boldsymbol{x}} = \boldsymbol{A}\boldsymbol{x} + \boldsymbol{h}(\boldsymbol{x}).$$

Hence the transformation of variables $\boldsymbol{x} = \boldsymbol{T}\boldsymbol{\eta}$ yields

$$\begin{aligned} \dot{\boldsymbol{\eta}} &= \boldsymbol{T}^{-1}\boldsymbol{A}\boldsymbol{T}\boldsymbol{\eta} + \boldsymbol{T}^{-1}\boldsymbol{h}(\boldsymbol{T}\boldsymbol{\eta}) \\ &\equiv \boldsymbol{T}^{-1}\boldsymbol{A}\boldsymbol{T}\boldsymbol{\eta} + \boldsymbol{g}(\boldsymbol{\eta}). \end{aligned} \tag{64}$$

With the aid of (63), (64) can be expressed as

$$\frac{d\eta_k}{dt} \equiv \dot{\eta}_k = \lambda_k\eta_k + \sum_{j>k} b_{kj}\eta_j + g_k(\boldsymbol{\eta}) \quad (k = 1, \ldots, n). \tag{65}$$

Let the Liapunov function $V(\boldsymbol{\eta}) = \frac{1}{2}\bar{\boldsymbol{\eta}}' \cdot \boldsymbol{\eta}$.[28] Noticing that

$$\frac{d(|\eta_k|^2)}{dt} = \bar{\eta}_k \frac{d\eta_k}{dt} + \eta_k \frac{d\bar{\eta}_k}{dt} \qquad (k = 1, \ldots, n),$$

we obtain

$$\begin{aligned} \dot{V}(\boldsymbol{\eta}) &= \operatorname{Re}\left[\sum_{k=1}^{n} \bar{\eta}_k\left(\lambda_k\eta_k + \sum_{j>k} b_{kj}\eta_j + g_k(\boldsymbol{\eta})\right)\right] \\ &= \sum_{k=1}^{n}\left[\operatorname{Re}(\lambda_k)|\eta_k|^2 + \operatorname{Re}\left(\bar{\eta}_k \sum_{j>k} b_{kj}\eta_j\right) + \operatorname{Re}(\bar{\eta}_k \cdot g_k(\boldsymbol{\eta}))\right]. \end{aligned}$$

Let α and β be respectively the largest of $\operatorname{Re}(\lambda_j), j = 1, \ldots, n$, and the greatest of $\sum_{j>k} |b_{kj}|, k = 1, \ldots, n$, and let $\|\boldsymbol{\xi}\| = \max_{1 \le j \le n}|\xi_j|$, where ξ_j denotes the jth element of $\boldsymbol{\xi}$. Then we easily see that

$$\dot{V}(\boldsymbol{\eta}) \leqq \sum_{k=1}^{n} \{\|\boldsymbol{\eta}\|^2(\alpha + \beta) + \|\boldsymbol{\eta}\| \cdot \|\boldsymbol{g}(\boldsymbol{\eta})\|\}. \tag{66}$$

From the assumptions imposed on $\boldsymbol{f}(\boldsymbol{x})$ it follows that, for any $\varepsilon > 0$ there exists a $\delta > 0$ such that if $\|\boldsymbol{\eta}\| < \delta$ then $\|\boldsymbol{g}(\boldsymbol{\eta})\|/\|\boldsymbol{\eta}\| < \varepsilon$. Since without loss of generality δ may be assumed to be less than ε, from the above conclusion and from (66) it follows that

$$\dot{V}(\boldsymbol{\eta}) \leqq n \cdot \|\boldsymbol{\eta}\|^2 \cdot (\alpha + \beta + \varepsilon).$$

Since $\alpha < 0$ and since, in view of Lemma 11, β can be assumed so small that $\alpha + \beta < 0$, by choosing ε sufficiently small we can be sure that

$$\dot{V}(\boldsymbol{\eta}) < 0 \quad \text{for all relevant } \boldsymbol{\eta} \neq \mathbf{0}.$$

Hence, by assertion (ii) of Theorem 21, $\boldsymbol{x}_e = \mathbf{0}$ is stable. □

5 Stability Theorems for Systems of Difference Equations

Dynamic economic analysis can be conducted in terms of systems of differential equations or in terms of systems of difference equations (or, indeed, in terms of general "mixed" systems). So far we have confined ourselves to systems of differential equations. We turn now to systems of difference equations.

Let $x_j(t)$ now stand for the value assumed by the jth variable during the tth time *interval*. Then we may consider the nonautonomous system of mth-order difference equations

$$F_j(\boldsymbol{x}(t+m), \boldsymbol{x}(t+m-1), \ldots, \boldsymbol{x}(t), t) = 0 \qquad (j = 1, \ldots, n) \tag{67}$$

in n variables, with $\boldsymbol{x}(t) \equiv (x_1(t), \ldots, x_n(t))$. However, if each F_j in (67) satisfies some sufficient condition for univalence,[29] we may solve (67) to obtain

$$x_j(t+m) = g_j(\boldsymbol{x}(t+m-1), \ldots, \boldsymbol{x}(t), t) \qquad (j = 1, \ldots, n)$$

or

$$\boldsymbol{x}(t+m) = \boldsymbol{g}(\boldsymbol{x}(t+m-1), \ldots, \boldsymbol{x}(t), t). \tag{67'}$$

Introducing the change of variables

$$\boldsymbol{x}(t+k) = \boldsymbol{y}_{k+1}(t) \qquad (k = 0, 1, \ldots, m-1)$$

we can then rewrite the mth-order system (67′) as the first-order system

$$\begin{aligned} \boldsymbol{y}_j(t+1) &= \boldsymbol{y}_{j+1}(t) \qquad (j = 1, \ldots, m-1), \\ \boldsymbol{y}_m(t+1) &= \boldsymbol{g}(\boldsymbol{y}_m(t), \boldsymbol{y}_{m-1}(t), \ldots, \boldsymbol{y}_1(t), t). \end{aligned}$$

Henceforth we shall confine our attention to systems which can be written in this form, that is, in the form

$$\boldsymbol{x}(t+1) = \boldsymbol{f}(\boldsymbol{x}(t), t) \tag{68}$$

where the f_j are functions.

Let $\boldsymbol{x}_0$ and t_0 be respectively a given point in $\mathbb{R}^n$ and the moment of time at which the motion of the system begins. Then a vector function $\boldsymbol{x}(t)$ which satisfies both (68) and $\boldsymbol{x}(t_0) = \boldsymbol{x}_0$ is called a *solution* of (68) satisfying the initial condition $(\boldsymbol{x}_0, t_0)$; if necessary, it can be written in the explicit form $\boldsymbol{x}(\boldsymbol{x}_0, t_0)$. Evidently a solution to (68) always exists; for, given $\boldsymbol{x}(t)$, the state of the system during the time interval $t + 1$ is given by (68). Moreover, since the f_j are functions, the solution is unique.

Turning to questions of stability, we find that, *mutatis mutandis*, the Liapunov stability theorems for systems of differential equations remain valid for systems of difference equations. Since time is now discrete we no longer can avail ourselves of the notion of the trajectory derivative of a Liapunov function. Instead of $\dot{V}(\boldsymbol{x}, t)$ we must work with

$$\Delta V(\boldsymbol{x}, t) \equiv V(\boldsymbol{f}(\boldsymbol{x}, t), t + 1) - V(\boldsymbol{x}, t), \tag{69}$$

that is, with the rate of change per unit time interval along the path defined by (68), starting at $\boldsymbol{x}$ and t. Omitting the proof, which is similar to those of Section 4, we simply state

Theorem 22. *With $\Delta V(\boldsymbol{x}, t)$ substituted for $\dot{V}(\boldsymbol{x}, t)$, Theorems* 19 *and* 20 *are valid for the nonautonomous system of difference equations* (68).

If $\boldsymbol{f}$ in (68) does not depend on the time variable t then the system is said to be *autonomous* and we may write

$$\boldsymbol{x}(t + 1) = \boldsymbol{f}(\boldsymbol{x}(t)). \tag{70}$$

The rate of change of V along the path defined by (70), starting at $\boldsymbol{x}$, can be defined as

$$\Delta V(\boldsymbol{x}) \equiv V(\boldsymbol{f}(\boldsymbol{x})) - V(\boldsymbol{x}). \tag{71}$$

Substituting $\Delta V(x)$ for $\dot{V}(\boldsymbol{x})$ in conditions (a′) and (a″) of Section 4, we then obtain the difference equation version of Theorem 21.

Theorem 23. *With $\Delta V(\boldsymbol{x})$ substituted for $\dot{V}(\boldsymbol{x})$ in conditions (a′) and (a″), Theorem* 21 *is valid for the autonomous system of difference equations* (70).

Remark Concerning Theorem 23. Recently Brock and Scheinkman [1975; Theorem 1] have described restrictions on the Liapunov function that ensure that condition (a″) in Theorem 23 is satisfied. In particular, they have shown that if

i. $V(\boldsymbol{f}(\boldsymbol{x})) = V(\boldsymbol{x})$ implies that $\boldsymbol{V}'(\boldsymbol{f}(\boldsymbol{x}))\boldsymbol{J}(\boldsymbol{x})\boldsymbol{x} - \boldsymbol{V}'(\boldsymbol{x})\boldsymbol{x} < 0$, for all $\boldsymbol{x} \neq \boldsymbol{0}$,
ii. $\boldsymbol{V}'(\boldsymbol{0}) \neq \boldsymbol{0}$ implies that $\boldsymbol{V}'(\boldsymbol{0})\boldsymbol{J}(\boldsymbol{0})\boldsymbol{x} - \boldsymbol{V}'(\boldsymbol{0})\boldsymbol{x} < 0$, for all $\boldsymbol{x} \neq \boldsymbol{0}$,
iii. $\boldsymbol{V}'(\boldsymbol{0}) = \boldsymbol{0}$ implies that $(\boldsymbol{V}''(\boldsymbol{0})\boldsymbol{J}(\boldsymbol{0})\boldsymbol{x})'\boldsymbol{J}(\boldsymbol{0})\boldsymbol{x} - \boldsymbol{x}'\boldsymbol{V}''(\boldsymbol{0})\boldsymbol{x} < 0$ for all $\boldsymbol{x} \neq \boldsymbol{0}$, and
iv. $V(\boldsymbol{x}) \to \infty$ as $\|\boldsymbol{x}\| \to \infty$,

then the equilibrium state $\boldsymbol{0}$ of (70) (if it exists) is globally stable. (Here $\boldsymbol{V}'(\boldsymbol{x})$ and $\boldsymbol{V}''(\boldsymbol{x})$ respectively denote the gradient and Hessian of V evaluated at $\boldsymbol{x}$, and $\boldsymbol{J}(\boldsymbol{x})$ is the Jacobian of $\boldsymbol{f}$ evaluated at $\boldsymbol{x}$.)

PROOF. We begin by showing that the Liapunov function is decreasing in t along every solution of (70). Corresponding to any given $\boldsymbol{x}$, we define $g(\lambda) \equiv \lambda(V(\boldsymbol{f}(\lambda\boldsymbol{x})) - V(\lambda\boldsymbol{x}))$, where the domain of λ is taken to be the closed interval $[0, 2]$. Then, obviously, $g(0) = 0$. Direct differentiation of $g(\lambda)$ yields

$$g'(\lambda) = V(\boldsymbol{f}(\lambda\boldsymbol{x}) - V(\lambda\boldsymbol{x}) + \lambda(\boldsymbol{V}'(\boldsymbol{f}(\lambda\boldsymbol{x}))\boldsymbol{J}(\lambda\boldsymbol{x})\boldsymbol{x} - \boldsymbol{V}'(\lambda\boldsymbol{x})\boldsymbol{x}).$$

Recalling that $\boldsymbol{0} = \boldsymbol{f}(\boldsymbol{0})$ and that $V(\boldsymbol{0}) = 0, g'(0) = 0$. For ease of calculation, let $\psi(\lambda) \equiv \boldsymbol{V}'(\boldsymbol{f}(\lambda\boldsymbol{x}))\boldsymbol{J}(\lambda\boldsymbol{x})\boldsymbol{x} - \boldsymbol{V}'(\lambda\boldsymbol{x})\boldsymbol{x}$. Then

$$g''(\lambda) = 2\psi(\lambda) + \lambda\psi'(\lambda),$$

$$g'''(\lambda) = 3\psi'(\lambda) + \lambda\psi''(\lambda).$$

Hence, from condition (ii), if $\boldsymbol{V}'(\boldsymbol{0}) \neq \boldsymbol{0}$ then

$$g''(0) = 2[\boldsymbol{V}'(\boldsymbol{0})\boldsymbol{J}(\boldsymbol{0})\boldsymbol{x} - \boldsymbol{V}'(\boldsymbol{0})\boldsymbol{x}] < 0;$$

and, from condition (iii), if $\boldsymbol{V}'(\boldsymbol{0}) = \boldsymbol{0}$ then

$$g'''(0) = 3\,(\boldsymbol{V}''(\boldsymbol{0})\boldsymbol{J}(\boldsymbol{0})\boldsymbol{x})'\boldsymbol{J}(\boldsymbol{0})\boldsymbol{x} - \boldsymbol{x}'\boldsymbol{V}''(\boldsymbol{0})\boldsymbol{x} < 0.$$

From the continuity of $g''(\lambda)$ (if $\boldsymbol{V}'(\boldsymbol{0}) \neq \boldsymbol{0}$) or of $g'''(\lambda)$ (if $\boldsymbol{V}'(\boldsymbol{0}) = \boldsymbol{0}$), therefore, there exists $\varepsilon > 0$ such that

$$g(\lambda) < 0 \quad \text{for all } \lambda \in (0, \varepsilon).$$

We next show that $g(1) < 0$. If $\varepsilon > 1$ there is nothing to prove. Suppose therefore that $\varepsilon \leqq 1$ and that $g(1) \geqq 0$. Next we notice that, if $\varepsilon = 1$, we can choose a new $\varepsilon < 1$ such that $g(\lambda) < 0$ for all $\lambda \in (0, \varepsilon)$ and that therefore there is no loss of generality in supposing that $\varepsilon < 1$. There then exists $\bar{\lambda} \in (\varepsilon, 1]$ such that $g(\bar{\lambda}) = 0$ and $g'(\bar{\lambda}) \geqq 0$, contradicting condition (i). Hence $g(1) < 0$ for any $\boldsymbol{x} \neq \boldsymbol{0}$, implying that the Liapunov function is decreasing in t along every solution of (70).

This fact, together with condition (iv), ensures that every solution of (70) is bounded above. Suppose the contrary. Then there exists a solution containing a subsequence $\{\boldsymbol{x}(t_j)\}$ such that $\|\boldsymbol{x}(t_j)\| \to \infty$ as $t_j \to \infty$. From condition (iv), therefore, $\lim_{t_j \to \infty} V(\boldsymbol{x}(t_j)) = \infty$, contradicting the monotone decreasing property of $V(\boldsymbol{x})$.

To establish global stability, we suppose that there exists a solution $\boldsymbol{x}(t)$ of (70) with the property that there exists an $\varepsilon > 0$ such that $\|\boldsymbol{x}(t)\| \geqq \varepsilon$ for sufficiently large t. Since this solution is bounded above, there exists $\beta > 0$ such that $\|\boldsymbol{x}(t)\| < \beta$ for all $t \geqq 0$. Let $C_\varepsilon \equiv \{\boldsymbol{x}: \varepsilon \leqq \|\boldsymbol{x}\| \leqq \beta\}$. Since C_ε is compact, a subsequence $\{\boldsymbol{x}(t_i)\}$ in C_ε converges to a point $\bar{\boldsymbol{x}} \in C_\varepsilon$. On the other hand, from the continuity of $V(\boldsymbol{x})$ it follows that

$$\begin{aligned} V(\boldsymbol{f}(\bar{\boldsymbol{x}})) - V(\bar{\boldsymbol{x}}) &= \lim_{t_i \to \infty} \{V(\boldsymbol{f}(\boldsymbol{x}(t_i)) - V(\boldsymbol{x}(t_i)\} \\ &= \lim_{t_i \to \infty} \{V(\boldsymbol{x}(t_{i+1}) - V(\boldsymbol{x}(t_i)\} = 0. \end{aligned}$$

Since $\bar{\boldsymbol{x}} \neq \boldsymbol{0}$ by definition, this is a contradiction. □

To this point we have been concerned with fairly general systems of difference equations. We now focus on the special case of linear equations. Consider first the linear homogeneous system

$$\boldsymbol{x}(t+1) - \boldsymbol{A}\boldsymbol{x}(t), \tag{72}$$

where $\boldsymbol{A}$ is an $n \times n$ matrix with constant coefficients. Since $\boldsymbol{A}^{t+1} = \boldsymbol{A} \cdot \boldsymbol{A}^t$, every column of $\boldsymbol{A}^t$ is a solution of system (72) and the set of all solutions of (72) forms a linear subspace. Let us assume for the time being that $\boldsymbol{A}$ is nonsingular. Then $\boldsymbol{A}^t$ is nonsingular for any t. Hence any solution $\boldsymbol{x}(t)$ of (72) can be expressed as a linear combination of all columns of $\boldsymbol{A}^t$, that is, there exists $\boldsymbol{\xi}(t)$ such that

$$\boldsymbol{x}(t) = \boldsymbol{A}^t \boldsymbol{\xi}(t). \tag{73}$$

Since $\boldsymbol{x}(t)$ is a solution of (72),

$$\boldsymbol{x}(t+1) = \boldsymbol{A} \cdot \boldsymbol{x}(t) = \boldsymbol{A}^{t+1} \cdot \boldsymbol{\xi}(t).$$

On the other hand, by definition,

$$\boldsymbol{x}(t+1) = \boldsymbol{A}^{t+1} \cdot \boldsymbol{\xi}(t+1).$$

And $\boldsymbol{A}^{t+1}$ is nonsingular. Hence

$$\boldsymbol{\xi}(t+1) = \boldsymbol{\xi}(t) \quad \text{for all } t,$$

and (73) can be written more simply as

$$\boldsymbol{x}(t) = \boldsymbol{A}^t \cdot \boldsymbol{\xi}. \tag{74}$$

Successive substitutions reveal that (74) can be viewed as a solution of (72), starting from a point $\boldsymbol{\xi}$ at $t_0 = 0$.

We now show that the assumption that $\boldsymbol{A}$ is nonsingular involves no loss of essential generality. If $\boldsymbol{A}$ is singular, it can be partitioned as

$$\begin{pmatrix} \boldsymbol{A}_{11} & \boldsymbol{A}_{12} \\ \boldsymbol{A}_{21} & \boldsymbol{A}_{22} \end{pmatrix}$$

where $\boldsymbol{A}_{11}$ is a nonsingular matrix of order equal to the rank of $\boldsymbol{A}$. Conformably with the partitioning of $\boldsymbol{A}$ we further partition $\boldsymbol{x}(t)$ as $(\boldsymbol{x}_1'(t), \boldsymbol{x}_2'(t))'$. From the definition of the rank of a matrix, there exists a matrix $\boldsymbol{Y}$ such that

$$(\boldsymbol{A}_{21}, \boldsymbol{A}_{22}) = \boldsymbol{Y}(\boldsymbol{A}_{11}, \boldsymbol{A}_{12}).$$

Hence $\boldsymbol{x}(t) = (\boldsymbol{x}_1'(t), \boldsymbol{x}_2'(t))'$ is a solution of (72) if and only if $\boldsymbol{x}_2(t) = \boldsymbol{Y}\boldsymbol{x}_1(t)$ for all t, where $\boldsymbol{x}_1(t)$ is a solution of

$$\boldsymbol{x}_1(t+1) = (\boldsymbol{A}_{11} + \boldsymbol{A}_{12}\boldsymbol{Y})\boldsymbol{x}_1(t). \tag{72'}$$

If $\boldsymbol{A}_{11} + \boldsymbol{A}_{12}\boldsymbol{Y}$ is still singular, finitely many repetitions of the above procedure lead to a situation in which the resultant coefficient matrix is nonsingular or a situation where it is a zero (scalar). Hence we can assume without loss of generality that the coefficient matrix $\boldsymbol{A}$ in (72) is nonsingular.

Let $\boldsymbol{J}$ be the Jordan canonical form of $\boldsymbol{A}$ and again let $\boldsymbol{P}^{-1}\boldsymbol{A}\boldsymbol{P} = \boldsymbol{J}$. Then the change of variables $\boldsymbol{y} = \boldsymbol{P}^{-1}\boldsymbol{x}$ transforms (74) into

$$\boldsymbol{y}(t) = \boldsymbol{J}^t\boldsymbol{\eta} = \begin{pmatrix} \boldsymbol{J}_1^t & \boldsymbol{0} \\ \boldsymbol{0} & \boldsymbol{J}_m^t \end{pmatrix}\boldsymbol{\eta},$$

where $\boldsymbol{\eta} \equiv \boldsymbol{P}^{-1}\xi$ and $\boldsymbol{J}_j$ is the jth Jordan diagonal block of $\boldsymbol{J}$. A tedious calculation of each element of $\boldsymbol{y}(t)$, together with the fact that $|\lambda_j^t| = |\lambda_j|^t = \exp(t \log|\lambda_j|)$, leads us to

Theorem 24. *The stationary solution* $\boldsymbol{x}(t) = \boldsymbol{0}$ *of* (72) *is uniformly stable in the large if and only if every eigenvalue of* $\boldsymbol{A}$ *is less than one in modulus.*

The difference equation version of Theorem 3 is

Theorem 25. *Every eigenvalue of an* $n \times n$ *complex matrix* $\boldsymbol{A}$ *is less than one in modulus if and only if for any* hnd $\boldsymbol{W}$ *there exists a unique* hpd $\boldsymbol{Q}$ *such that* $\boldsymbol{A}^*\boldsymbol{Q}\boldsymbol{A} - \boldsymbol{Q} = \boldsymbol{W}$.

PROOF (*Sufficiency*). Define the Liapunov function associated with (72) by $V(\boldsymbol{x}) = \boldsymbol{x}^*\boldsymbol{Q}\boldsymbol{x}$. Then the rate of change of V along the trajectory $\boldsymbol{x}(t)$, starting from any given initial condition, is

$$\Delta V(\boldsymbol{x}(t)) = (\boldsymbol{x}(t))^*(\boldsymbol{A}^*\boldsymbol{Q}\boldsymbol{A} - \boldsymbol{Q})\boldsymbol{x}(t) < 0,$$

and the assertion follows directly from Theorem 23.

(*Necessity*). For any hnd $\boldsymbol{W}$, define $\boldsymbol{x}(t) = -(\boldsymbol{A}^*)^t\boldsymbol{W}\boldsymbol{A}^t$. Then

$$-\boldsymbol{W} = \boldsymbol{x}(0),$$

$$\lim_{t\to\infty} \boldsymbol{x}(t) = \boldsymbol{0},$$

and

$$\boldsymbol{x}(t) - \boldsymbol{x}(0) = \sum_{\tau=0}^{t-1} (\boldsymbol{A}^*\boldsymbol{x}(\tau)\boldsymbol{A} - \boldsymbol{x}(\tau)). \tag{75}$$

Defining

$$\boldsymbol{Q} = \lim_{t\to\infty} \sum_{\tau=0}^{t} \boldsymbol{x}(\tau)$$

and letting t go to infinity in (75), we obtain

$$\boldsymbol{W} = \boldsymbol{A}^*\boldsymbol{Q}\boldsymbol{A} - \boldsymbol{Q}.$$

The uniqueness and hermitian positive definiteness of $\boldsymbol{Q}$ are easy to verify. □

As a straightforward corollary of Theorem 25 we list

Corollary 11 (Samuelson [1945; Theorem 3, p. 438]). *If* $\boldsymbol{A}^*\boldsymbol{A} - \boldsymbol{I}$ *is* hnd *then every eigenvalue of* $\boldsymbol{A}$ *is less than one in modulus.*

Notes to Chapter 4

1. $d(e^{At})/dt$ is the matrix consisting of the time derivatives of the elements of e^{At}.
2. See Definition 4 in Appendix A.
3. Let $\boldsymbol{B}(t)$ be any matrix with $b_{ij}(t)$ as its (ij)th element. Then $\int_\alpha^\beta B(t)dt$ is the matrix $(\int_\alpha^\beta b_{ij}(t)dt)$, where α and β are arbitrary constants.
4. See Lemma 2 of Appendix A.
5. Orthogonality of the basis vectors can always be achieved. Here, we briefly explain the so-called Gram–Schmidt method of orthogonalization. Let $\boldsymbol{x}_i$ $(i = 1, \ldots, k)$ be linearly independent complex vectors of arbitrary order, and by $(\boldsymbol{x}, \boldsymbol{y})$ denote the innerproduct of vectors $\boldsymbol{x}$ and $\boldsymbol{y}$. Define $\boldsymbol{y}_1 = \boldsymbol{x}_1$ and let $\boldsymbol{y}_2 = \boldsymbol{x}_2 - \alpha_{21}\boldsymbol{y}_1$. Then

$$(\boldsymbol{y}_2, \boldsymbol{y}_1) - (\boldsymbol{x}_2, \boldsymbol{y}_1) - \alpha_{21}(\boldsymbol{y}_1, \boldsymbol{y}_1).$$

Hence, choosing $\alpha_{21} = (\boldsymbol{x}_2, \boldsymbol{y}_1)/(\boldsymbol{y}_1, \boldsymbol{y}_1)$, $\boldsymbol{y}_2$ becomes orthogonal to $\boldsymbol{y}_1$ and $\boldsymbol{y}_2 \neq \boldsymbol{0}$, for otherwise $\boldsymbol{x}_1, \ldots, \boldsymbol{x}_k$ would be linearly dependent. In general, define $\boldsymbol{y}_j = \boldsymbol{x}_j - \sum_{i=1}^{j-1}\alpha_{ji}\boldsymbol{y}_i$, $j = 3, \ldots, k$, and let $\alpha_{ji} = (\boldsymbol{x}_j, \boldsymbol{y}_i)/(\boldsymbol{y}_i, \boldsymbol{y}_i)$, $i = 1, \ldots, j-1$, $j = 3, \ldots, k$. Then it is easy to see that the transformed vectors $\boldsymbol{y}_1, \ldots, \boldsymbol{y}_k$ are mutually perpendicular. In matrix notation, this procedure can be represented as:

$$(\boldsymbol{x}_1, \ldots, \boldsymbol{x}_k) = (\boldsymbol{y}_1, \ldots, \boldsymbol{y}_k)\begin{pmatrix} 1 & \alpha_{21} & \alpha_{31} & \cdots & \alpha_{k1} \\ 0 & 1 & \alpha_{32} & \cdots & \alpha_{k2} \\ & 0 & 1 & \ddots & \vdots \\ & & 0 & \ddots & \\ \vdots & \vdots & \vdots & \ddots & \alpha_{kk-1} \\ 0 & 0 & 0 & \cdots & 1 \end{pmatrix}.$$

Furthermore, defining $\boldsymbol{u}_j = \boldsymbol{y}_j/(\boldsymbol{y}_j, \boldsymbol{y}_j)$, $j - 1, \ldots, k$, $\boldsymbol{u}_j$ $(j - 1, \ldots, k)$ becomes a normal orthogonal set of vectors, that is, $(\boldsymbol{u}_i, \boldsymbol{u}_j) = \delta_{ij}$, $i, j = 1, \ldots, k$, where δ_{ij} is, as usual, Kronecker's delta. Employing this procedure, any basis for $\mathbb{R}^n$ can be transformed into an orthogonal basis. Hence we can depict, without loss of generality, the motions of solutions of (3) in a space with perpendicular axes.
6. In Chapter 3, an n $\times$ n matrix $\boldsymbol{A}$ with the property $(-1)^{\#J} \cdot \det \boldsymbol{A}_{JJ} > 0$ for every subset set J of $\{1, \ldots, n\}$ was said to be an NP-matrix. Here, an NP-matrix means a complex matrix with the same property.
7. An $n \times n$ matrix $\boldsymbol{A}$ with complex entries is said to be *normal* and *unitary*, respectively, if $\boldsymbol{A}^*\boldsymbol{A} = \boldsymbol{A}\boldsymbol{A}^*$ and if $\boldsymbol{A}\boldsymbol{A}^* = \boldsymbol{I}$. A necessary and sufficient condition for A to be normal is that corresponding to $\boldsymbol{A}$ there exists a unitary matrix $\boldsymbol{U}$ such that $\boldsymbol{A} = \boldsymbol{U}\Lambda\boldsymbol{U}^*$ where Λ is a diagonal matrix consisting of all eigenvalues of A. An elegant proof of the above assertion can be found in Gantmacher [1959 (Volume 1); Chapter IX, Sections 9 and 10].
8. See Subsection 2.3 of Appendix B.
9. Since $\boldsymbol{z}(s)$ is differentiable on [0, 1], the so-called "chain rule of differentiation" is applicable. Precisely stated, the rule is as follows. Let $\boldsymbol{f}$ be an m-dimensional vector function defined on an open set G of $\mathbb{R}^n$ and let g be a real-valued function defined on $\mathbb{R}^m$. Then the composite function $\psi(\boldsymbol{x}) \equiv g(\boldsymbol{f}(\boldsymbol{x}))$ is differentiable on G if $\boldsymbol{f}$ and g are differentiable on the relevant domains, and

$$\nabla\psi(\boldsymbol{x}) = \nabla g(\boldsymbol{f}(\boldsymbol{x})) \cdot J_{\boldsymbol{f}}(\boldsymbol{x}),$$

where

$$J_{\boldsymbol{f}}(\boldsymbol{x}) = \left(\frac{\partial f_i}{\partial x_j}\bigg|_{\boldsymbol{x}}\right) \qquad (j = 1, \ldots, n).$$

10. Geometrically, this is equivalent to saying that the graph of $\boldsymbol{x}^*(t)$ belongs to Γ.
11. In view of the remark concerning Definition 5(8) in Appendix A, Π is clearly bounded, and its closedness follows from Theorem 6 in Appendix A.
12. To verify the continuity of the norm, let ε be any positive number and note that

$$|\|\boldsymbol{x}\| - \|\boldsymbol{y}\|| \leqq \|\boldsymbol{x} - \boldsymbol{y}\|.$$

Then, choosing $\delta = \varepsilon$, it is immediate that for any $\varepsilon > 0$ there exists $\delta > 0$ such that $\|\boldsymbol{x} - \boldsymbol{y}\| < \delta$ implies $|\|\boldsymbol{x}\| - \|\boldsymbol{y}\|| < \varepsilon$, which is simply the desired continuity of the norm.
13. Since no bound on the values of $\partial f_i/\partial \tau$, $i = 1, \ldots, n$, has been assumed, Lemma 9 apparently is inapplicable. However, we are comparing $\boldsymbol{f}(\boldsymbol{\psi}(\tau), \tau)$ with $\boldsymbol{f}(\boldsymbol{x}(\tau), \tau)$ at the same point of time τ; hence (35) still holds *mutatis mutandis* and Lemma 9 is in fact applicable.
14. That the solution of (1) is continuous with respect to t, was shown in the proof of (i).
15. Since the definition of a closed set in Appendix A (Definition 5(3)) is apparently different from that used here, we prove this assertion briefly. Let E be any subset of a metric space. Suppose that E is not closed. Then, by Definition 5(3), there exists a limit point $\boldsymbol{p}$ of E which is not in E. Let $\varepsilon_i = 1/i$, $i = 1, 2, \ldots$. Then, from the definition of a limit point, there exist $\boldsymbol{p}_i \in (B(\boldsymbol{p}, \varepsilon_i) \cap E)$ such that $\boldsymbol{p}_i \neq \boldsymbol{p}$, $i = 1, 2, \ldots$, and the sequence $\{\boldsymbol{p}_i\}$ converges to $\boldsymbol{p}$ as i tends to infinity. Taking the contrapositive of the above, the assertion in the text is immediate.
16. See Definition 2 in Appendix A.
17. From (i), the continuity of $V(\boldsymbol{x}, t)$ is immediate. For the detail, see Definition 14 in Chapter 1.
18. Here, we implicitly postulate that the origin $\mathbf{0}$ is a stationary solution of (1).
19. Since we are interested in local stability, the Liapunov function need not be defined on the whole space $\mathbb{R}^{n+1}$.
20. The inequality $\alpha(\varepsilon_0) \leqq V(\boldsymbol{x}(t_1), t_1)$ is obtained in the following way. Let $\|\boldsymbol{x}(t_1)\| \equiv \varepsilon'$. Then, clearly, $\varepsilon' \geqq \varepsilon_0$. Since $\alpha(\xi)$ is nondecreasing,

$$0 < \alpha(\varepsilon_0) \leqq \alpha(\varepsilon') \leqq V(\boldsymbol{x}(t_1), t_1).$$

21. If $r(R) < \delta(\mu)$ then by letting $\delta(\mu)$ be sufficiently small we can still maintain the nonnegativity of T, for the reduction in $\delta(\mu)$ does not violate (57).
22. Since $V(\boldsymbol{x})$ is continuous and $V(\mathbf{0}) = 0$, for $\eta_\varepsilon > 0$ there exists a $\delta' > 0$ such that if $\|\boldsymbol{x}\| < \delta'$ then $V(\boldsymbol{x}) < \eta_\varepsilon$. Hence δ_ε can be chosen so that $0 < \delta_\varepsilon < \min(\delta', \varepsilon)$.
23. Define $W(\boldsymbol{x}) = V(\boldsymbol{x}) - \eta_\varepsilon$. Then $W(\boldsymbol{x})$ is continuous in $\boldsymbol{x}$. Hence for $\mu = W(\boldsymbol{y}) > 0$ there exists a $\rho > 0$ such that if $\|\boldsymbol{y} - \boldsymbol{z}\| < \rho$ then $|W(\boldsymbol{y}) - W(\boldsymbol{z})| < \mu$, from which it follows that

$$0 = W(\boldsymbol{y}) - \mu < W(\boldsymbol{z}) = V(\boldsymbol{z}) - \eta_\varepsilon.$$

24. The final inequality follows from the assertion (i) above.
25. In this context, it must be noted that we implicitly assume the continuity of each $f_j(\boldsymbol{x})$. As was seen in Theorem 18, this is almost inevitable for the existence of solutions of (1′). Thus, recalling that $V(\boldsymbol{x})$ is assumed to have continuous partial derivatives, $\dot{V}(\boldsymbol{x})$ becomes a continuous function. Hence it surely attains its maximum v' on a compact set $\{\boldsymbol{x}: \delta(c/2) \leqq \|\boldsymbol{x}\| \leqq \varepsilon\}$. Moreover, that $v' < 0$ is clear, because $\dot{V}(\boldsymbol{x}) < 0$ for all $\boldsymbol{x} \neq \mathbf{0}$. Thus defining $v = -v'$, we see that

$$\dot{V}(\boldsymbol{x}) \leq -v < 0 \quad \text{for all } \boldsymbol{x} \in \left\{\boldsymbol{x}: \delta\left(\frac{c}{2}\right) \leqq \|\boldsymbol{x}\| \leqq \varepsilon\right\}.$$

26. The symbol $\boldsymbol{df}(\tau\boldsymbol{x})/\boldsymbol{d}\tau$ simply denotes a vector of which the kth $df_k(\tau\boldsymbol{x})/d\tau = \dot{V}f_k(\tau\boldsymbol{x}) \cdot \boldsymbol{x}$.

27. Let $\xi_1, \ldots, \xi_k$ $(k < n)$ be k linearly independent vectors of order n. Then there exist $\xi_{k+1}, \ldots, \xi_n$ such that $\{\xi_1, \ldots, \xi_n\}$ constitute a basis for an n-dimensional vector space $\mathscr{V}_n$. Here we give a brief outline of the proof. Since the set of vectors $\{\xi_1, \xi_2, \ldots, \xi_k\}$ spans a linear subspacc properly contained in $\mathscr{V}_n$, we can choose a ξ_{k+1} that is not in the linear subspace spanned by $\{\xi_1, \ldots, \xi_k\}$, which we denote by $L\{\xi_1, \ldots, \xi_k\}$. Hence $\xi_1, \ldots, \xi_k, \xi_{k+1}$ are certainly linearly independent, for otherwise, $\xi_{k+1} \in L\{\xi_1, \ldots, \xi_k\}$ a contradiction. Repeating the above procedure, we can find $\xi_{k+1}, \ldots, \xi_n$ such that $\xi_1, \ldots, \xi_n$ are linearly independent and hence constitute a basis for $\mathscr{V}_n$. Needless to say, the matrix $(\xi_1, \ldots, \xi_n)$ is nonsingular.
28. Noticing that the region in which the linear approximation by Taylor expansion holds is limited, our Liapunov function is considered to be defined on a bounded set.
29. See, for example, Gale and Nikaido [1965] and Inada [1971].

5 Neoclassical Economies—Statics

In this chapter we consider some of the properties of a particular class of multisectoral economies, *viz*., those based on "neoclassical" technologies and an arbitrary number of "primary" factors of production. Such economies have been studied in some depth by specialists in the theories of international trade and public finance. However, some of the questions we shall study are more typical of capital theory.

Throughout the chapter our attention is restricted to steady states, that is, states of the economy which, except possibly for a common exponential change of quantities, can be repeated through time. When therefore we speak of the effects of a change in this or that parameter, we must be understood to be comparing alternative steady states, not as describing the path of transition from one steady state to another. The task of describing such transitional paths is reserved for Section 3 of Chapter 6.

1 Neoclassical Technologies

It is supposed that commodities are produced by applying stocks of primary and secondary factors of production. By a *primary factor of production* we mean a stock with a rate of accumulation or decumulation which is entirely independent of human decisions. The definition does not exclude the possibility of *natural* replenishment or decay, where the rates of change depend only on position in time and the current stocks of the primary factors. In fact we shall suppose that all primary factors have the same nonnegative rate of natural change. By a secondary factor of production we mean any non-primary factor. Evidently the class of secondary factors is quite heterogeneous, including stocks of commodities which are producible (machines, buildings, forests, fisheries) and stocks which are depletable (mineral deposits, machines, forests, fisheries). However, we shall confine our attention to those secondary factors which are nondepletable. Thus the distinction between primary and secondary factors reduces to one between producible and nonproducible factors. Henceforth we shall follow convention in contrasting *primary* and *producible* factors. Altogether there are m primary factors and n producible factors.

With each commodity there is associated a production relationship

$$\begin{aligned} x_j &\leqq f^j(v_{1j}, \ldots, v_{mj}; V_{1j}, \ldots, V_{nj}) \\ &= f^j(\boldsymbol{v}^j; \boldsymbol{V}^j) \qquad (j = 1, \ldots, n), \end{aligned} \tag{1}$$

where x_j is the gross output of the jth commodity (gross in that no provision is made for the equipment of the increment of primary-factor endowment), v_{ij} is the amount of the ith primary factor allocated to the production of the jth commodity, and V_{ij} is the amount of the ith producible factor allocated to the production of the jth commodity. The function f^j is supposed to possess continuous second derivatives, to be positively homogeneous of degree one, and to be strictly quasi-concave in its positive inputs.[1] In view of the homogeneity of f^j, if x_j is positive then (1) may be written as

$$\begin{aligned} 1 &\leqq f^j(a_{1j}, \ldots, a_{mj}; A_{1j}, \ldots, A_{nj}) \\ &\equiv f^j(\boldsymbol{a}^j, \boldsymbol{A}^j), \end{aligned} \tag{2}$$

where $a_{ij} \equiv v_{ij}/x_j$ and $A_{ij} \equiv V_{ij}/x_j$. Any vector $(\boldsymbol{a}^j; \boldsymbol{A}^j)$ which satisfies (2) is said to be *feasible* and defines a particular *process* or *activity* in the jth industry. Any set of n processes $(\boldsymbol{a}^j; \boldsymbol{A}^j)$, one for each j constitutes a *technique* of production. Any technique can be expressed as an $(n + m) \times n$ matrix

$$\begin{pmatrix} a_{11} & \cdots & a_{1n} \\ \vdots & & \vdots \\ a_{m1} & \cdots & a_{mn} \\ A_{11} & \cdots & A_{1n} \\ \vdots & & \vdots \\ A_{n1} & \cdots & A_{nn} \end{pmatrix} \equiv \begin{pmatrix} \boldsymbol{a} \\ \boldsymbol{A} \end{pmatrix}. \tag{3}$$

The set of all available techniques constitutes the *technology* of the economy.

Each technique matrix $\binom{\boldsymbol{a}}{\boldsymbol{A}}$ is necessarily nonnegative. It is not required that the production of each commodity absorb a positive amount of each of the $n + m$ factors of production; some of the a_{ij} and A_{ij} may be zero. However, it is required that in each technique the production of each commodity use, directly or indirectly, some of at least one primary factor of production; that is, it is required that for each commodity j there exist a primary factor i such that there can be found a finite chain "production of commodity j requires some of commodity k, production of commodity k requires some of commodity $l, \ldots,$ production of commodity q requires some of primary factor i." It is required also that in each technique each primary factor be needed in the production of at least one commodity.

2 Mapping from Factor-Rentals and Primary-Factor Endowments to Commodity Prices and Outputs

Price-rental relationships

Producers may either purchase or rent the inputs required. Let us suppose that all factors are rented. Then each producer of the jth commodity chooses a process in the light of the prevailing rentals $(w_1, \ldots, w_m; W_1, \ldots, W_n) \equiv (\boldsymbol{w}; \boldsymbol{W})$, where w_i is the rental per unit of time of the ith primary factor of production and W_i is the rental of the ith producible factor. If the technique $\binom{\boldsymbol{a}}{\boldsymbol{A}}$ is chosen at the rentals $(\boldsymbol{w}; \boldsymbol{W}) \geqq \boldsymbol{0}$ then the cost inequality

$$(\boldsymbol{w}; \boldsymbol{W})\begin{pmatrix} \boldsymbol{a}^j \\ \boldsymbol{A}^j \end{pmatrix} \leqq (\boldsymbol{w}; \boldsymbol{W})\begin{pmatrix} \hat{\boldsymbol{a}}^j \\ \hat{\boldsymbol{A}}^j \end{pmatrix} \tag{4}$$

must be satisfied for all alternative processes

$$\begin{pmatrix} \hat{\boldsymbol{a}}^j \\ \hat{\boldsymbol{A}}^j \end{pmatrix}$$

and for all $j, j = 1, \ldots, n$. If $(\boldsymbol{w}; \boldsymbol{W}) > \boldsymbol{0}$ then the inequality (4) is strict. From (4) and the assumed smoothness of f^j,

$$(\boldsymbol{w}; \boldsymbol{W})\begin{pmatrix} \Delta \boldsymbol{a} \\ \Delta \boldsymbol{A} \end{pmatrix} \geqq \boldsymbol{0}, \tag{5}$$

with strict inequality in the jth place if $(\boldsymbol{w}; \boldsymbol{W}) > \boldsymbol{0}$ and

$$\begin{pmatrix} \Delta \boldsymbol{a}^j \\ \Delta \boldsymbol{A}^j \end{pmatrix} \neq \boldsymbol{0}.$$

The cost-minimizing technique will be written sometimes in the explicit form

$$\begin{pmatrix} \boldsymbol{a}(\boldsymbol{w}; \boldsymbol{W}) \\ \boldsymbol{A}(\boldsymbol{w}; \boldsymbol{W}) \end{pmatrix}.$$

Under competitive conditions, with complete freedom of entry, the price p_j of the jth commodity is not greater than the average cost of production π_j; that is,

$$\boldsymbol{p} \leqq \boldsymbol{\pi}(\boldsymbol{w}; \boldsymbol{W}) \equiv (\boldsymbol{w}; \boldsymbol{W})\begin{pmatrix} \boldsymbol{a}(\boldsymbol{w}; \boldsymbol{W}) \\ \boldsymbol{A}(\boldsymbol{w}; \boldsymbol{W}) \end{pmatrix}, \tag{6}$$

where $\boldsymbol{p} \equiv (p_1, \ldots, p_n)$ and $\boldsymbol{\pi} \equiv (\pi_1, \ldots, \pi_n)$. If $(\boldsymbol{w}; \boldsymbol{W}) \geqq \boldsymbol{0}$ then $\boldsymbol{\pi}(\boldsymbol{w}; \boldsymbol{W}) \geqq \boldsymbol{0}$ and is unique. Moreover, since the production of each commodity requires, directly or indirectly, something of at least one primary factor,

$$\boldsymbol{\pi}(\boldsymbol{w}; \boldsymbol{W}) > \boldsymbol{0} \quad \text{if } (\boldsymbol{w}; \boldsymbol{W}) > \boldsymbol{0}. \tag{7a}$$

Finally,

$$\frac{\partial \pi(\boldsymbol{w};\boldsymbol{W})}{\partial w_j} \geq \mathbf{0} \quad (j = 1, \ldots, m),$$

$$\frac{\partial \pi(\boldsymbol{w};\boldsymbol{W})}{\partial W_j} \geqq \mathbf{0} \qquad (j = 1, \ldots, n), \tag{7b}$$

where the vector $\partial\boldsymbol{\pi}/\partial w_j$ is semipositive (rather than merely nonnegative) since each primary factor is needed in the production of some commodity. If our standing assumptions are strengthened by the requirement that each primary factor is needed in the production of each commodity then $\partial\boldsymbol{\pi}/\partial w_j > \mathbf{0}$; if they are strengthened by the requirement that each producible factor be needed, directly or indirectly, in the production of each commodity then $\partial\boldsymbol{\pi}/\partial w_j > \mathbf{0}$ and $\partial\pi/\partial W_j > \mathbf{0}$. Sometimes, when it is understood that all n commodities are produced, the notation $\boldsymbol{p}(\boldsymbol{w};\boldsymbol{W})$ will be employed.

The own-rate of return to owners of the jth producible factor is

$$\rho_j(\boldsymbol{w};\boldsymbol{W}) \equiv \frac{W_j}{p_j(\boldsymbol{w};\boldsymbol{W})} \qquad (j = 1, \ldots, n). \tag{8}$$

Of course, in steady-state equilibrium all own-rates of return are equal:

$$\rho_1(\boldsymbol{w};\boldsymbol{W}) = \cdots = \rho_n(\boldsymbol{w};\boldsymbol{W}) = \rho(\boldsymbol{w};\boldsymbol{W}). \tag{9}$$

For the time being, however, we choose to ignore (9). Now suppose that there are given not $\boldsymbol{w}$ and $\boldsymbol{W}$ but $\boldsymbol{w}$ and $\boldsymbol{\rho}(\boldsymbol{w};\boldsymbol{W})$. Does $(\boldsymbol{w};\boldsymbol{\rho}(\boldsymbol{w};\boldsymbol{W}))$ yield the price vector $\boldsymbol{p}(\boldsymbol{w};\boldsymbol{W})$? The answer is: Yes.[2] Thus we may equivalently take $(\boldsymbol{w};\boldsymbol{W})$ or $(\boldsymbol{w};\boldsymbol{\rho}(\boldsymbol{w};\boldsymbol{W}))$ as given. However, this conclusion must be interpreted with care. While we can take an arbitrary rental vector $(\boldsymbol{w};\boldsymbol{W}) \geqq \mathbf{0}$ and expect it to determine a unique price vector $\boldsymbol{p}(\boldsymbol{w};\boldsymbol{W}) \geqq \mathbf{0}$, we cannot take an arbitrary vector $(\boldsymbol{w};\boldsymbol{\rho})$ of primary-factor rentals and own-rates of return and expect it to determine a unique price vector $\boldsymbol{p}(\boldsymbol{w};\boldsymbol{\rho}) \geqq \mathbf{0}$. The chosen rates of return may be so high that the existing technology cannot support them. What follows from the preceding analysis is that given an arbitrary $\boldsymbol{w} \geqq \mathbf{0}$ there exists a nonempty set $\Gamma(\boldsymbol{w})$ of nonnegative own-rate of return vectors such that, for any $\boldsymbol{\rho} \in \Gamma(\boldsymbol{w})$, the system

$$\boldsymbol{p} = \boldsymbol{w}\boldsymbol{a}(\boldsymbol{w};\boldsymbol{p}[\rho]) + \boldsymbol{p}[\rho]\boldsymbol{A}(\boldsymbol{w};\boldsymbol{p}[\rho]) \tag{10}$$

has a unique and nonnegative solution $\boldsymbol{p}(\boldsymbol{w};\boldsymbol{\rho})$. (In (10), $[\rho]$ is the diagonal matrix with ρ_i in the ith diagonal place.) If the given $\boldsymbol{w}$ is positive then, for any $\boldsymbol{\rho} \in \Gamma(\boldsymbol{w})$, the solution $\boldsymbol{p}(\boldsymbol{w};\boldsymbol{\rho})$ also is positive.

In the special case in which all own-rates of return are constrained to be equal, so that $\rho_1 = \cdots = \rho_n = \rho$, say, it is possible to be more specific about $\Gamma(\boldsymbol{w})$: It is a nondegenerate interval in $\mathbb{R}_+$. The following proof follows Burmeister and Kuga [1970]. Consider the sets

$$\Omega \equiv \{\boldsymbol{p}: 0 \leqq p_i \leqq 1, i = 1, \ldots, n\}$$

and

$$\bar{\Gamma} \equiv \{\rho: \rho > 0 \text{ and there exists } \boldsymbol{w} > \boldsymbol{0} \text{ such that } \pi_i(\boldsymbol{w}; \rho\boldsymbol{p}) \leqq 1$$

for all $i = 1, \ldots, n$ and all $\boldsymbol{p} \in \Omega\}$.

Now suppose that $\boldsymbol{w} > \boldsymbol{0}$ and $\rho > 0$ are given. Then, since π_i is an increasing function of p_i, $i = 1, \ldots, n$,

$$\pi_i(\boldsymbol{w}; \rho\boldsymbol{p}) \leqq \pi_i(\boldsymbol{w}; \rho(1, \ldots, 1)) \equiv \bar{\pi}_i(\boldsymbol{w}; \rho)$$

Let $\lambda \equiv 1/\max_i \bar{\pi}_i > 0$ and $(\boldsymbol{w}', \rho') \equiv \lambda \cdot (\boldsymbol{w}, \rho)$. Then since π_i is homogeneous of degree one,

$$\pi_i(\boldsymbol{w}'; \rho'\boldsymbol{p}) = \lambda\pi_i(\boldsymbol{w}; \rho\boldsymbol{p}) \leqq \lambda\bar{\pi}_i \leqq 1 \quad \text{for all } \boldsymbol{p} \in \Omega \text{ and all } i = 1, \ldots, n.$$

Thus $\bar{\Gamma}$ is not empty. Now for any choice of $\rho \in \bar{\Gamma}$ and a suitable choice of $\boldsymbol{w} > \boldsymbol{0}$, $\pi_i(\boldsymbol{w}; \rho\boldsymbol{p})$, $i = 1, \ldots, n$, defines a mapping from Ω into itself. Moreover, the π_i are continuous functions defined on the convex set Ω. Hence, from Brouwer's fixed point theorem, there exists a fixed point $\boldsymbol{p}^*$ such that

$$p_i^* = \pi_i(\boldsymbol{w}; \rho\boldsymbol{p}^*) \qquad (i = 1, \ldots, n).$$

Again recalling the homogeneity of the π_i, we conclude that $\Gamma(\boldsymbol{w})$ contains a positive element. Finally, from the preceding argument, if ρ is in $\Gamma(\boldsymbol{w})$ then so is any number in the closed interval $[0, \rho]$.

In concluding this subsection we note briefly the even more special case in which all own-rates of return are equal and in which $m = 1$. Suppose that w, the rental of the sole primary factor, is positive. Then, clearly, the set Γ is independent of w. We conclude that for any $w > 0$ and any ρ less than some positive value ρ^* (which may be infinite) there exists a uniquely determined, positive, equilibrium vector of commodity prices.

Generalized factor-price frontiers

Confronted by $\boldsymbol{w} > \boldsymbol{0}$ and $\boldsymbol{\rho} \in \Gamma(\boldsymbol{w})$, producers will choose feasible input-output matrices $\boldsymbol{a}$ and $\boldsymbol{A}$ that, component by component, minimize

$$\begin{aligned} \boldsymbol{p} &= \boldsymbol{\pi}(\boldsymbol{w}; \boldsymbol{\rho}) \\ &= \boldsymbol{wa}(\boldsymbol{I} - [\rho]\boldsymbol{A})^{-1} \\ &= \boldsymbol{wG}(\boldsymbol{w}; \boldsymbol{\rho}), \end{aligned} \tag{11}$$

say. The equilibrium real rental of the ith primary factor in terms of the jth commodity is then

$$\frac{w_i}{\pi_j(\boldsymbol{w}; \boldsymbol{\rho})} \equiv \omega_{ij}(\boldsymbol{w}; \boldsymbol{\rho}) \qquad (i = 1, \ldots, m; j = 1, \ldots, n). \tag{12}$$

The functions ω_{ij} provide a generalization of the familiar factor-price frontier of two-sector models. (See Samuelson [1957; 1962].) What are the properties of the generalized frontier? Let $\boldsymbol{\omega}_j \equiv (\omega_{1j}, \ldots, \omega_{mj})$. It will be shown that *the frontier* (in $(\boldsymbol{p}, \boldsymbol{\omega}_j)$-space) *is nonpositively sloped in all directions and that, for given $\boldsymbol{p}$, it is quasi-concave in $\boldsymbol{\omega}_j$*. (This proposition was first stated by Samuelson [1975; 1976]. The following proof is based on Burmeister [1975].)

From the strict quasi-concavity of the individual production functions and the fact that each primary factor is used in the production of some commodity, we may deduce that, for any $\boldsymbol{w}, \boldsymbol{w}' \geqq \boldsymbol{0}$,

$$\boldsymbol{\pi}(\boldsymbol{w}; \boldsymbol{p}) = \boldsymbol{w}\boldsymbol{G}(\boldsymbol{w}; \boldsymbol{p}) \leq \boldsymbol{w}\boldsymbol{G}(\boldsymbol{w}'; \boldsymbol{P}) \text{ if } \boldsymbol{w} \neq \boldsymbol{w}'.$$

Hence, for $0 < \lambda < 1$,

$$\begin{aligned}
&\lambda\boldsymbol{\pi}(\boldsymbol{w}; \boldsymbol{p}) + (1 - \lambda)\boldsymbol{\pi}(\boldsymbol{w}'; \boldsymbol{p}) \\
&= \lambda\boldsymbol{w}\boldsymbol{G}(\boldsymbol{w}; \boldsymbol{p}) + (1 - \lambda)\boldsymbol{w}'\boldsymbol{G}(\boldsymbol{w}'; \boldsymbol{p}) \\
&\leq \lambda\boldsymbol{w}\boldsymbol{G}(\lambda\boldsymbol{w} + (1 - \lambda)\boldsymbol{w}'; \boldsymbol{p}) + (1 - \lambda)\boldsymbol{w}'\boldsymbol{G}(\lambda\boldsymbol{w} + (1 - \lambda)\boldsymbol{w}'; \boldsymbol{p}) \\
&= (\lambda\boldsymbol{w} + (1 - \lambda)\boldsymbol{w}')\boldsymbol{G}(\lambda\boldsymbol{w} + (1 - \lambda)\boldsymbol{w}'; \boldsymbol{p}) \\
&= \boldsymbol{\pi}(\lambda\boldsymbol{w} + (1 - \lambda)\boldsymbol{w}'; \boldsymbol{p}).
\end{aligned}$$

That is, given $\boldsymbol{p}$, $\boldsymbol{\pi}$ is a concave function of $\boldsymbol{w}$. Now consider the equilibrium condition

$$p_i = \pi_i(\boldsymbol{w}; \boldsymbol{p}) \qquad (i = 1, \ldots, n). \tag{13}$$

Holding constant all variables except w_j and w_k, where the jth primary factor is essential to the production of the ith commodity, we obtain, first,

$$\frac{dw_j}{dw_k} = -\frac{\partial \pi_i}{\partial w_k}\bigg/\frac{\partial \pi_i}{\partial w_j} \tag{14}$$

and then,

$$\frac{d^2 w_j}{dw_k^2} = -\left(\frac{\partial \pi_i}{\partial w_j}\right)^{-1}\begin{pmatrix}\dfrac{dw_j}{dw_k} & 1\end{pmatrix}\begin{pmatrix}\dfrac{\partial^2 \pi_i}{\partial w_j^2} & \dfrac{\partial^2 \pi_i}{\partial w_j \partial w_k} \\ \dfrac{\partial^2 \pi_i}{\partial w_k \partial w_j} & \dfrac{\partial^2 \pi_i}{\partial w_k^2}\end{pmatrix}\begin{pmatrix}\dfrac{dw_j}{dw_k} \\ 1\end{pmatrix}. \tag{15}$$

Now $\partial \pi_i/\partial w_k \geqq 0$ and $\partial \pi_i/\partial w_j > 0$; moreover, π_i is a concave function of w_j and w_k, so that the Hessian matrix on the right-hand side of (15) is nonpositive definite. Hence

$$\frac{dw_j}{dw_k} \leqq 0 \quad \text{and} \quad \frac{d^2 w_j}{dw_k^2} \geqq 0. \tag{16}$$

If our standing assumptions are strengthened by the condition that either each primary factor be required in the production of every commodity or each producible factor be required, directly or indirectly, in the production of every commodity (so that $\boldsymbol{A}$ is power-positive of degree not greater than n)

then the weak inequalities of (16) become strict. It remains only to consider the slopes of the frontier in other directions. From (13),

$$dp_i = \sum_j \frac{\partial \pi_i}{\partial w_j} dw_j + \sum_j \frac{\partial \pi_i}{\partial \rho_k} d\rho_k. \tag{17}$$

Equating to zero both dp_i and all but two of the dw_j and $d\rho_k$, and noting that $\partial\pi_i/\partial w_j \geqq 0$, $\partial\pi_i/\partial\rho_k \geqq 0$, it may be verified that all slopes are nonpositive. (It does not seem possible without additional assumptions to say anything about terms like $d^2\rho_j/d\rho_k^2$.)

Before leaving the generalized factor-price frontiers we briefly examine the manner in which they respond to simple kinds of technical improvement. To this end, let us rewrite $\boldsymbol{a}$ and $\boldsymbol{A}$ as functions of $\boldsymbol{w}$, $\boldsymbol{W}$, and α, where α is a shift parameter whose increases indicate technical improvements. Differentiating (10) with respect to α, holding $\boldsymbol{w}$ and $\boldsymbol{W}$ constant, we obtain, after applying (5) and collating terms involving $d\boldsymbol{p}$,

$$d\boldsymbol{p}(\boldsymbol{I} - [\rho]\boldsymbol{A}) = \left(\boldsymbol{w}\frac{\partial \boldsymbol{a}}{\partial\alpha} + \boldsymbol{p}[\rho]\frac{\partial \boldsymbol{A}}{\partial\alpha}\right)d\alpha, \tag{18}$$

where $\partial\binom{\boldsymbol{a}}{\boldsymbol{A}}/\partial\alpha \leq \boldsymbol{0}$. Evidently $d\boldsymbol{p}/d\alpha \leq \boldsymbol{0}$ so that, in a weak sense, every real rental increases. If only a single element of $\binom{\boldsymbol{a}}{\boldsymbol{A}}$, say a_{ij}, is affected by the change in α, implying that $a_{ij} > 0$, then (18) reduces to

$$d\boldsymbol{p}(\boldsymbol{I} - [\rho]\boldsymbol{A}) = \left(0, \ldots, 0, w_i\left(\frac{\partial a_{ij}}{\partial\alpha}\right), 0, \ldots, 0\right)d\alpha. \tag{19}$$

It follows that $dp_j/d\alpha < 0$ and that if the production of each commodity requires something of each producible factor then $d\boldsymbol{p}/d\alpha < \boldsymbol{0}$. Similarly, if only A_{ij} is affected by the change in α, implying that $A_{ij} > 0$, then (18) reduces to

$$d\boldsymbol{p}(\boldsymbol{I} - [\rho]\boldsymbol{A}) = \left(0, \ldots, 0, p_i\rho_i\left(\frac{\partial A_{ij}}{\partial\alpha}\right), 0, \ldots, 0\right)d\alpha, \tag{20}$$

whence $dp_j < 0$ and, if the production of each commodity requires something of each producible factor, $d\boldsymbol{p}/d\alpha < \boldsymbol{0}$. Of course, if $d\boldsymbol{p} < \boldsymbol{0}$ then each real factor-rental unambiguously increases.

Relationships between endowments and net outputs

Let us continue to take $\boldsymbol{w}$ and $\boldsymbol{W}$ as given, so that $\boldsymbol{a}(\boldsymbol{w}; \boldsymbol{W})$ and $\boldsymbol{A}(\boldsymbol{w}; \boldsymbol{W})$ also are given. If $\boldsymbol{w}$ and $\boldsymbol{W}$ are consistent with the full employment of all primary factors, we have

$$\begin{aligned} \boldsymbol{a}(\boldsymbol{w}; \boldsymbol{W})\boldsymbol{x} &= \boldsymbol{v}, \\ \boldsymbol{x} &\geq \boldsymbol{0}, \\ \boldsymbol{v} &> \boldsymbol{0}, \end{aligned} \tag{21}$$

where $\boldsymbol{x} \equiv (x_1, \ldots, x_n)$ is the vector of gross outputs and $\boldsymbol{v} \equiv (v_1, \ldots, v_m)$ is the vector of available amounts (endowments) of the primary factors. In a steady state, $\boldsymbol{x}$ and $\boldsymbol{v}$ grow at the same rate $g \geqq 0$. Moreover, the stocks of producible factors are adjusted to requirements, determined in turn by $\boldsymbol{w}$ and $\boldsymbol{W}$ and by g; thus

$$A(\boldsymbol{w}; \boldsymbol{W})\boldsymbol{x} = \boldsymbol{V}, \tag{22}$$

where $\boldsymbol{V} \equiv (V_1, \ldots, V_m)$ is the vector of stocks of producible inputs and grows at the same rate as $\boldsymbol{x}$ and $\boldsymbol{v}$. The net-output (or, in a closed economy, the consumption) vector $\boldsymbol{y} = (y_1, \ldots, y_n)$ is given by

$$\boldsymbol{y} = \boldsymbol{x} - g\boldsymbol{V}. \tag{23}$$

Finally, from (22) and (23),

$$\boldsymbol{x} = (\boldsymbol{I} - gA)^{-1}\boldsymbol{y}$$

so that, in view of (21),

$$\boldsymbol{v} = \boldsymbol{a}(\boldsymbol{w}; \boldsymbol{W})\boldsymbol{x} = \boldsymbol{a}(\boldsymbol{w}; \boldsymbol{W})(\boldsymbol{I} - gA(\boldsymbol{w}; \boldsymbol{W}))^{-1}\boldsymbol{y}. \tag{24}$$

From Theorem 2 of Chapter 2, $(\boldsymbol{I} - gA)^{-1} \geq \boldsymbol{0}$ if and only if $g < (1/\lambda^*)$, where λ^* is the Frobenius root of A; under the same condition, therefore, $\boldsymbol{a}(\boldsymbol{I} - gA)^{-1} \geqq \boldsymbol{0}$. Now $\boldsymbol{a}(\boldsymbol{I} - gA)^{-1}$ is the matrix of direct plus indirect primary-factor requirements per unit of output; and the production of each commodity requires, directly or indirectly, something of at least one primary factor. Hence $\boldsymbol{a}(\boldsymbol{I} - gA)^{-1}$ contains at least one positive element in each column, with the reassuring but not surprising implication that an increase in the net output of any commodity requires an increase in the endowment of at least one primary factor. If we impose the strong additional assumption that the production of each commodity requires something of each producible factor then $\boldsymbol{a}(\boldsymbol{w}; \boldsymbol{W})(\boldsymbol{I} - gA(\boldsymbol{w}; \boldsymbol{W})) > \boldsymbol{0}$, implying that an increase in the net output of any commodity requires an increase in the endowment of every primary factor.

It may or may not be possible to solve (21) uniquely for $\boldsymbol{x}$, given $\boldsymbol{w}$ and $\boldsymbol{W}$. However that may be, we can consider the implications for outputs of a change in primary-factor endowments. Suppose that initially it is not possible to express $\boldsymbol{v}$ as a positive linear combination of less than s, $1 < s \leqq \min\{m, n\}$, columns of $\boldsymbol{a}(\boldsymbol{w}; \boldsymbol{W})$. Then, after a small increase in the endowment of any primary factor, say the ith, we have

$$d\boldsymbol{v} = \boldsymbol{a}(\boldsymbol{w}; \boldsymbol{W})\Delta\boldsymbol{x}, \tag{25a}$$

$$\boldsymbol{x} + \Delta\boldsymbol{x} \quad \geq \boldsymbol{0}, \tag{25b}$$

where $d\boldsymbol{v} \equiv (0, \ldots, 0, dv_i, 0, \ldots, 0)$ and (25b) contains at least s strict inequalities. Since $dv_i > 0$ at least one output increases. Moreover, if in some member of (25a) other than the ith at least one output that increases appears with a positive coefficient then at least one output must contract; in particular, if $\boldsymbol{a} > \boldsymbol{0}$ then at least one output contracts.

In the special case in which $m = n$ and $s = n$, (25) contains exactly n equations, $\boldsymbol{a}^{-1}$ exists, and

$$d\boldsymbol{x} = \boldsymbol{a}^{-1}d\boldsymbol{v}. \tag{26}$$

(Notice that $\Delta \boldsymbol{x}$ is now "small.") To this special case one can apply Theorems 6 and 7 of Chapter 3. Thus if Condition [3] (or Condition [4]) of Chapter 3 is imposed on $\boldsymbol{a}(\boldsymbol{w}; \boldsymbol{W})$ then there exists a numbering of products and primary factors such that an increase in the endowment of the ith primary factor causes the output of the ith commodity to increase or remain unchanged (respectively, decrease or remain unchanged) and all other outputs to decrease or remain unchanged (respectively, increase or remain unchanged). (When $m = n = s = 2$ this proposition is known as the Samuelson–Rybczynski Theorem. See Stolper and Samuelson [1941], Samuelson [1953], and Rybczynski [1955].) Thus, for the price of Condition [3] (or Condition [4]), there is achieved a very considerable sharpening of earlier conclusions. However, when $m = n = s > 2$, Conditions [3] and [4] are quite strict. For example, Condition [3] requires that for *any* nontrivial J and *any* set of positive outputs $\bar{x}_i$, $i \in \bar{J}$, there exist a set of positive outputs x_i, $i \in J$, such that more of the jth primary factor, $j \in J$, and less of the jth primary factor, $j \in \bar{J}$, be used in producing the composite good J than in producing the composite good $\bar{J}$. (For more detailed analysis and interpretation the reader is referred to Uekawa, Kemp, and Wegge [1973]. He also may consult Uekawa [1971], Ethier [1974], and Kemp and Wan [1976].)

Before leaving Equation (21) we briefly note the implications of small autonomous changes in the rentals of primary factors and of small technical improvements in the use of primary factors. From (21),

$$d\boldsymbol{a} \cdot \boldsymbol{x} + \boldsymbol{a} \cdot \Delta \boldsymbol{x} = \boldsymbol{0}. \tag{27}$$

If the disturbance consists in an increase in the rental of the ith primary factor, while all other factor-rentals and the technology remain unchanged, then the ith row of $d\boldsymbol{a}$ is seminegative; moreover, at least one negative term must be associated with a positive element of $\boldsymbol{x}$, so that $(d\boldsymbol{a})_i \cdot \boldsymbol{x} < 0$. It follows that $(\boldsymbol{a} \cdot \Delta \boldsymbol{x})_i > 0$, so that at least one output must increase. If, on the other hand, the disturbance consists in an autonomous technical improvement in the use of primary factors of production, we have $d\boldsymbol{a} = (\partial \boldsymbol{a}/\partial \alpha)d\alpha \leq \boldsymbol{0}$. If at least one of the negative entries in $\partial \boldsymbol{a}/\partial \alpha$ is associated with a positive output then $d\boldsymbol{a} \cdot \boldsymbol{x} \leq \boldsymbol{0}$ and, not surprisingly, at least one output must increase. In the special case in which the improvement is confined to the jth industry, (27) reduces to

$$d\boldsymbol{a}^j \cdot x_j + \boldsymbol{a} \cdot \Delta \boldsymbol{x} = \boldsymbol{0}. \tag{28}$$

From (28), if some output other than the jth increases then some other output must decrease.

Let us turn to Equation (24), and suppose that $g < (1/\lambda^*)$. In general outline, the analysis of Equation (21) can be repeated. Thus it may or may not be possible to solve (24) uniquely for $\boldsymbol{y}$. However, it is always true that an increase in dv_i gives rise to an increase in the net output of at least one commodity; and, if $\boldsymbol{a}(\boldsymbol{w}; \boldsymbol{W})(\boldsymbol{I} - g\boldsymbol{A}(\boldsymbol{w}; \boldsymbol{W}))^{-1}$ is positive (as when the production of each commodity requires something of each producible factor) then the net output of at least one commodity must fall. In the special case in which $m = n = s$,

$$d\boldsymbol{y} = (\boldsymbol{I} - g\boldsymbol{A})\boldsymbol{a}^{-1}\, d\boldsymbol{v}. \tag{29}$$

Then Corollary 5 of Chapter 3 can be applied, for $\boldsymbol{I} - g\boldsymbol{A}$ (with $g < 1/\lambda^*$) is a Leontief matrix. The details are left to the reader.

3 Mapping from Commodity Prices and Own-Rates of Return to Primary-Factor Rentals

In Section 2 we took as given the vector of factor-rentals and the vector of primary-factor endowments and explored the implications for commodity prices of variations in factor-rentals and for outputs of variations in endowments. In the present section we turn the procedure about, taking commodity prices as given and setting free the rentals of primary factors. Thus the given data now include commodity prices, the rentals of producible factors, and the endowments of primary factors; and the dependent variables include the rentals of primary factors.

We consider first the implications for primary-factor rentals of changes in commodity prices. Of course, we cannot hold constant both $\boldsymbol{W}$ and $\boldsymbol{\rho}$ while $\boldsymbol{p}$ is manipulated; a choice must be made. Throughout our calculations, $\boldsymbol{\rho}$ is held constant, implying that $\boldsymbol{W}$ adjusts to variations in $\boldsymbol{p}$. It is supposed that, given $\boldsymbol{p}$ and $\boldsymbol{\rho}$, there exists a solution $\boldsymbol{w}^0 > \boldsymbol{0}$ to the system

$$\boldsymbol{p} \leqq \boldsymbol{w}\boldsymbol{a}(\boldsymbol{w}; \boldsymbol{p}[\rho]) + \boldsymbol{p}[\rho]\boldsymbol{A}(\boldsymbol{w}; \boldsymbol{p}[\rho]) \tag{30}$$

with strict equalities everywhere; it may or may not be the only such solution. And it is supposed that after some small change in commodity prices there exists a solution $\boldsymbol{w}^0 + \Delta\boldsymbol{w}^0 > \boldsymbol{0}$ such that at least s, $s > 1$, commodities continue in production and $\Delta\boldsymbol{w}^0$ is not necessarily small. From (30), noting (4) and dropping the superscript from $\boldsymbol{w}$,

$$d\boldsymbol{p} \leqq \Delta\boldsymbol{w} \cdot \boldsymbol{a}(\boldsymbol{w}; \boldsymbol{p}[\rho]) + d\boldsymbol{p} \cdot [\rho] \cdot \boldsymbol{A}(\boldsymbol{w}; \boldsymbol{p}[\rho]) \tag{31}$$

with at least s strict equalities. From (31)

$$d\boldsymbol{p} \leqq \Delta\boldsymbol{w} \cdot \boldsymbol{a}(\boldsymbol{w}; \boldsymbol{p}[\rho])(\boldsymbol{I} - [\rho]\boldsymbol{A}(\boldsymbol{w}; \boldsymbol{p}[\rho]))^{-1} \tag{32}$$

since $(\boldsymbol{I} - [\rho]\boldsymbol{A})^{-1} \geq \mathbf{0}$. Now let us consider the special case in which only the jth commodity price changes, with $dp_j > 0$. Then the jth inequality (31) holds as a strict equality and we can be sure that

$$\frac{\Delta w_k}{w_k} \geqq \frac{dp_j}{p_j} > 0 \quad \text{for some } k. \tag{33}$$

Moreover, since at least one other member of (31) holds as a strict equality,

$$\frac{\Delta w_l}{w_l} \leqq 0 \quad \text{for some } l. \tag{34}$$

Thus at least one primary-factor rental increases in terms of some commodity and decreases in terms of none; and at least one primary-factor rental falls in terms of one commodity and increases in terms of none. These conclusions can be strengthened if it is supposed that each primary factor is used, directly or indirectly, in the production of every commodity, so that

$$\boldsymbol{a}(\boldsymbol{I} - [\rho]\boldsymbol{A})^{-1} > \mathbf{0}.$$

For then (33) implies that the inequality of (34) is strict, and a strict inequality in (34) implies that the inequality of (33) must be strict.

In the special case in which $m = n = s$, $\boldsymbol{w}^0$ is locally unique, all members of (31) hold as strict equalities, and (31) can be solved uniquely:

$$d\boldsymbol{w} = d\boldsymbol{p}(\boldsymbol{I} - [\rho]\boldsymbol{A}(\boldsymbol{w}; \boldsymbol{p}[\rho]))\boldsymbol{a}(\boldsymbol{w}; \boldsymbol{p}[\rho])^{-1}. \tag{35}$$

(Notice that $\Delta \boldsymbol{w}$ is now small). In this case, Corollary 5 of Chapter 3 can be applied, for $\boldsymbol{I} - [\rho]\boldsymbol{A}$ is a Leontief matrix. The details are left to the reader.

Let us turn now to the implications for primary-factor rentals of small changes in the vector of own-rates of return or, equivalently, in the vector of rentals for producible factors. From (30)

$$\mathbf{0} \leqq \Delta \boldsymbol{w} \cdot \boldsymbol{a}(\boldsymbol{w}; \boldsymbol{p}[\rho]) + d\boldsymbol{W} \cdot \boldsymbol{A}(\boldsymbol{w}; \boldsymbol{p}[\rho]), \tag{36}$$

with strict equalities in at least s places. Suppose that W_j (or ρ_j) increases, with all other W's constant. If the jth producible factor is used in producing not more than $n - s$ commodities, the only effect of the change may be to drive out those industries which hire the jth producible factor, with no change in the vector of primary-factor rentals. Suppose then that the jth producible factor is used in at least $n - s$ industries. At least one of those industries remains active in the new equilibrium and is associated with a strict equality in (36). It follows that the nominal rental of at least one primary factor must fall, implying that the real rental of that factor, in terms of whatever *numéraire*, must also fall. It seems to be impossible to advance beyond this rather obvious conclusion without imposing extremely strong restrictions on the technology or retreating to special cases in which n and/or m are small. (See, for example, Kemp [1976, Chapter 5.])

4 Mapping from Commodity Prices and Own-Rates of Return to Gross and Net Outputs

We have examined some of the properties of mappings from prices to prices and from physical quantities to physical quantities. In the present section we turn our attention to "mixed" mappings from prices to quantities—in particular, from commodity prices to gross and net outputs. At the same time, we examine some of the properties of the loci of short-run and long run (or, gross and net) production possibilities.

With any vector of primary-factor endowments $\boldsymbol{v} > \mathbf{0}$ and any vector of producible factors $V > \mathbf{0}$ there may be associated a short-run locus of production possibilities defined by

$$\begin{aligned} y_1 + \dot{V}_1 = x_1 &= F(x_2, \ldots, x_n; V_1, \ldots, V_n; v_1, \ldots, v_m) \\ &= F(y_2 + \dot{V}_2, \ldots, y_n + \dot{V}_n; V_1, \ldots, V_n; v_1, \ldots, v_m), \end{aligned} \tag{37}$$

where F $(= x_1 = y_1 + \dot{V}_1)$ is the largest attainable output of the first commodity, given the (feasible) outputs $x_i = y_i + \dot{V}_i$ $(i = 2, \ldots, n)$ of the remaining commodities. As is well known,[3] the function F is concave, nonpositively sloped with respect to outputs and nonnegatively sloped with respect to inputs, and homogeneous of the first degree; moreover, if both the ith and jth industries are active then, under competitive conditions,

$$\frac{dx_i}{dx_j} = -\frac{\partial F/\partial x_j}{\partial F/\partial x_i} = -\frac{p_j}{p_i} \qquad (i, j = 1, \ldots, n). \tag{38}$$

In general, the function F is only weakly concave. Necessary and sufficient conditions for strict concavity have been provided by Khang [1971]; see also Khang and Uekawa [1973]. We proceed to a general analysis, the outcome of which will be necessary and sufficient conditions for any degree of (local) flatness of the locus of short-run production possibilities. The analysis is based on that of Kemp, Khang and Uekawa [1975].

A preliminary remark is needed. In the short run, with stocks of producible factors given, the distinction between primary and producible factors of production evaporates. We therefore may dispense with the distinction, treat all factors as primary, and write the jth production function simply as

$$x_j = g^j(\boldsymbol{v}^j) \qquad (j = 1, \ldots, n). \tag{1'}$$

Under this convention the number of primary factors m necessarily exceeds the number of producible commodities n. We can now state

Theorem 1. *Let $\mathbf{x}^*$ be a point in the locus of short-run production possibilities with n^0, $n^0 \leqq n$, industries active. Then the locus contains an $(n^0 - r)$-dimensional flat embracing $\mathbf{x}^*$ if and only if at $\mathbf{x}^*$ there are exactly r linearly independent vectors of primary-factor inputs.*

Remark. The plausibility of the proposition can be established by reflecting on the following familiar cases. (i) When $n = 2$ and $m = 1$ (the Torrens–Ricardo case), the locus is a straight line in $\mathbb{E}^2$, that is, the locus is a flat of dimension $n - r = n - m = 1$. (ii) When $n = 3$ and $m = 1$, the locus is a plane in $\mathbb{E}^3$, that is, it is a flat of dimension $n - r = n - m = 2$. (iii) When $n = 3$ and $m = 2$, and when at no point in the locus do the three industries share a common ratio of factor inputs, the locus is ruled or lined, that is, consists of flats of dimension $n - r = n - m = 1$. (See, for example, Kemp [1964, Chapter 7] and Melvin [1968].)

PROOF (*Sufficiency*). We have

$$\boldsymbol{x}^* \equiv (x_1^*, \ldots, x_{n^0}^*; x_{n^0+1}^*, \ldots, x_n^*) \equiv (\boldsymbol{x}_{\mathrm{I}}^*; \boldsymbol{x}_{\mathrm{II}}^*),$$

where $x_j^* = f^j(\boldsymbol{v}^{j*})$ and $\boldsymbol{v}^{j*} \equiv (v_{1j}^*, \ldots, v_{mj}^*)$, $j = 1, \ldots, n$, and where $f^j(\boldsymbol{v}^{j*})$ is positive for $1 \leqq j \leqq n^0$ and zero for $n^0 < j \leqq n$. Let

$$\boldsymbol{Z}^* \equiv (\boldsymbol{v}^{1*}, \ldots, \boldsymbol{v}^{n^0*})$$

have rank r, $r \leqq n^0 < n < m$. Now consider the system of homogeneous linear equations

$$\boldsymbol{Z}^*\boldsymbol{c} = \boldsymbol{0}. \tag{39}$$

The solution space in $\mathbb{E}^{n^0}$ is spanned by $n^0 - r$ independent vectors, say $\boldsymbol{c}^1, \ldots, \boldsymbol{c}^{n^0-r}$. Corresponding to any one of these vectors, say $\boldsymbol{c}^h$ (with jth component c_j^h), we may define the new input vectors

$$\boldsymbol{v}^{jh} \equiv (1 + \delta_h c_j^h)\boldsymbol{v}^{j*} \qquad (j = 1, \ldots, n^0), \tag{40}$$

where the scalar δ_h is so chosen that $1 + \delta_h c_j^h > 0$ and, therefore, $\boldsymbol{v}^{jh} \geqq \boldsymbol{0}$. Corresponding to the new input vectors are the outputs

$$\begin{aligned} x_j^h &= f^j(\boldsymbol{v}^{jh}) \\ &= (1 + \delta_h c_j^h) f^j(\boldsymbol{v}^{j*}) \quad \text{[from (40) and the homogeneity of } f^j] \\ &= x_j^* + \delta_h c_j^h x_j^* \qquad (j = 1, \ldots, n^0). \end{aligned}$$

In matrix notation,

$$\boldsymbol{x}_{\mathrm{I}}^h = \boldsymbol{x}_{\mathrm{I}}^* + \delta_h [x_{\mathrm{I}}^*] \boldsymbol{c}^h, \tag{41}$$

where $[x_I^*]$ is the $n^0 \times n^0$ diagonal matrix with x_j^* in the jth diagonal place, $j = 1, \ldots, n^0$. We note that

$$\sum_{j=1}^{n^0} \boldsymbol{v}^{jh} = \sum_{j=1}^{n^0} (1 + \delta_h c_j^h)\boldsymbol{v}^{j*} + \delta_h \sum_{j=1}^{n^0} c_j^h \boldsymbol{v}^{j*} = \boldsymbol{v},$$

implying that $\boldsymbol{x}^h$ is feasible.

Since the production set is convex, there is associated with $\boldsymbol{x}_{\mathrm{I}}^*$ a vector $\boldsymbol{p}_{\mathrm{I}}^*$ of shadow prices such that

$$0 \geqq \boldsymbol{p}_{\mathrm{I}}^*(\boldsymbol{x}^h - \boldsymbol{x}^*) = \delta_h \boldsymbol{p}_{\mathrm{I}}^*[x_{\mathrm{I}}^*]\boldsymbol{c}^h.$$

Since δ_h can be of either sign, $\boldsymbol{p}_{\mathrm{I}}^* \boldsymbol{x}_{\mathrm{I}}^h = \boldsymbol{p}^* \boldsymbol{x}_{\mathrm{I}}^*$. Thus the line segment $\overline{\boldsymbol{x}^* \boldsymbol{x}^h}$ traced by δ_h, $1 + \delta_h c_j^h > 0$, lies in the locus of production possibilities. Since h can take any of the values $1, \ldots, n^0 - r$, we have generated $n^0 - r$ such line segments. To complete the proof of sufficiency we must show that these segments are linearly independent or, equivalently, that the $n^0 - r$ vectors $(\boldsymbol{x}^h - \boldsymbol{x}^*)$ are linearly independent.

From (41),

$$\boldsymbol{x}_{\mathrm{I}}^h - \boldsymbol{x}_{\mathrm{I}}^* = \delta_h[x_{\mathrm{I}}^*]\boldsymbol{c}^h \qquad (h = 1, \ldots, n^0 - r), \tag{42}$$

that is, $(\boldsymbol{x}_{\mathrm{I}}^h - \boldsymbol{x}_{\mathrm{I}}^*)$ is a linear transformation of $\boldsymbol{c}^h$. Evidently $[x_{\mathrm{I}}^*]$ is of full rank. Moreover, the $\boldsymbol{c}^h$, $h = 1, \ldots, n^0 - r$, are linearly independent. Hence the $(\boldsymbol{x}_{\mathrm{I}}^h - \boldsymbol{x}_{\mathrm{I}}^*)$, and therefore the $(\boldsymbol{x}^h - \boldsymbol{x}^*)$, $h = 1, \ldots, n^0 - r$, are linearly independent.

(*Necessity*). Consider any point $\boldsymbol{x}^* \equiv (\boldsymbol{x}_{\mathrm{I}}^*, \boldsymbol{x}_{\mathrm{II}}^*)$ in an $(n^0 - r)$-dimensional flat of the transformation surface. Let $\boldsymbol{Z}^* = (\boldsymbol{v}_1^*, \ldots, \boldsymbol{v}_{n^0}^*)$ be the matrix of primary-factor inputs to the n^0 active industries. Suppose that the rank of $\boldsymbol{Z}^*$ is s, $s \neq r$. Then, from the sufficiency part of the theorem, the dimension of the flat must be $n^0 - s$, a contradiction. □

That concludes our analysis of the locus of *short-run* production possibilities, defined for given $\boldsymbol{v}$ and given V. We turn our attention now to a companion concept, the locus of steady-state or long-run net outputs. In a steady state, $\boldsymbol{V}$ grows at a constant relative rate g, $g \geqq 0$, so that (37) takes the form

$$y_1 + gV_1 = F(y_2 + gV_2, \ldots, y_n + gV_n; V_1, \ldots, V_n; v_1, \ldots, v_m). \tag{37'}$$

Moreover, $\boldsymbol{V}$ is not arbitrary but satisfies

$$\frac{\partial F}{\partial V_1} - \rho_1 \frac{\partial F}{\partial x_1} = 0,$$

$$\frac{\partial F}{\partial V_j} + \rho_j \frac{\partial F}{\partial x_j} = 0 \qquad (j = 2, \ldots, n^0), \tag{43}$$

where $\partial F/\partial x_1 \equiv 1$ and n^0 $(n^0 \leqq n)$ is the number of active industries. If (as we suppose) all commodities can be traded internationally, y_j may be of either sign. We shall be concerned, then, with the locus of points $\boldsymbol{y}$ in $\mathbb{E}^n$ compatible with (37′) and (43).

We consider first the special case in which $\rho_j = g, j = 1, \ldots, n^0$, so that (43) becomes

$$\frac{\partial F}{\partial V_1} - g\frac{\partial F}{\partial x_1} = 0,$$

$$\frac{\partial F}{\partial V_j} + g\frac{\partial F}{\partial x_j} = 0 \qquad (j = 2, \ldots, n^0). \tag{43'}$$

Differentiating (37′) with respect to y_k $(1 < k \leqq n^0)$, holding constant the net outputs $y_2, \ldots, y_{k-1}, y_{k+1}, \ldots, y_{n^0}$, the gross outputs $x_{n^0+1}, \ldots, x_n$, and the capital stocks $V_{n^0+1}, \ldots, V_n$, we obtain

$$\begin{aligned}\frac{dy_1}{dy_k} &= \frac{\partial F}{\partial x_k} + \sum_{i=1}^{n^0}\frac{\partial F}{\partial V_i}\frac{dV_i}{dy_k} + g\left[\sum_{i=2}^{n^0}\frac{\partial F}{\partial x_i}\frac{dV_i}{dy_k} - \frac{dV_1}{dy_k}\right]\\ &= \frac{\partial F}{\partial x_k} \qquad \text{[from (43′)]}\\ &\leqq 0.\end{aligned} \tag{44}$$

Differentiating again, we obtain, after applying (43′),

$$\begin{aligned}\frac{d^2y_1}{dy_k^2} &= \frac{\partial^2 F}{\partial x_k^2} + 2\sum_{i=1}^{n^0}\frac{\partial^2 F}{\partial x_k\,\partial V_i}\frac{dV_i}{dy_k} + \sum_{i,j=1}^{n^0}\frac{\partial^2 F}{\partial V_i\,\partial V_j}\frac{dV_i}{dy_k}\frac{dV_j}{dy_k}\\ &\quad + 2g\sum_{i=2}^{n^0}\frac{\partial^2 F}{\partial x_i\,\partial x_k}\frac{dV_i}{dy_k} + 2g\sum_{\substack{i=1\\ j=2}}^{n^0}\frac{\partial^2 F}{\partial V_i\,\partial x_j}\frac{dV_i}{dy_k}\frac{dV_j}{dy_k}\\ &\quad + g^2\sum_{i,j=2}^{n^0}\frac{\partial^2 F}{\partial x_i\,\partial x_j}\frac{dV_i}{dy_k}\frac{dV_j}{dy_k}\\ &= \frac{\boldsymbol{q}\boldsymbol{H}\boldsymbol{q}'}{(dy_k)^2},\end{aligned} \tag{45}$$

where

$$\boldsymbol{q} \equiv (g\,dV_2, \ldots, g\,dV_k + dy_k, \ldots, g\,dV_{n^0}, dV_1, \ldots, dV_{n^0}) \tag{46}$$

and

$$\boldsymbol{H} \equiv \left(\begin{array}{c|c}\dfrac{\partial^2 F}{\partial x_i\,\partial x_j} & \dfrac{\partial^2 F}{\partial x_i\,\partial V_j}\\ \hline \dfrac{\partial^2 F}{\partial V_i\,\partial x_j} & \dfrac{\partial^2 F}{\partial V_i\,\partial V_j}\end{array}\right) \tag{47}$$

is the relevant Hessian matrix derived from F. From the concavity of F, $\boldsymbol{H}$ is nonpositive definite and

$$\frac{d^2y_1}{dy_k^2} \leqq 0. \tag{48}$$

In the special stationary case, in which $g = 0$, $\boldsymbol{q}$ reduces to

$$\boldsymbol{q} = (0, \ldots, 0, dy_k, 0, \ldots, 0) \tag{46'}$$

and (45) becomes

$$\frac{d^2 y_1}{dy_k^2} = (dy_k, dV_1, \ldots, dV_{n^0}) \begin{pmatrix} \dfrac{\partial^2 F}{\partial y_k^2} & \dfrac{\partial^2 F}{\partial y_k \, \partial V_1} & \cdots & \dfrac{\partial^2 F}{\partial y_k \, \partial V_{n^0}} \\ \dfrac{\partial^2 F}{\partial V_1 \, \partial y_k} & \dfrac{\partial^2 F}{\partial V_1^2} & \cdots & \dfrac{\partial^2 F}{\partial V_1 \, \partial V_{n^0}} \\ \vdots & \vdots & & \vdots \\ \dfrac{\partial^2 F}{\partial V_{n^0} \, \partial y_k} & \dfrac{\partial^2 F}{\partial V_{n^0} \, \partial V_1} & \cdots & \dfrac{\partial^2 F}{\partial V_{n^0}^2} \end{pmatrix} \begin{pmatrix} dy_k \\ dV_1 \\ \vdots \\ dV_{n^0} \end{pmatrix}. \tag{45'}$$

Thus, when $\rho_j = g$, $j = 1, \ldots, n^0$, the locus of steady-state net-output points is nonpositively sloped and concave; that is, it has the same properties as the locus of production possibilities defined by (37) for given V. And it remains true, of course, that at interior points of the locus the slopes are equal in everything but sign to the commodity price ratios.

We can go further and develop an analogue of Theorem 1.

Theorem 2. *Let $\boldsymbol{y}^*$ be a point in the locus of steady-state net-output possibilities, with n^0 industries active. Then the locus contains an $(n^0 - r)$-dimensional flat embracing $\boldsymbol{y}^*$ if and only if at $\boldsymbol{y}^*$ there are exactly r linearly independent vectors of primary-factor inputs.*

Proof. The proof runs parallel to that of Theorem 1 but differs from the latter in detail.

(*Sufficiency*). We have $\boldsymbol{y}^* \equiv (y_1^*, \ldots, y_{n^0}^*, y_{n^0+1}^*, \ldots, y_n^*) \equiv (\boldsymbol{y}_{\mathrm{I}}^*, \boldsymbol{y}_{\mathrm{II}}^*)$, where $y_j^* = f^j(\boldsymbol{v}^{j*}, \boldsymbol{V}^{j*}) - gV_j$, $\boldsymbol{v}^{j*} \equiv (v_{1j}^*, \ldots, v_{mj}^*)$ and $\boldsymbol{V}^{j*} \equiv (V_{1j}^*, \ldots, V_{nj}^*)$, and where $f^j(\boldsymbol{v}^{j*}, \boldsymbol{V}^{j*})$ is positive for $1 \leqq j \leqq n^0$ and zero for $n^0 < j \leqq n$. Suppose that $\boldsymbol{Z} \equiv (\boldsymbol{v}^{1*}, \ldots, \boldsymbol{v}^{n^0*})$ has rank r, $r \leqq \min\{m, n^0\}$. Now consider the system of homogeneous linear equations

$$\boldsymbol{Z}^*\boldsymbol{c} = \boldsymbol{0}. \tag{49}$$

The solution space in $\mathbb{E}^{n^0}$ is spanned by $n^0 - r$ independent vectors, say $\boldsymbol{c}^1, \ldots, \boldsymbol{c}^{n^0 - r}$. Corresponding to any one of these vectors, say $\boldsymbol{c}^h$ (with jth component c_j^h), we may define the new input vectors

$$\left.\begin{aligned} \boldsymbol{v}^{jh} &\equiv (1 + \delta_h c_j^h)\boldsymbol{v}^{j*} \\ \boldsymbol{V}^{jh} &\equiv (1 + \delta_h c_j^h)\boldsymbol{V}^{j*} \end{aligned}\right\} \quad (j = 1, \ldots, n^0) \tag{50}$$

where the scalar δ_h is so chosen that $1 + \delta_h c_j^h > 0$ and, therefore, $\boldsymbol{v}^{jh}, \boldsymbol{V}^{jh} \geqq \boldsymbol{0}$.

Corresponding to the new input vectors are the outputs

$$\begin{aligned} y_j^h &= f^j(\boldsymbol{v}^{jh}, \boldsymbol{V}^{jh}) - gV_j^h \\ &= (1 + \delta_h c_j^h) f^j(\boldsymbol{v}^{j*}, \boldsymbol{V}^{j*}) - gV_j^h \quad \text{[from (50) and the homogeneity of } f^j\text{]} \\ &= y_j^* + \delta_h c_j^h x_j^* - g(V_j^h - V_j^*). \end{aligned}$$

In matrix notation,

$$\boldsymbol{y}^h = \boldsymbol{y}^* + \delta_h \boldsymbol{Q}^* \boldsymbol{c}^h - g(\boldsymbol{V}^h - \boldsymbol{V}^*) \tag{51}$$

where

$$\boldsymbol{Q}^* \equiv \begin{pmatrix} [x_{\mathrm{I}}^*] - \boldsymbol{A}_{\mathrm{I}}^* \\ -\boldsymbol{A}_{\mathrm{II}}^* \end{pmatrix},$$

$$\boldsymbol{A}_{\mathrm{I}}^* \equiv \begin{pmatrix} V_{11}^* & \cdots & V_{1n^0}^* \\ \vdots & & \vdots \\ V_{n^01}^* & \cdots & V_{n^0n^0}^* \end{pmatrix}, \qquad \boldsymbol{A}_{\mathrm{II}}^* \equiv \begin{pmatrix} V_{n^0+1,1}^* & \cdots & V_{n^0+1,n^0}^* \\ \vdots & & \vdots \\ V_{n1}^* & \cdots & V_{nn^0}^* \end{pmatrix}$$

and where $[x_{\mathrm{I}}^*]$ is the diagonal matrix of order n^0 with x_i^* in the ith diagonal place, $i = 1, \ldots, n^0$. Now

$$\begin{aligned} \boldsymbol{V}^h &= \sum_{j=1}^{n^0} \boldsymbol{V}^{jh} \\ &= \sum_{j=1}^{n^0} (1 + \delta_h c_j^h) \boldsymbol{V}^{j*} \\ &= \boldsymbol{V}^* + \delta_h \begin{pmatrix} \boldsymbol{A}_{\mathrm{I}} \\ \boldsymbol{A}_{\mathrm{II}} \end{pmatrix} \boldsymbol{c}^h \end{aligned} \tag{52}$$

Hence, substituting in (51),

$$\boldsymbol{y}^h = \boldsymbol{y}^* + \delta_h \left\{ \boldsymbol{Q}^* - g \begin{pmatrix} \boldsymbol{A}_{\mathrm{I}} \\ \boldsymbol{A}_{\mathrm{II}} \end{pmatrix} \right\} \boldsymbol{c}^h \tag{53}$$

We note that

$$\begin{aligned} \sum_{j=1}^{n^0} \boldsymbol{v}^{jh} &= \sum_{j=1}^{n^0} (1 + \delta_h c_j^h) \boldsymbol{v}^{j*} \\ &= \sum_{j=1}^{n^0} \boldsymbol{v}^{j*} + \delta_h \sum_{j=1}^{n^0} c_j^h \boldsymbol{v}^{j*} \\ &= \boldsymbol{v}, \end{aligned}$$

implying that $\boldsymbol{y}^h$ is feasible.

Since the steady-state production set is convex, there is associated with $\boldsymbol{y}^*$ a (shadow price) vector $\boldsymbol{p}^* \equiv (\boldsymbol{p}_{\mathrm{I}}^*, \boldsymbol{p}_{\mathrm{II}}^*)$ such that $\boldsymbol{p}_{\mathrm{I}}^* > \boldsymbol{0}$ and

$$0 \geqq \boldsymbol{p}^*(\boldsymbol{y}^h - \boldsymbol{y}^*) = \delta_h \boldsymbol{p}^* \left\{ \boldsymbol{Q}^* - g \begin{pmatrix} \boldsymbol{A}_{\mathrm{I}}^* \\ \boldsymbol{A}_{\mathrm{II}}^* \end{pmatrix} \right\} \boldsymbol{c}^h.$$

Since δ_h can be of either sign, $p^* y^h = p^* y^*$. Thus the line segment $\overline{y^* y^h}$ traced by δ_h, $1 + \delta_h c_j^h > 0$, lies in the locus of steady-state production possibilities. Since h can take any of the values $1, \ldots, n^0 - r$, we have generated $n^0 - r$ such segments. To complete the proof of sufficiency we must show that these segments are linearly independent or, equivalently, that the $n^0 - r$ vectors $(y^h - y^*)$ are linearly independent.

From (53),

$$y^h - y^* = \delta_h \begin{pmatrix} [x_{\text{I}}^*] - (1 + g)A_{\text{I}}^* \\ -(1 + g)A_{\text{II}}^* \end{pmatrix} c^h \qquad (h = 1, \ldots, n^0 - r),$$

that is, $y^h - y^*$ is a linear transformation of c^h. Since the production of each commodity requires, directly or indirectly, some of at least one primary factor, $p_{\text{I}}([x_{\text{I}}^*] - A_{\text{I}}^*) > \mathbf{0}$, implying that $[x_{\text{I}}^*] - A_{\text{I}}$ has a dominant diagonal and therefore is of full rank. (See Theorem 6 of Chapter 1.) Moreover, the c^h, $h = 1, \ldots, n^0 - r$, are linearly independent. Hence the $y^h - y^*$, $h = 1, \ldots, n^0 - r$, are linearly independent.

(*Necessity*). Consider any point $y^* \equiv (y_{\text{I}}^*, y_{\text{II}}^*)$ in an $(n^0 - r)$-dimensional flat of the locus of steady-state net-output possibilities. Let

$$Z^* \equiv (v^{1*}, \ldots, v^{n^0 *})$$

be the matrix of primary-factor inputs to the n^0 active industries. Suppose that the rank of Z^* is s, $s \neq r$. Then, from the sufficiency part of the proposition, the dimension of the flat must be $n^0 - s$, a contradiction. □

When $\rho_j \neq g$, all of this changes. Then it is impossible to prove, on the basis of (37′) and (43) alone, either that the locus of long run production possibilities is everywhere nonpositively sloped or that it is uniformly concave. In fact Metcalfe and Steedman [1972] have produced a two-commodities example (with $\rho_1 = \rho_2 > g$) in which parts of the locus are positively sloped and parts strictly convex. They note also that, in examples of the kind they have constructed, the locus may be not everywhere continuous.[4] Moreover, when $\rho_j \neq g$ then, even at interior points of the locus, the slopes are only in singular cases proportional to commodity price ratios. Finally, Metcalfe and Steedman have shown that the response of steady-state production to changes in commodity prices may be "perverse," that is, an increase in price may be associated with a decline in production.[5] Whether there exist interesting conditions whose satisfaction ensures that the locus has the conventional concavity property, and whether there exist interesting conditions that ensure that price-production responses are normal, are open questions. Indeed, the possibility that concavity implies or is implied by the normality of price-output responses remains largely unexplored. (For conditions that, in the familiar case of two primary factors and two products, rule out perversity of price-output responses the reader may refer to Kemp and Khang [1974].)

It remains to consider the manner in which variations in the own-rates of return shift the locus of long-run production possibilities. Only one special case has been at all carefully studied, *viz.*, the case in which there is only one primary factor of production and in which all own-rates of return are both equal and not less than the growth rate g. (See, for example, Samuelson [1961], Morishima [1964], and Bruno *et al.* [1966].) Then (10) and (24) reduce to

$$\boldsymbol{p} = w\boldsymbol{a} + \rho\boldsymbol{p}\boldsymbol{A} \tag{10'}$$

and

$$\boldsymbol{a}(\boldsymbol{I} - g\boldsymbol{A})^{-1}\boldsymbol{y} = v \tag{24'}$$

respectively, where w, ρ (the rate of interest), and v are scalars and $\boldsymbol{a}$ is a row vector. Setting $w = 1$, and expressing the solution to (10′) somewhat elaborately as $\boldsymbol{p}(\rho; \boldsymbol{a}(\rho), \boldsymbol{A}(\rho))$, we have, from (10′),

$$\boldsymbol{p}(\rho; \boldsymbol{a}(\rho), \boldsymbol{A}(\rho)) = \boldsymbol{a}(\rho)(\boldsymbol{I} - \rho\boldsymbol{A}(\rho))^{-1}. \tag{54}$$

Hence (24′) can be rewritten as

$$\boldsymbol{p}(g; \boldsymbol{a}(\rho), \boldsymbol{A}(\rho)) \cdot \boldsymbol{y} = v. \tag{24''}$$

Thus, in the special case under study, the locus of long-run production possibilities is a plane, with slopes proportional to those prices appropriate to a rate of interest equal to the growth rate and the technique actually chosen when the rate of interest is ρ, $\rho \geqq g$. It is easy to verify that

$$\boldsymbol{p}(g; \boldsymbol{a}(\rho), \boldsymbol{A}(\rho)) \geq \boldsymbol{p}(g; \boldsymbol{a}(g), \boldsymbol{A}(g)) \quad \text{if } \rho > g,$$

so that the locus appropriate to ρ, $\rho > g$, lies unambiguously "inside" the locus associated with $\rho = g$. The locus associated with $\rho = g$ is sometimes called the "golden rule" locus. On the other hand, when $\rho > g$ it is not the case that the locus necessarily shrinks monotonically as ρ increases. Thus, when $m = 1$ and $\rho_1 = \cdots = \rho_n = \rho \geqq g$,

a. the locus of long-run production possibilities is a hyperplane,
b. the locus associated with any ρ, $\rho > g$, lies unambiguously inside the golden-rule locus defined by $\rho = g$,
c. no two loci intersect, and
d. for $\rho > g$, the locus does not necessarily shrink monotonically as the rate of interest increases

(For a verification of (d) the reader may refer to Bruno *et al.* [1966].)

When it is not required that $\rho_1 = \cdots = \rho_n$, conclusions (b)–(d) must be modified in fairly obvious ways. Then

b′. the locus of long-run production possibilities associated with any $\boldsymbol{\rho}$, $[\rho] \geq [g]$, lies unambiguously inside the golden-rule locus,
c′. the loci associated with the own-rate vectors $\boldsymbol{\rho}^{(1)}$ and $\boldsymbol{\rho}^{(2)}$, with $[\rho^{(i)}] \geq [g]$ and $\boldsymbol{\rho}^{(1)}$ not a positive multiple of $\boldsymbol{\rho}^{(2)}$, may possibly intersect, and

d′. for $\boldsymbol{\rho}$, with $[\rho] \geq [g]$, the locus does not necessarily shrink monotonically as the elements of $\boldsymbol{\rho}$ increase in uniform proportion.

When $m > 1$, conclusion (a) must be abandoned. As we have seen, when $m > 1$ and not all own-rates of return are equal to the growth rate, one cannot even be sure that the locus is concave and downward sloping. Presumably, properties (b′)–(d′) continue to hold.

Notes to Chapter 5

1. This assumption is slightly weaker than that of strict concavity. Strict concavity implies strict quasi-concavity, but the latter does not imply strict concavity.
2. Suppose that we are given $\boldsymbol{w}$ and $\boldsymbol{\rho} = \boldsymbol{\rho}(\boldsymbol{w}; \boldsymbol{W})$. Then $\boldsymbol{\rho}(\boldsymbol{w}; \boldsymbol{W})$ is an equilibrium price vector; for $\boldsymbol{W} = \boldsymbol{\rho}(\boldsymbol{w}; \boldsymbol{W})[\rho]$, where $[\rho]$ is the diagonal matrix with ρ_j in the jth diagonal place, and (6) may be written in the equivalent form

$$\boldsymbol{p}(\boldsymbol{w}; \boldsymbol{W}) = \boldsymbol{w}\boldsymbol{a}(\boldsymbol{w}; \boldsymbol{W}) + \boldsymbol{p}(\boldsymbol{w}; \boldsymbol{W})[\rho]A(\boldsymbol{w}; \boldsymbol{W}) \tag{6′}$$

Is it possible that there exists a second equilibrium price vector, say $\boldsymbol{p}^*$, with $\boldsymbol{p}^* \neq \boldsymbol{p}(\boldsymbol{w}; \boldsymbol{W})$? Now $\boldsymbol{p}^*$, if it exists, is either a positive multiple of $\boldsymbol{p}(\boldsymbol{w}; \boldsymbol{W})$ or it is not. Suppose that $\boldsymbol{p}^*$ is λ times $\boldsymbol{p}(\boldsymbol{w}; \boldsymbol{W})$, with $0 < \lambda \neq 1$. Then $\boldsymbol{W}^* \equiv \boldsymbol{p}^*[\rho]$ must be the same multiple of $\boldsymbol{W}$. But each primary factor is required for the production of at least one commodity, and the rentals of primary factors are unchanged. It follows that not all average costs can change in the proportion λ, a contradiction. Consider therefore the alternative possibility, that there does not exist a positive λ, $\lambda \neq 1$, such that $\boldsymbol{p}^* = \lambda \boldsymbol{p}(\boldsymbol{w}; \boldsymbol{W})$. Suppose that $p_i^* > p_i(\boldsymbol{w}; \boldsymbol{W})$ for some i and that $(p_i^* - p_i(\boldsymbol{w}; \boldsymbol{W}))/p_i(\boldsymbol{w}; \boldsymbol{W})$ is largest for $i = k$, so that $(W_i^* - W_i)/W_i$ also is largest for $i = k$; and suppose that k is unique. Now producers of the kth commodity use, directly or indirectly, some of at least one primary factor. Hence they must buy at least one factor other than the kth producible factor. But all other factor rentals have risen in smaller proportion than W_k. Hence the average cost of producing the kth commodity cannot have risen in the proportion $(p_k^* - p_k(\boldsymbol{w}; \boldsymbol{W}))/p_k(\boldsymbol{w}; \boldsymbol{W})$, a contradiction. If there is no i such that $p_i^* > p_i(\boldsymbol{w}; \boldsymbol{W})$, a similar argument can be developed in terms of that commodity for which $(p_i^* - p_i(\boldsymbol{w}; \boldsymbol{W}))/p_i(\boldsymbol{w}; \boldsymbol{W})$ is smallest. If k is not unique the argument merely takes a slightly more elaborate form. Hence we may equivalently take $(\boldsymbol{w}; \boldsymbol{W})$ or $(\boldsymbol{w}; \boldsymbol{\rho}(\boldsymbol{w}; \boldsymbol{W}))$ as given.
3. The properties of F to be listed can be verified by considering the nonlinear programming problem

$$\max_{v_{ij}, V_{ij}} \sum p_j f^j(v_{1j}, \ldots, v_{mj}; V_{1j}, \ldots, V_{nj})$$

$$\text{subject to} \sum_j v_{ij} \leqq v_i \qquad (i = 1, \ldots, m),$$

$$\sum_j V_{ij} \leqq V_i \qquad (i = 1, \ldots, n),$$

$$v_{ij} \geqq 0 \qquad (i = 1, \ldots, m; j = 1, \ldots, n),$$

$$V_{ij} \geqq 0 \qquad (i, j = 1, \ldots, n),$$

and recalling the concavity-homogeneity properties of f^j. The reader should carry out the verification in detail, after refreshing his memory of Section 4 of Chapter 1.

4. Metcalfe and Steedman have noted also the possibility of multiple steady-state equilibrium in their example. And Samuelson [1975] has shown that, when $r > 0$ and $m > 1$, there may be many steady-state equilibria even when community preferences are homothetic.
5. When $\rho_j \neq g$, as when $\rho_j = g$, the locus of steady-state production possibilities may contain flats of dimension greater than zero, implying that outputs are, within limits, indeterminate. Evidently the statement of the text applies only to situations in which outputs are determinate.

Neoclassical Economies— Dynamics 6

In Chapter 5 we studied questions concerning the existence and properties of steady-state equilibria in neoclassical economies, in particular, concerning the sensitivity of steady-state equilibria to changes in parameters. Conspicuously lacking was any discussion of the nonsteady-state behavior of such economics and, more specifically, of the dynamic stability of such equilibria. To these matters we now turn.

1 Introduction

Traditionally, two types of stability question have been considered in economics. On the one hand, economists have wanted to know whether market economies will reach or approach a state of *market clearance*, that is, a state for which, in each market, either demand equals supply, or the price is zero, or both. The answer to such a question depends on the assumed restrictions on the initial prices and quantities and on the rule or process by which nonmarket-clearing prices and quantities adjust through time. In the most intensively studied adjustment processes, the so-called *tâtonnement* processes, binding contracts are entered into only when market-clearing prices and quantities are discovered. And in most *tâtonnement* processes so far studied all demand and supply functions are stable over time, suggesting that all preferences, production sets, and asset holdings are constant. In Section 2 we discuss a market adjustment process of *tâtonnement* type and provide conditions for both local and global stability of the economy. Because we want our analysis to be applicable to economies that produce as well as exchange, we proceed without the usual assumption of gross substitutability. Our analysis is based on that of Mukherji [1974].

On the other hand, taking for granted market equilibrium, economists have asked whether the economy will approach a steady state. In particular, they have studied economies that are always in market equilibrium but in which either (i) individual agents adjust sluggishly to market data, or (ii) preferences, technical knowledge, or assets change over time. In Section 3 we offer an analysis of an economy of type (i). It is, in fact, a dynamical and otherwise generalized version of the Heckscher–Ohlin model familiar to students of international trade and public finance. The analysis builds on that of Kemp, Kimura, and Okuguchi [1977].

All realistic market adjustment processes take place in time, and as they work themselves out the resources and technology of the economy are changing. Similarly, the development of the economy through time depends to some extent on the characteristics of the market adjustment process. Evidently we should aspire to a general dynamic analysis, with market adjustment and long-run development intermingled. However, such an analysis would be exceedingly complicated and is far beyond our aim in this chapter.

2 Stability of Market Equilibrium

As in Chapter 5, we consider an economy that produces n goods with the aid of stock of those n goods and of m primary factors. Let $N \equiv 2n + m$. Varying our earlier notation slightly, we denote by p_i, $i = 1, \ldots, n, n + 1, \ldots, 2n, 2n + 1, \ldots, N$, the price of the ith commodity, with p_{n+i}, $i = 1, \ldots, n$, the rental of the produced input and with p_{2n+i}, $i = 1, \ldots, m$, the rental of the ith primary factor. And we denote by $E_i(\boldsymbol{p})$ the excess demand for the ith commodity where, of course, $\boldsymbol{p} \equiv (p_1, \ldots, p_N)$.

The excess demand functions will be subject to the following restrictions.

i. *There exists a positive equilibrium price vector* $\boldsymbol{p}^*$, *implying that* $E_i(\boldsymbol{p}^*) = 0$, $i = 1, \ldots, N$.
ii. *The excess demand functions are positively homogeneous of degree zero*; *that is,* $E_i(\lambda \boldsymbol{p}) = E_i(\boldsymbol{p})$ *for all positive* λ *and for all* $i = 1, \ldots, N$.
iii. *For any* $\boldsymbol{p} \geq \boldsymbol{0}$, $\boldsymbol{p}\boldsymbol{E}(\boldsymbol{p}) = 0$, *where* $\boldsymbol{E} \equiv (E_1, \ldots, E_N)$; *that is, Walras's Law holds.*
iv. *Excess demand functions are single-valued and continuous, with continuous partial derivatives for all* $\boldsymbol{p} > \boldsymbol{0}$.

In view of assumption (ii), if $p_k \neq 0$ then the N excess demand functions may be written $E_i(\boldsymbol{q})$, $i = 1, \ldots, N$, where

$$\boldsymbol{q} \equiv (q_1, q_{k-1}, 1, q_{k+1}, \ldots, q_N), \tag{1}$$

$$q_i \equiv \frac{p_i}{p_k}.$$

Evidently

$$\frac{\partial E_i(\boldsymbol{p})}{\partial p_j} = \left(\frac{1}{p_k}\right)\left(\frac{\partial E_i(\boldsymbol{q})}{\partial q_j}\right) \qquad (i, j \neq k). \tag{2}$$

However, normalization of the price vector may give rise to analytical difficulties. For, in the course of adjustment, some elements of $\boldsymbol{q}$ may go to infinity even though the excess demand for the *numéraire* is positive (but finite). To avoid this complication, the excess demand functions are subjected to an additional restriction.

v. *There exists at least one commodity, say the* kth, *with the property that if* $\|\boldsymbol{q}\| \to \infty$ *then* $E_k(\boldsymbol{q}) \to \infty$.

An auctioneer or market secretary is charged with the task of finding an equilibrium normalized price vector $\boldsymbol{q}^* \geq \boldsymbol{0}$ such that $q_i^* E_i(\boldsymbol{q}^*) = 0$, $i = 1, \ldots, N$. He announces a nonnegative price vector $\boldsymbol{q}$, asks each market participant to declare his excess demand for the ith commodity, computes E_i, then revises the ith price according to the rule (adjustment process)

$$\begin{aligned} \dot{q}_i &= h_i(E_i(\boldsymbol{q})) \equiv H_i(\boldsymbol{q}) \qquad (i \neq k), \\ q_k &= 1, \end{aligned} \tag{3}$$

where h_i is a differentiable, sign preserving function, that is,

$$\left.\begin{aligned} \text{sign } h_i(E_i(\boldsymbol{q})) &= \text{sign } E_i(\boldsymbol{q}) \\ h_i(0) &= 0 \end{aligned}\right\} \qquad (i \neq k). \tag{4}$$

The process of revision (or *tâtonnement*) continues until an equilibrium price vector is obtained.

Although the initial price vector is nonnegative, (3) and (4) do not by themselves ensure the nonnegativity of each revised price vector; for, so far, we have not excluded the possibility that there exists an index i such that $E_i(\boldsymbol{q}) < 0$ for some $\boldsymbol{q} \geqq \boldsymbol{0}$ with $q_i = 0$. To ensure that the price vector is positive throughout the process of revision is the purpose of our next assumption.[1]

vi. *There exists an* $e > 0$ *such that if* $q_i \leqq e$ *then* $E_i(\boldsymbol{q}) > 0$.

We first examine the *local* stability of $\boldsymbol{q}^*$ under *tâtonnement* (3). For this specific purpose we introduce the following assumption.

vii. *Let* $\boldsymbol{A}(\boldsymbol{p}^*)$ *be the Jacobian matrix of the excess demand functions* $E_i(\boldsymbol{p})$, *evaluated at the nonnormalized equilibrium price vector* $\boldsymbol{p}^*$, *and let* $\boldsymbol{X} \equiv \boldsymbol{A}(\boldsymbol{p}^*) + \boldsymbol{A}'(\boldsymbol{p}^*)$. *Then there exists a nonsingular* $\boldsymbol{S}$ *such that* $\boldsymbol{SXS}^{-1}$ *is indecomposable and Metzlerian and such that* $\boldsymbol{Sp}^* > \boldsymbol{0}$.

Lemma 1. *Under assumptions* (i)–(iv) *and* (vii), *every principal submatrix of* $\boldsymbol{A}(\boldsymbol{p}^*)$ *of order* $N - 1$ *is quasi-negative definite.*

PROOF. By virtue of assumptions (i)–(iii), condition (a) of Theorem 16 of Chapter 4 is satisfied; and, by virtue of assumption (vii), conditions (b) and (c) of that theorem also are satisfied. Let λ^* be the Frobenius root of $\boldsymbol{SXS}^{-1}$. Then Theorem 16 of Chapter 4, together with the assumed indecomposability of $\boldsymbol{SXS}^{-1}$, implies that $\lambda^* = 0$ is a simple root of $\boldsymbol{X}$. Moreover, $\boldsymbol{X}$ is symmetrical; hence, from Theorem 16(ii) of Chapter 4, $\boldsymbol{X}$ contains a principal submatrix of order $N - 1$ that is negative definite and every nonsingular principal submatrix $\boldsymbol{X}$ of order $N - 1$ is negative definite. Again by virtue of assumptions (i)–(iii), Lemma 11 of Chapter 1 assures us that every principal submatrix of $\boldsymbol{X}$ of order $N - 1$ is nonsingular and hence negative definite. The assertion follows immediately. □

With the aid of Lemma 1 we can establish the local stability of $\boldsymbol{q}^*$ under *tâtonnement* (3).

Theorem 1. *Given assumptions* (i)–(vii), *the normalized equilibrium price vector* $\boldsymbol{q}^*$ *is locally stable under tâtonnement* (3) *for all possible combinations of positive adjustment speeds* $h_i'(0), i \neq k$, *and for all possible choices of numéraire; in other words,* $\boldsymbol{q}^*$ *is totally stable for all possible choices of numéraire.*

PROOF. In view of (2) and assumption (i), the assertion follows from Lemma 1 and Theorem 10 of Chapter 4. □

Remark Concerning Theorem 1. From assumption (i) it is clear that $\boldsymbol{q}^* > \boldsymbol{0}$. Hence there exists a neighborhood $B(\boldsymbol{q}^*, \varepsilon)$ of $\boldsymbol{q}^*$ consisting of positive vectors. Since $\boldsymbol{q}^*$ is locally stable, by choosing the initial price vector $\boldsymbol{q}_0$ both positive and sufficiently close to $\boldsymbol{q}^*$ we can ensure that the solution of (3) starting from $\boldsymbol{q}_0$ will always remain in $B(\boldsymbol{q}^*, \varepsilon)$. (See the proof of Theorem 21 of Chapter 4.) For such a choice of $\boldsymbol{q}_0$, therefore, we can dispense with assumptions (v) and (vi) and Theorem 1 takes the stronger form of

Theorem 1′. *Given assumptions* (i)–(iv) *and* (vii), *and if* $\boldsymbol{q}_0$ *is sufficiently close to* $\boldsymbol{q}^*$, *then* $\boldsymbol{q}^*$ *is totally stable for any choice of numéraire.*

It is implausible that the auctioneer can infallibly find an initial price vector so close to $\boldsymbol{q}^*$, which is unknown to him, that assumptions (v) and (vi) are redundant. Thus, to the extent that we are concerned with the stability of normalized prices, assumptions like (v) and (vi) seem inevitable.

We now turn our attention to the global properties of *tâtonnement* (3). To ensure global stability an additional assumption is introduced.

viii. *Let* $\boldsymbol{J}(\boldsymbol{q})$ *be the Jacobian matrix of* $\boldsymbol{H}(\boldsymbol{q}) = (H_1(\boldsymbol{q}), \ldots, H_{k-1}(\boldsymbol{q}), H_{k+1}(\boldsymbol{q}), \ldots, H_N(\boldsymbol{q}))'$. *Then, for any* $\boldsymbol{q} > \boldsymbol{0}$, *there exist* $\rho > 0$ *and a nonsingular matrix* $\boldsymbol{S}$ *such that* $\boldsymbol{W} = \boldsymbol{S}(\boldsymbol{J}(\boldsymbol{q}) + \boldsymbol{J}'(\boldsymbol{q}))\boldsymbol{S}^{-1} + \rho \boldsymbol{I}$ *is a power positive matrix of* jth *degree with the property that* $\rho > \lambda^* > 0$; *moreover,*

$$0 > -\varepsilon = \sup_{\substack{\boldsymbol{x} \neq \boldsymbol{0} \\ \boldsymbol{q} > \boldsymbol{0}}} \boldsymbol{x}'\boldsymbol{J}(\boldsymbol{q})\boldsymbol{x}, \tag{5}$$

where λ^* *is the greatest absolute eigenvalue of* W^2.

We can now state our main theorem.

Theorem 2. *If assumptions* (i)–(vi) *and* (viii) *hold then* $\boldsymbol{q}^*$ *is unique and is globally stable under tâtonnement* (3).

PROOF. Let $V(\boldsymbol{q}) \equiv \frac{1}{2}\boldsymbol{H}'(\boldsymbol{q}) \cdot \boldsymbol{H}(\boldsymbol{q})$. Then $V(\boldsymbol{q})$ is a Liapunov function associated with (3). The trajectory derivative $\dot{V}(\boldsymbol{q})$ of $V(\boldsymbol{q})$ is given by

$$\dot{V}(\boldsymbol{q}) = \boldsymbol{H}'(\boldsymbol{q}) \cdot \boldsymbol{J}(\boldsymbol{q}) \cdot \boldsymbol{H}(\boldsymbol{q}) = \tfrac{1}{2}[\boldsymbol{H}'(\boldsymbol{q}) \cdot (\boldsymbol{J}(\boldsymbol{q}) + \boldsymbol{J}'(\boldsymbol{q})) \cdot \boldsymbol{H}(\boldsymbol{q})].$$

Since a symmetric matrix $\boldsymbol{J}(\boldsymbol{q}) + \boldsymbol{J}'(\boldsymbol{q})$ is, by definition, similar to a stable matrix $\boldsymbol{W} - \rho \boldsymbol{I}$, $\boldsymbol{J}(\boldsymbol{q})$ is quasi-negative definite and hence $\dot{V}(\boldsymbol{q}) < 0$ for any $\boldsymbol{q} \neq \boldsymbol{q}^*$.

To verify the uniqueness of $\boldsymbol{q}^*$, suppose that there exists $\hat{\boldsymbol{q}} \neq \boldsymbol{q}^*$ for which $\boldsymbol{H}(\hat{\boldsymbol{q}}) = \boldsymbol{0}$. Then $\boldsymbol{q}$ can never be proportionate to $\boldsymbol{q}^*$; for, otherwise, there would exist $\lambda > 0$ satisfying $\boldsymbol{q}^* = \lambda \hat{\boldsymbol{q}}$. Consequently

$$\hat{q}_k = 1 = q_k^* = \lambda \hat{q}_k,$$

from which it follows that $\lambda = 1$, contradicting the hypothesis.

Define $\boldsymbol{q}(\tau) = \tau \hat{\boldsymbol{q}} + (1 - \tau)\boldsymbol{q}^*$ for any $\tau \in [0, 1]$. Then, in view of the fact that $q_k(\tau) = 1$ for all $\tau \in [0, 1]$, direct differentiation of $\boldsymbol{H}(\boldsymbol{q}(\tau))$ with respect to τ yields

$$\begin{aligned} \frac{d\boldsymbol{H}}{d\tau} &\equiv \left(\frac{dH_1(\boldsymbol{q}(\tau))}{d\tau}, \ldots, \frac{dH_{k-1}(\boldsymbol{q}(\tau))}{d\tau}, \frac{dH_{k+1}(\boldsymbol{q}(\tau))}{d\tau}, \ldots, \frac{dH_N(\boldsymbol{q}(\tau))}{d\tau}\right)' \\ &= \boldsymbol{J}(\boldsymbol{q}(\tau))\bar{\boldsymbol{x}}, \end{aligned} \tag{6}$$

where

$$\bar{\boldsymbol{x}} = (\hat{q}_1 - q_1^*, \ldots, \hat{q}_{k-1} - q_{k-1}^*, \hat{q}_{k+1} \quad q_{k+1}, \ldots, \hat{q}_N - q_N^*)'.$$

Integrating both sides of (6) from $\tau = 0$ to $\tau = 1$, and recalling that $\boldsymbol{H}(\hat{\boldsymbol{q}}) = \boldsymbol{H}(\boldsymbol{q}^*) = \boldsymbol{0}$, we obtain

$$\boldsymbol{0} = \boldsymbol{H}(\hat{\boldsymbol{q}}) - \boldsymbol{H}(\boldsymbol{q}^*) = \int_0^1 \boldsymbol{J}(\boldsymbol{q}(\tau))\bar{\boldsymbol{x}} \cdot d\tau. \tag{7}$$

Since $\boldsymbol{J}(\boldsymbol{q}(\tau))$ is quasi-negative definite, as was shown above, it follows from (7) that

$$0 = \int_0^1 \bar{\boldsymbol{x}}'\boldsymbol{J}(\boldsymbol{q}(\tau))\bar{\boldsymbol{x}}\, d\tau < 0,$$

which is a self-contradiction.

We finally prove that $V(\boldsymbol{q}) \to \infty$ as $\|\boldsymbol{q}\| \to \infty$, where $\|\boldsymbol{q}\|$ denotes the Euclidean norm of $\boldsymbol{q}$. For any given $\bar{\boldsymbol{q}} > \boldsymbol{0}$ and $\tau > 0$, define $\boldsymbol{q}(\tau) = \tau \cdot \bar{\boldsymbol{q}} + \boldsymbol{q}^*$. Then, integrating both sides of $d\boldsymbol{H}/d\tau = \boldsymbol{J}(\boldsymbol{q}(\tau))\boldsymbol{x}'$ from $\tau = 0$ to $\tau = \alpha$, an arbitrarily given positive number, and proceeding as in the proof of uniqueness, the assumed inequality (5) enables us to assert that

$$\boldsymbol{x} \cdot \boldsymbol{H}(\boldsymbol{q}(\alpha)) = \int_0^\alpha \boldsymbol{x} \cdot \boldsymbol{J}(\boldsymbol{q}(\tau))\boldsymbol{x}'\, d\tau \leq -\varepsilon\alpha,^3 \tag{8}$$

where

$$\boldsymbol{x} = (\bar{q}_1, \ldots, \bar{q}_{k-1}, \bar{q}_{k+1}, \ldots, \bar{q}_N).$$

Since $\boldsymbol{x}$ is fixed and since $-\boldsymbol{x} \cdot \boldsymbol{H}(\boldsymbol{q}(\alpha))$ becomes infinitely large as α tends to infinity, there exists an $i \in \{1, \ldots, k-1, k+1, \ldots, N\}$ such that

$$|H_i(\boldsymbol{q}(\alpha))| \to \infty$$

as $\alpha \to \infty$. Again noticing that $\bar{\boldsymbol{q}}$ is fixed arbitrarily, it is easily seen that $\lim_{\|\boldsymbol{q}\| \to \infty} V(\boldsymbol{q}) = \infty$, as desired.[4] □

Our examination of global stability has proceeded to this point without any assumptions concerning the relationships of complementarity and substitutability between commodities. We now show that by introducing such an assumption it is possible to achieve a considerable sharpening of conclusions. In particular, we suppose that all commodities are *strong gross substitutes*, in the sense that

ix. $\partial E_i(\boldsymbol{p})/\partial p_j > 0$ *for any* $i \neq j$ *and for any* $\boldsymbol{p} > \boldsymbol{0}$,

and examine the implications of the assumption in the context of the tâtonnement process

$$\dot{p}_i = h_i(E_i(\boldsymbol{p})) \equiv H_i(\boldsymbol{p})) \qquad (i = 1, \ldots, N), \tag{3'}$$

where h_i, $i = 1, \ldots, N$, is a differentiable, sign-preserving function and where prices are now in nonnormalized form.

Remarks Concerning the Assumption of Gross Substitutability. (1) The definition of gross substitutability has been expressed in differential form. Alternatively, one might have concentrated on finite price changes and defined the N commodities to be gross substitutes if, for any two price vectors $\boldsymbol{p}^1$ and $\boldsymbol{p}^2$ such that $p_{k_0}^1 > p_{k_0}^2$ for some $k_0 \in \{1, \ldots, N\}$ and such that $p_i^1 = p_i^2$ for all $i \neq k_0$, $E_i(\boldsymbol{p}^1) > E_i(\boldsymbol{p}^2)$ for every $i \neq k_0$. It can be shown (with the aid of the Mean Value Theorem) that if every $E_i(\boldsymbol{q})$ is continuous on the nonnegative orthant then differential gross substitutability implies finite gross substitutability. (2) We have assumed strong gross substitutability. However, the conclusions that we shall reach can be obtained also with the assumption of *weak* gross substitutability $[\partial E_i(\boldsymbol{p})/\partial p_j \geqq 0, i \neq j, \boldsymbol{p} > \boldsymbol{0}]$ if, in addition, the matrix of partial derivatives $(\partial E_i/\partial p_j)$ is indecomposable. (See McKenzie [1960].)

We begin by listing and verifying some of the remarkable properties of "gross substitute" economies.

Lemma 2. *If assumptions* (ii), (iii), *and* (ix) *are satisfied and if, in addition,*

i′. *every market is cleared by some price vector* $\boldsymbol{p}^* \geq \boldsymbol{0}$, *that is,* $\boldsymbol{E}(\boldsymbol{p}^*) = \boldsymbol{0}$,

then:

1. $\boldsymbol{p}^*$ *in* (i′) *is in fact positive*;[5]
2. *for any positive price vector* $\boldsymbol{p}$ *and any* $j \in \{1, \ldots, N\}$, $E_j(\boldsymbol{p}) \to \infty$ *as* $p_j \to 0$;

3. *any solution* $\boldsymbol{p}(t)$ *of* (3′) *with initial state* $\boldsymbol{p}(0) > \boldsymbol{0}$ *remains in the non-negative orthant;*
4. *if, for any semipositive* $\boldsymbol{p}$ *that is not proportional to* $\boldsymbol{p}^*$, $\max_j \{p_j/p_j^*\} = p_k/p_k^*$ *and* $\min_j \{p_j/p_j^*\} = p_r/p_r^*$ *then*

$$E_k(\boldsymbol{p}) < 0 \tag{9.1}$$

and

$$E_r(\boldsymbol{p}) > 0, \tag{9.2}$$

whence

$$\dot{p}_k = H_k(\boldsymbol{p}) < 0 \tag{9.1'}$$

and

$$\dot{p}_r = H_r(\boldsymbol{p}) > 0; \tag{9.2'}$$

5. *the equilibrium price vectors are unique up to a positive multiple.*

PROOF. (1) Let $p_j^* = 0$ for some j and let $\boldsymbol{p}^j = (p_1^j, \ldots, p_N^j)$ be such that $p_i^j > p_i^*$ for all $i \neq j$ and such that $p_j^j = p_j^* = 0$. Then, from (i′), (ix), and the first remark concerning (ix), $E_j(\boldsymbol{p}^j) > E_j(\boldsymbol{p}^*) = 0$. Let $\boldsymbol{p}$ be the vector obtained from $\boldsymbol{p}^j$ by replacing its jth element with $p_j > 0$ so small that $\boldsymbol{p}$ is sufficiently near to $\boldsymbol{p}^j$. Then

$$\begin{aligned} 0 = p_j^* E_j(\boldsymbol{p}^j) &= -\sum_{i \neq j} p_i^j E_i(\boldsymbol{p}^j) && \text{[from (iii)]} \\ &> -\sum_{i \neq j} p_i^j E_i(\boldsymbol{p}) && \text{[from (ix) and the first remark concerning (ix)]} \\ &= p_j E_j(\boldsymbol{p}) && \text{[from (iii) again].} \end{aligned}$$

Since $p_j > 0$, $E_j(\boldsymbol{p}) = E_j(p_1^j, \ldots, p_{j-1}^j, p_j, p_{j+1}^j, \ldots, p_N^j) < 0$. However, this inequality contradicts the continuity of E_j since $\boldsymbol{p}$ can be taken so close to $\boldsymbol{p}^j$ that $\operatorname{sgn} E_j(\boldsymbol{p}) = \operatorname{sgn} E_j(\boldsymbol{p}^j)$. Hence, $\boldsymbol{p}^*$ is positive.

(2) Since $\boldsymbol{p}^* > \boldsymbol{0}$ and since we are interested in the behavior of $E_j(\boldsymbol{p})$ as p_j tends to zero, it is harmless to assume that $0 < p_j < p_j^*$. Moreover, by hypothesis, $\boldsymbol{p} > \boldsymbol{0}$; hence there can be found a positive λ such that $\lambda p_i > p_i^*$ for all $i \neq j$. Define $\boldsymbol{p}(\lambda) = (\lambda p_1, \ldots, \lambda p_{j-1}, p_j^*, \lambda p_{j+1}, \ldots, \lambda p_N)$ and let $\boldsymbol{p}'(\lambda)$ be the vector obtained from $\boldsymbol{p}(\lambda)$ by replacing its jth element p_j^* by p_j. Proceeding as in the proof of assertion (1), we then obtain

$$\begin{aligned} 0 < p_j^* E_j(\boldsymbol{p}(\lambda)) &= -\sum_{i \neq j} \lambda p_i E_i(\boldsymbol{p}(\lambda)) \\ &< -\sum_{i \neq j} \lambda p_i E_i(\boldsymbol{p}'(\lambda)) && \text{[from (ix), the first remark concerning (ix), and the hypothesis that } p_j^* > p_j > 0] \\ &= p_j E_j(\boldsymbol{p}'(\lambda)). \end{aligned}$$

Therefore

$$\frac{(p_j^* E_j(\boldsymbol{p}(\lambda)))}{p_j} < E_j(\boldsymbol{p}'(\lambda)) \quad \text{for any } p_j \in (0, p_j^*).$$

Letting $p_j \to 0$, and noticing that the numerator of the left-hand side of the above inequality is a positive constant independent of p_j, assertion (2) follows at once.

(3) This is a straightforward consequence of the previous assertion, whence the proof is omitted.

(4) Since, by virtue of (1) above, $\boldsymbol{p}^*$ is positive, and since $\boldsymbol{p}$ is semipositive, every ratio p_i/p_i^* is well defined and p_k is surely positive. Furthermore, it is evident that

$$\frac{p_k}{p_k^*} > \frac{p_r}{p_r^*} \tag{10}$$

for, otherwise, $\boldsymbol{p} = (p_k/p_k^*) \cdot \boldsymbol{p}^* \; (= (p_r/p_r^*) \cdot \boldsymbol{p}^*)$, contradicting the hypothesis that $\boldsymbol{p}$ is not proportionate to $\boldsymbol{p}^*$. Defining $\boldsymbol{p}^1 = (p_k/p_k^*) \cdot \boldsymbol{p}^*$, we have

$$\begin{aligned} p_k^1 &= p_k \\ p_j^1 &= \left(\frac{p_k}{p_k^*}\right) \cdot p_j^* \geq \left(\frac{p_j}{p_j^*}\right) \cdot p_j^* = p_j \qquad (j \neq k). \end{aligned} \tag{11.1}$$

In particular, in view of (10),

$$p_r^1 > p_r. \tag{11.2}$$

Hence

$$\begin{aligned} 0 = E_k(\boldsymbol{p}^*) &= E_k(\boldsymbol{p}^1) \qquad [\text{from the definition of } \boldsymbol{p}^1 \text{ and (ii)}] \\ &> E_k(\boldsymbol{p}) \qquad [\text{from (11.1), (11.2), and (ix)}]. \end{aligned}$$

This verifies (9.1).

In view of (2), (9.2) is immediate if $p_r = 0$. Hence, without loss, we henceforth assume that $p_r > 0$. Define now $\boldsymbol{p}^2 = (p_r/p_r^*) \cdot \boldsymbol{p}^*$. Then, as a matter of definition, we get

$$\begin{aligned} \boldsymbol{p}^2 &\leq \boldsymbol{p}, \\ p_r^2 &= p_r, \end{aligned} \tag{12}$$

and

$$p_k^2 < p_k.$$

Proceeding as before, it can be seen that

$$0 = E_r(\boldsymbol{p}^*) = E_r(\boldsymbol{p}^2) < E_r(\boldsymbol{p}).$$

(9.1′) and (9.2′) are now obvious since H_i is assumed to be sign-preserving. The following two-dimensional diagram (with $p_2/p_2^* < p_1/p_1^*$) will facilitate the intuitive understanding of the proof.

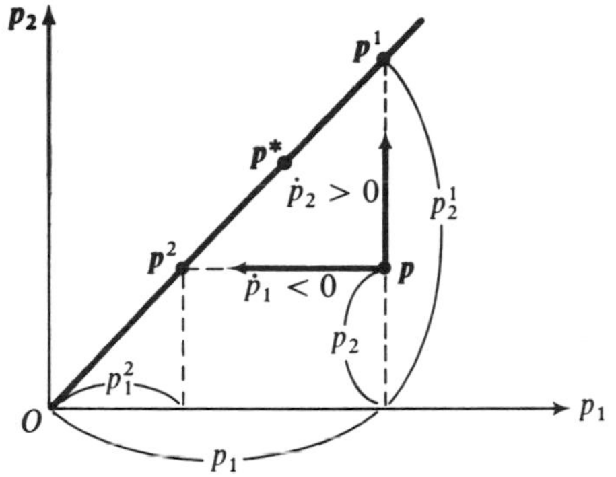

Figure 6.1

(5) Let $\boldsymbol{p}^*$ and $\bar{\boldsymbol{p}}$ be two positive vectors such that $\mathbf{0} = \boldsymbol{E}(\boldsymbol{p}^*) = \boldsymbol{E}(\bar{\boldsymbol{p}})$ and such that $\bar{\boldsymbol{p}} \neq \lambda \boldsymbol{p}^*$ for all positive λ. Then, from the foregoing arguments,

$$E_r(\bar{\boldsymbol{p}}) > 0 \quad \text{for any } r \text{ such that } \min_j \left\{\frac{\bar{p}_j}{p_j^*}\right\} = \frac{\bar{p}_r}{p_r^*},$$

contradicting the definition of $\bar{\boldsymbol{p}}$. □

Remark Concerning Lemma 2. Consider the normalized *tâtonnement* process (3) and let $\boldsymbol{q} > \mathbf{0}$ be a nonequilibrium normalized price vector. By l denote an index such that $\max\{\max_{j \neq k}(q_j/q_j^*), 1\} = q_l/q_l^*$. Then, from the fact that $q_j/q_j^* = (p_j/p_j^*)/(p_k/p_k^*)$, $j = 1, \ldots, N$, it immediately follows that $p_l/p_l^* = \max_j (p_j/p_j^*)$. Hence, the assertion 4 of the lemma, together with the assumed homogeneity, guarantees that $E_l(\boldsymbol{q}) < 0$. Similarly, for an index h satisfying $\min\{\min_{j \neq k}(q_j/q_j^*), 1\} = q_h/q_h^*$, $E_h(\boldsymbol{q}) > 0$. Moreover, either index h or l does exist according as $\max_{j \neq k}(q_j/q_j^*) \leqq 1$ or $\min_{j \neq k}(q_j/q_j^*) \geqq 1$. Thus, it is clear that the stabilizing force worked in the nonnormalized *tâtonnement* process is still in effect in the normalized *tâtonnement* process.

Before entering upon our discussion of global stability under strong gross substitutability, we provide an essential piece of mathematical information.

Definition 1. Let $f(\boldsymbol{x})$ be a real-valued function defined on an open set X of a metric space and let $\boldsymbol{a}$ be an arbitrary but fixed point of X. Then

$$\limsup_{\boldsymbol{x} \to \boldsymbol{a}} f(\boldsymbol{x}) \equiv \lim_{n \to \infty}\left(\sup_{\boldsymbol{x} \in B_n} f(\boldsymbol{x})\right)$$

and

$$\liminf_{\boldsymbol{x} \to \boldsymbol{a}} f(\boldsymbol{x}) \equiv \lim_{n \to \infty}\left(\inf_{\boldsymbol{x} \in B_n} f(\boldsymbol{x})\right)$$

where $B_n = B(\boldsymbol{a}, 1/n)$, $n = 1, 2, \ldots$.

Lemma 3. *Let $f(\boldsymbol{x})$ be as in Definition* 1. *Then there exists a sequence $\{\boldsymbol{x}_n\}$ with the properties that $\{\boldsymbol{x}_n\}$ converges to $\boldsymbol{a}$ as n tends to infinity and that*

$$\limsup_{\boldsymbol{x}\to\boldsymbol{a}} f(\boldsymbol{x})\left(\liminf_{\boldsymbol{x}\to\boldsymbol{a}} f(\boldsymbol{x})\right) = \lim_{n\to\infty} f(\boldsymbol{x}_n).$$

Proof. It suffices to verify the lemma for $\limsup_{\boldsymbol{x}\to\boldsymbol{a}} f(\boldsymbol{x})$. Let a_n be $\sup_{\boldsymbol{x}\in B_n} f(\boldsymbol{x})$. Then the numerical sequence $\{a_n\}$ is clearly monotone decreasing. For the time being let us assume that this sequence is bounded below.[6] Then, in view of Theorem 1 of Appendix A, $\inf A \equiv \alpha$ surely exists, where A denotes the set consisting of all terms of $\{a_n\}$.

To any $\varepsilon > 0$ there therefore corresponds an a_{n_ε} such that $a_{n_\varepsilon} < \alpha + \varepsilon$. Since $\{a_n\}$ is monotone decreasing, the above inequality further implies that

$$\alpha + \varepsilon > a_n \quad \text{for any } n \text{ such that } n \geqq n_\varepsilon.$$

Combining this inequality with the definition of α, it is easy to see that with any $\varepsilon > 0$ there is associated an n_ε such that if $n \geqq n_\varepsilon$ then $|\alpha - a_n| < \varepsilon$. In other words, we have obtained

$$\lim_{n\to\infty} a_n = \alpha.$$

Since $a_n = \sup_{\boldsymbol{x}\in B_n} f\boldsymbol{x})$, for any $\varepsilon > 0$ and for $n = 1, 2, \ldots$ we can find $\boldsymbol{x}_n$ of B_n such that $a_n \geqq f(\boldsymbol{x}_n) > a_n - \varepsilon$. From the definition of α, it again follows that $a_{n_\varepsilon} < \alpha + \varepsilon$ and that $\alpha \leqq a_n$ for all n. Proceeding as before with the monotone decreasing property of $\{a_n\}$ in mind, it immediately follows that $\lim_{n\to\infty} f(\boldsymbol{x}_n) = \alpha \equiv \limsup_{\boldsymbol{x}\to\boldsymbol{a}} f(\boldsymbol{x})$. Moreover, from the construction of $\{\boldsymbol{x}_n\}$ it is also evident that $\boldsymbol{x}_n \to \boldsymbol{a}$ as $n \to \infty$.

To complete the proof, suppose that $\lim_{n\to\infty} a_n = -\infty$. In this case, any sequence $\{\boldsymbol{x}_n\}$ constructed by picking up one $\boldsymbol{x}_n$ from each B_n possesses the desired property, since then we have

$$a_n \geqq f(\boldsymbol{x}_n) \quad \text{for } n = 1, 2, \ldots. \qquad \square$$

That completes the preliminaries. We can now state and prove the global stability of *tâtonnement* (3′).

Theorem 3. *Suppose that assumptions* (i′)–(iv) *and* (ix) *hold and let $\boldsymbol{p}(t)$ be the solution of* (3′) *starting from any given initial state $\boldsymbol{p}^0 > \boldsymbol{0}$. Then $\boldsymbol{p}(t)$ approaches a point $\bar{\boldsymbol{p}}$ on the ray $E = \{\lambda \boldsymbol{p}^*, \lambda > 0\}$ as time t tends to infinity.*

Proof. If $\boldsymbol{p}^0$ is proportionate to $\boldsymbol{p}^*$, which is positive by virtue of Lemma 2(1), there remains nothing to verify. We therefore assume that for any $\lambda > 0$, $\boldsymbol{p}^0 \neq \lambda \boldsymbol{p}^*$. By a and b respectively denote $\min_j \{p_j^0/p_j^*\}$ and $\max_j \{p_j^0/p_j^*\}$; let $v_M(t) = \max_j \{p_j(t)/p_j^*\}$ and let $v_m(t) = \min_j \{p_j(t)/p_j^*\}$. Then, by definition, $a = v_m(0)$ and $v_M(0) = b$. Define further $q_i(t) = p_i(t)/p_i^*$, $i = 1, \ldots, N$. Moreover, unless otherwise specified, we henceforth use the index $k(r)$ to mean $q_k(t) = v_M(t)$ $(q_r(t) = v_m(t))$. The proof is completed by demonstrating successively the following three assertions:

a. The solution $\boldsymbol{p}(t)$ always remains in the N-cell

$$Q = \{\boldsymbol{p}: ap_i^* \leqq p_i \leqq b \cdot p_i^*, i = 1, \ldots, N\};$$

b. $\{v_M(t)\}(\{v_m(t)\})$ is a bounded monotone decreasing (respectively, increasing) sequence;
c. $v_M(t)$ and $v_m(t)$ converge to a common limit, say $\bar{\lambda}$, as t tends to infinity.

(*Proof of* (a)). In view of Theorem 18 of Chapter 4, (3′) has a solution $\boldsymbol{p}(t)$ defined on some time interval, say $[0, v]$. We first show that $\boldsymbol{p}(t)$ remains in Q throughout this time interval. Arguing by contradiction, we suppose this false. Then, for some $j \in \{1, \ldots, N\}$ and for some $t_1 \in [0, v]$, $p_j(t_1) = p_j^* b$ because $\boldsymbol{p}(t)$ must pass through the boundary of Q before it emerges from the cell. Since we need no further proof if $\boldsymbol{p}(t_1) = b \cdot \boldsymbol{p}^*$, we can suppose without loss of generality that

$$p_i(t_1) < b \cdot p_i^* \quad \text{for } i \neq j.^7 \tag{13}$$

It follows from Lemma 2(4) that $\dot{p}_j(t_1)$ is negative. This, in conjunction with (13), implies that there exists a u positive and so small that $p_i(t) \leqq bp_i^*$ for all i and for all $t \in [t_1, t_1 + u] \subseteq [0, v]$. Repeating the same reasoning, it is clear that $p_i(t)$ cannot exceed bp_i^*. Similarly, $p_i(t) \geqq ap_i^*$ for any i and for all $t \in [0, v]$. Consider next the solution of (3′) starting from $\boldsymbol{p}(v)$.[8] Then by analogous reasoning the solution also remains in

$$Q' = \{\boldsymbol{p}: a'p_i^* \leqq p_i \leqq b'p_i^*, i = 1, \ldots, N\}$$

throughout the time interval on which it is defined, where $a' = \min_j \{q_j(v)\}$ and $b' = \max_j \{q_j(v)\}$. Since $\boldsymbol{p}(v) \in Q$, it is clear that Q' is contained in Q. The repetition of this kind of argument establishes assertion (a).[9]

(*Proof of* (b)). In view of assertion (a) just established, the boundedness of $\{v_M(t)\}$ and $\{v_m(t)\}$ is obvious. To show that $\{v_M(t)\}$ is monotone decreasing,[10] consider $\limsup_{h \to 0} ((v_M(t + h) - v_M(t))/h)$, where t is an arbitrary but fixed time point. Applying Lemma 3, there exists a sequence $\{h_n\}$ satisfying

$$\lim_{n \to \infty} h_n = 0 \tag{14.1}$$

and

$$\limsup_{h \to 0} \left(\frac{(v_M(t + h) - v_M(t))}{h} \right) = \lim_{n \to \infty} \left(\frac{(v_M(t + h_n) - v_M(t))}{h_n} \right). \tag{14.2}$$

By definition of $v_M(t)$, it is possible to choose an index k_n of $\{1, \ldots, N\}$ such that

$$v_M(t + h_n) = q_{k_n}(t + h_n) \qquad (n = 1, 2, \ldots). \tag{15}$$

Since k_n lies in a finite set $\{1, 2, \ldots, N\}$ for all values of n, $n = 1, 2, \ldots$, at least one index, say k, of the set must be represented infinitely many times. In other words, the sequence $\{n\}$ contains a subsequence $\{n_i\}$ such that

$$k_{n_i} = k \quad \text{for all } n_i. \tag{16}$$

Moreover, taking (14.1) into account, we can further assert that

$$\lim_{n_i \to \infty} h_{n_i} = 0.^{11} \tag{17}$$

From the continuity of $p_k(t)$ and $v_M(t)$, together with (15), (16), and (17), it follows that

$$v_M(t) = q_k(t).^{12} \tag{18}$$

Again by virtue of assertion (a), $(v_M(t + h_n) - v_M(t))/h_n$ surely converges to a finite limit as n tends to infinity,[13] whence so does $(v_M(t + h_{n_i}) - v_M(t))/h_{n_i}$, as $n_i \to \infty$. Therefore

$$\begin{aligned}
\limsup_{h \to 0}\left(\frac{(v_M(t + h) - v_M(t))}{h}\right) &= \lim_{h_{n_i} \to 0}\left(\frac{(v_M(t + h_{n_i}) - v_M(t))}{h_{n_i}}\right) \\
&= \lim_{h_{n_i} \to 0}\left(\frac{(q_k(t + h_{n_i}) - q_k(t))}{h_{n_i}}\right) \\
&= \dot{q}_k(t) \\
&= \left(\frac{1}{p_k^*}\right) \cdot H_k(\boldsymbol{p}(t)) < 0.
\end{aligned} \tag{19}$$

The final inequality of course comes from Lemma 2(4).

Similarly, we have

$$\liminf_{h \to 0}\left(\frac{(v_m(t + h) - v_m(t))}{h}\right) = \dot{q}_r(t) = \left(\frac{1}{p_r^*}\right) \cdot H_r(\boldsymbol{p}(t)) > 0. \tag{20}$$

(*Proof of* (c)). From the argument so far, it is obvious that $v_M(t)$ and $v_m(t)$ are convergent as t tends to infinity. It therefore suffices to prove that $\lim_{t \to \infty} (v_M(t) - v_m(t)) = 0$; since then, in view of the fact that $v_M(t) \geqq (p_i(t)/p_i^*) \geqq v_m(t)$ for all t and i, $v_M(t)$, $q_i(t)$, and $v_m(t)$ approach a common limit, say $\bar{\lambda}$, or, equivalently, $\boldsymbol{p}(t) \to \bar{\boldsymbol{p}} \equiv \bar{\lambda}\boldsymbol{p}^*$ as $t \to \infty$.

Let $g(t) = (v_M(t) - v_m(t))$. Evidently $g(t)$ is nonnegative for all $t \geqq 0$ and is monotone decreasing in t. To obtain a contradiction, suppose that $g(t)$ is not convergent to zero, so that there exists an $e > 0$ such that

$$v_M(t) \geqq v_m(t) + e \quad \text{for all nonnegative } t. \tag{21}$$

Making use of the fact that $g(t)$ is monotone decreasing in t, we get

$$\begin{aligned}
g(0) &\geqq g(0) - g(t) \\
&\geqq \int_0^t -\left(\frac{dg(u)}{du}\right) du \qquad \text{[by Kestelman [1937; Theorem 258, p. 178]]} \\
&= \int_0^t \left(\left(\frac{1}{p_k^*}\right) \cdot |H_k(\boldsymbol{p}(u))| + \left(\frac{1}{p_r^*}\right) \cdot |H_r(\boldsymbol{p}(u))|\right) du \qquad \text{[from (19) and (20)].}
\end{aligned}$$

In view of assertion (a) and (21), each $|H_i(\cdot)|$ can be viewed as a continuous function defined on a compact set on which it never takes the value zero. Hence a suitable choice of $\delta > 0$ enables us to assert that

$$g(0) \geqq \int_0^t \left(\left(\frac{1}{p_k^*} \right) \cdot |H_k(\boldsymbol{p}(u))| + \left(\frac{1}{p_r^*} \right) \cdot |H_r(\boldsymbol{p}(u))| \right) \cdot du \geqq \delta \cdot t.$$

This yields a self-contradiction for t sufficiently large. □

Remark Concerning Theorem 3. (i) This theorem concerns the non-normalized *tâtonnement* process. Given the assumption of gross substitutability, the global stability of a *tâtonnement* of normalized prices, such as our (3), can be verified *mutatis mutandis*. (See Arrow, Block and Hurwicz [1959; Lemma 7, pp. 100–101].) (ii) Inspection of the proof of Theorem 3 reveals that assumption (iii) (that is, Walras's Law) is unnecessary if the equilibrium price vector is positive.

3 A Dynamic Heckscher–Ohlin Model

We turn now to an examination of a dynamic version of a model economy that is rather special but has played a decisive role in the development of some branches of economics. We have in mind the small, open, two-by-two economy; that is, the neoclassical economy that has just two nonjoint products, two primary inputs and no produced inputs, and that is so small in relation to the rest of the world that commodity prices can be treated as independent of its decisions. The comparative statics of general neoclassical economies have been described in Chapter 5. Here we concentrate on the stability and monotonicity properties of a particular dynamic generalization of the static 2×2 model. In fact, several of the properties of the dynamic 2×2 model are valid whatever the number of primary factors. We therefore begin with the larger $m \times 2$ economy, retaining the additional dimensions until we are compelled to abandon them.

The ith production relationship is now written

$$\begin{aligned} x_j &\leqq f^j(v_{1j}, \ldots, v_{mj}) \\ &\equiv f^j(\boldsymbol{v}^j) \qquad (j = 1, 2). \end{aligned} \tag{22}$$

(The reader should compare (22) with Equation (1) of Chapter 5, in which produced inputs are accommodated.) The function f^j is supposed to possess continuous second derivatives and, if $\boldsymbol{v}^j > \boldsymbol{0}$, positive first derivatives; and it is assumed to be positively homogeneous of degree one. Moreover, it is assumed that inputs are strongly substitutable, in the sense that

$$f_{is}^j > 0 \quad \text{if } i \neq s;\ i, s = 1, \ldots, m;\ j = 1, 2. \tag{23}$$

Then, from (23) and the assumed homogeneity of f^j, it further follows that on the positive orthant

$$f^j_{ii} < 0 \quad (i = 1, \ldots, m; j = 1, 2). \tag{23'}$$

Each factor earns the value of its marginal product in the industry in which it is employed. Thus, expressing earnings in terms of the first commodity, the reward of the ith factor is $f^1_i \equiv \partial f^1/\partial v_{i1}$ in the first industry and $pf^2_i \equiv (p_2/p_1)(\partial f^2/\partial v_{i2})$ in the second industry. It is assumed moreover that the locus of production possibilities is strictly concave and hence that with any positive p, $p \in (\underline{p}, \bar{p})$, there is associated a unique production equilibrium, with unique factor inputs $\boldsymbol{v}^j > \mathbf{0}$, $j = 1, 2$, such that

$$\boldsymbol{v}^1 + \boldsymbol{v}^2 = \boldsymbol{v} \tag{24}$$

and

$$pf^2_i(\boldsymbol{v}^2) = f^1_i(\boldsymbol{v}^1), \tag{25}$$

where $\boldsymbol{v} \equiv (v_1, \ldots, v_m)$ is the vector of primary factor inputs. (Here $1/\underline{p}$ and $1/\bar{p}$ denote the minimum and maximum values of the slopes (absolute values of the tangents) of the locus of production possibilities.) The solutions to (24) and (25) are written $\boldsymbol{v}^j(p, \boldsymbol{v})$, $j = 1, 2$. From Theorem 1 of Chapter 5, the locus of production possibilities is strictly concave if and only if $\boldsymbol{v}^1$ and $\boldsymbol{v}^2$ are linearly independent for all $p \in (\underline{p}, \bar{p})$.

However, a factor does not necessarily receive the same reward in both industries. Each factor moves sluggishly towards the industry in which it is relatively well paid. The proposed dynamic extension of the static $m \times 2$ model consists then of (24) and the equations of adjustment or interindustrial migration,

$$\frac{d}{dt} v_{i1} \equiv \dot{v}_{i1} = \phi_i(f^1_i - pf^2_i) \qquad (i = 1, \ldots, m), \tag{26}$$

where ϕ_i is a differentiable, sign-preserving function with $\phi_i(0) = 0$ and $\phi'_i(0) > 0$. Setting $\dot{v}_{i1} = 0$, $i = 1, \ldots, m$, Equations (24) and (26) can be solved for the unique stationary inputs $v_{ij}(p, \boldsymbol{v})$.

It will be shown that the system [(24), (26)] is globally stable and moreover that the motion of $\boldsymbol{v}^j$ is globally monotone. It will be shown also that, if $m = 2$, real factor rewards are monotone near their stationary values.

Our approach to these propositions will be indirect. First we provide some definitions and primary results. From the homogeneity of f^j,

$$\left(\frac{1}{v_{1j}}\right) f^j(v_{1j}, v_{2j}, \ldots, v_{mj}) = f^j\left(1, \frac{v_{2j}}{v_{1j}}, \ldots, \frac{v_{mj}}{v_{1j}}\right).$$

We therefore define $k_{ij} \equiv v_{ij}/v_{1j}$ and write

$$f^j(1, k_{2j}, \ldots, k_{mj}) \equiv g^j(k_{2j}, \ldots, k_{mj}) \equiv g^j(\boldsymbol{k}^j).$$

Writing $f^j_{is} \equiv \partial^2 f^j/\partial v_{ij}\, \partial v_{sj}$, *etc.*, we next define the $m \times m$ and $(m - 1) \times (m - 1)$ matrices

$$\boldsymbol{F}^j \equiv (f^j_{is}) \qquad (j = 1, 2)$$

and

$$G^j \equiv (g^j_{is}) \qquad (j = 1, 2)$$

respectively. In view of the assumed strict substitutability of factors, $\boldsymbol{F}^j$ is strictly Metzlerian.

Next we introduce a useful mathematical result.

Lemma 4. *If a real-valued function $f(\boldsymbol{x})$ defined on $\mathbb{R}^n$ is twice continuously differentiable and positively homogeneous of degree one, and if $f_{ij}(\boldsymbol{x}) \equiv \partial^2 f(\boldsymbol{x})/\partial x_i\, \partial x_j$ is positive for all $i \neq j$ and all $\boldsymbol{x} > \boldsymbol{0}$, then the $n \times n$ matrix $F(\boldsymbol{x}) \equiv (f_{ij}(\boldsymbol{x}))$ is nonpositive definite for all $\boldsymbol{x} > \boldsymbol{0}$, whence $f(\boldsymbol{x})$ is concave in the positive orthant.*

PROOF. By virtue of Euler's theorem on homogeneous functions,

$$\boldsymbol{F}(\boldsymbol{x}) \cdot \boldsymbol{x} = \boldsymbol{0},$$

implying that every principal submatrix of $\boldsymbol{F}$ of order $n - 1$ has a ndd. On the other hand, since $f(\boldsymbol{x})$ is twice continuously differentiable, each such submatrix is symmetric. In view of Theorem 14 of Chapter 4, therefore, each such submatrix is negative definite.

Since det $F(\boldsymbol{x}) = 0$, the concavity of $f(\boldsymbol{x})$ follows from Lemmas 4 and 4″ of Chapter 4 and Lemma 14(b) of Chapter 1. □

With the aid of Lemma 4 we can describe some of the properties of the matrices $\boldsymbol{F}^j$ and $\boldsymbol{G}^j$. First, from the homogeneity of f^j it follows by a direct but tedious calculation that

(P1)
$$\boldsymbol{F}^j = \frac{1}{v_{ij}} \begin{pmatrix} \boldsymbol{k}^j \cdot \boldsymbol{G}^j \cdot (\boldsymbol{k}^j)' & -\boldsymbol{k}^j \cdot \boldsymbol{G}^j \\ -\boldsymbol{G}^j \cdot (\boldsymbol{k}^j)' & \boldsymbol{G}^j \end{pmatrix}.$$

Second, recalling the homogeneity of f^j and the strict Metzlerian property of $\boldsymbol{F}^j$, we deduce from Lemma 4 that

P2a. $\boldsymbol{F}^j$ is nonpositive definite for all $\boldsymbol{v}^j > \boldsymbol{0}$, and f^j is concave on the positive orthant.

However, from the proof of Lemma 4,

P2b. $\boldsymbol{G}^j$ is negative definite for all $\boldsymbol{k}^j > \boldsymbol{0}$.

Finally, we list

P3. $\boldsymbol{F}^1 + p\boldsymbol{F}^2$ is (i) Metzlerian and (ii) negative definite on the positive orthant.

Proof of (P3). (i) The assertion follows from (23) and the positivity of $\boldsymbol{v}^1(p, \boldsymbol{v})$.

(ii) Since $\boldsymbol{F}^j$ is nonpositive definite for any $\boldsymbol{v}^j > \boldsymbol{0}$, so is $\boldsymbol{F}^1 + p\boldsymbol{F}^2$. It therefore suffices to verify the nonsingularity of $\boldsymbol{F}^1 + p\boldsymbol{F}^2$. The possibility that $\boldsymbol{k}^1 = \boldsymbol{k}^2 = \boldsymbol{k}$ has been assumed away. We therefore confine our attention to the case $\boldsymbol{k}^1 \neq \boldsymbol{k}^2$. Bearing (P1) and (P2b) in mind, a direct computation yields

$$N(\boldsymbol{F}^j) \equiv \{\xi : \boldsymbol{F}^j\xi = \boldsymbol{0}\} = \left\{\begin{pmatrix} 1 \\ \boldsymbol{k}^j \end{pmatrix} t, t \in \mathbb{R}\right\} \qquad (j = 1, 2).$$

Hence

$$N(\boldsymbol{F}^1) \cap N(\boldsymbol{F}^2) = \{\boldsymbol{0}\} \tag{27}$$

for, otherwise, there would exist $t_j \neq 0$, $j = 1, 2$, such that

$$\begin{pmatrix} 1 \\ \boldsymbol{k}^1 \end{pmatrix} t_1 = \begin{pmatrix} 1 \\ \boldsymbol{k}^2 \end{pmatrix} t_2,$$

implying that $t_1 = t_2$ and contradicting the hypothesis that $\boldsymbol{k}^1 \neq \boldsymbol{k}^2$. Suppose that $\boldsymbol{F}^1 + p\boldsymbol{F}^2$ is singular. Then $N(\boldsymbol{F}^1) \cap N(\boldsymbol{F}^2)$ contains a nonzero vector, contradicting (27). □

As a final preliminary, we prove

Lemma 5. *For any* $p \in (\underline{p}, \bar{p})$,

$$\left(\frac{\partial v_{11}(p, \boldsymbol{v})}{\partial p}, \ldots, \frac{\partial v_{m1}(p, \boldsymbol{v})}{\partial p}\right)' \equiv \frac{\partial \boldsymbol{v}^1(p, \boldsymbol{v})}{\partial p} < \boldsymbol{0}.$$

Proof. Totally differentiating (12) with respect to p, and recalling (11), we obtain

$$(\boldsymbol{F}^1 + p\boldsymbol{F}^2)\left(\frac{\partial \boldsymbol{v}^1(p, \boldsymbol{v})}{\partial p}\right) = \boldsymbol{f}^2,$$

where $\boldsymbol{f}^2 \equiv (f_1^2, \ldots, f_m^2)'$ and $f_i^2 \equiv \partial f^2/\partial v_{i2}$ and where the derivatives are evaluated at the equilibrium values $\boldsymbol{v}^2(p, \boldsymbol{v})$ and $\boldsymbol{v}^1 = \boldsymbol{v} - \boldsymbol{v}^2(p, \boldsymbol{v})$. Since $\boldsymbol{F}^1 + p\boldsymbol{F}^2$ is Metzlerian and since $\boldsymbol{f}^2 > \boldsymbol{0}$, we may infer from Theorem 7 of Chapter 1 that

$$\frac{\partial \boldsymbol{v}^1(p, \boldsymbol{v})}{\partial p} = (\boldsymbol{F}^1 + p\boldsymbol{F}^2)^{-1}\boldsymbol{f}^2 < \boldsymbol{0}. \qquad \square$$

Preliminary matters out of the way, we can proceed to state and verify two global properties of the $m \times 2$ economy.

Theorem 4 (Global stability). *If* $p \in (\underline{p}, \bar{p})$ *then the adjustment process* (13) *is globally stable.*

PROOF. The marginal products f_i^j are positively homogeneous of degree zero in $\boldsymbol{v}^j$; $\boldsymbol{F}^1 + p\boldsymbol{F}^2$ is Metzlerian; and, since $p \in (\underline{p}, \bar{p})$, $\boldsymbol{v}^1(p, \boldsymbol{v})$ is positive. The assertion then follows from Theorem 3 and the remark concerning Theorem 3(ii).

Theorem 5 (Global Monotonicity). *Suppose that the relative commodity price changes from p^α to p^β, with both p^α and p^β in $(\underline{p}, \bar{p})$. Then:*

a. $\boldsymbol{v}^1(p^\alpha, \boldsymbol{v}) \gtrless \boldsymbol{v}^1(p^\beta, \boldsymbol{v})$ *as* $p^\alpha \lesseqgtr p^\beta$; *and*
b. $\dot{\boldsymbol{v}}^1 \lesseqgtr 0$ *as* $p^\alpha \lesseqgtr p^\beta$ *if the economy is initially in stationary equilibrium.*

PROOF. (a) Applying the Mean Value Theorem to each $v_{i1}(p, \boldsymbol{v})$, there exists $\gamma_i \in (0, 1)$ such that

$$v_{i1}(p^\beta, \boldsymbol{v}) - v_{i1}(p^\alpha, \boldsymbol{v}) = (p^\beta - p^\alpha)\,\frac{\partial v_{i1}(\gamma_i p^\beta + (1 - \gamma_i)p^\alpha)}{\partial p} \qquad (i = 1, \ldots, m).$$

Moreover, $\gamma_i p^\beta + (1 - \gamma_i)p^\alpha$ is in $(\underline{p}, \bar{p})$. The assertion then follows from Lemma 5.

(b) Suppose without loss that $p^\alpha < p^\beta$. In view of (12), the interindustrial migration of factors is described by

$$\begin{aligned} \dot{v}_{i1} &= \phi_i(f_i^1(\boldsymbol{v}^1) - p^\beta f_i^2(\boldsymbol{v} - \boldsymbol{v}^1)) \\ &\equiv \phi_i(g_i(\boldsymbol{v}^1)) \qquad (i = 1, \ldots, m). \end{aligned} \tag{28}$$

In view of Theorem 1 and of the assumed properties of ϕ_i, it suffices to show that $g_i(\boldsymbol{v}^1) < 0$ for all i and all $t > 0$.

Since $p^\alpha < p^\beta$, $g_i(\boldsymbol{v}^1(p^\alpha, \boldsymbol{v})) < 0$, $i = 1, \ldots, m$. From the continuity of each $g_i(\boldsymbol{v}^1)$ there therefore exists a nonempty set

$$S \equiv \{\boldsymbol{v}^1 : \boldsymbol{g}(\boldsymbol{v}^1) \equiv (g_1(\boldsymbol{v}^1), \ldots, g_m(\boldsymbol{v}^1))' < \boldsymbol{0}\}.$$

Let ∂S be the boundary of S. It will be shown that if $\boldsymbol{v}^1 \in \partial S$ then both $\boldsymbol{g}(\boldsymbol{v}^1) \leqq \boldsymbol{0}$ and $g_i(\boldsymbol{v}^1) = 0$ for some i. Thus suppose that $\boldsymbol{v}^1 \in \partial S$. Then $\boldsymbol{g}(\boldsymbol{v}^1) \not< \boldsymbol{0}$; for otherwise, again from the continuity of $\boldsymbol{g}(\boldsymbol{v}^1)$, $\boldsymbol{v}^1$ would be an interior point of S. Hence the set $I(\boldsymbol{v}^1)$ of indices for which $g_i(\boldsymbol{v}^1) \geqq 0$ is nonempty. In fact $g_i(\boldsymbol{v}^1) = 0$ for all $i \in I(\boldsymbol{v}^1)$. Suppose the contrary, that $g_{i_0}(\boldsymbol{v}^1) > 0$ for some $i_0 \in I(\boldsymbol{v}^1)$. Then, from the continuity of $g_{i_0}(\boldsymbol{v}^1)$, there exists a δ-neighborhood $B(\boldsymbol{v}^1, \delta)$ of $\boldsymbol{v}^1$ such that $g_{i_0}(\boldsymbol{x}) > 0$ for any $\boldsymbol{x} \in B(\boldsymbol{v}^1, \delta)$, contradicting the hypothesis.

Turning to the heart of the proof, let $\boldsymbol{v}^1(t, \boldsymbol{v}^1(p^\alpha, \boldsymbol{v}))$ be that solution to (28) which passes through the initial point $\boldsymbol{v}^1(p^\alpha, \boldsymbol{v})$. Since the solution is a continuous function of time and since $\boldsymbol{v}^1(0, \boldsymbol{v}^1(p^\alpha, \boldsymbol{v}))$ is in S, the solution must pass through ∂S before it reaches a point exterior to S. Suppose then that $\boldsymbol{v}^1(t_1, \boldsymbol{v}^1(p^\alpha, \boldsymbol{v}))$ is in ∂S. From the argument of the preceding paragraph, $I(\boldsymbol{v}^1(t_1, \boldsymbol{v}^1(p^\alpha, \boldsymbol{v})))$ is nonvacuous. If $I(\boldsymbol{v}^1(t_1, \boldsymbol{v}^1(p^\alpha, \boldsymbol{v})))$ consists of all m indices, there is nothing left to prove; suppose then that it consists of the first m_0

indices, $m_0 < m$. We have already shown that $\partial g_i/\partial v_{k1} = (\partial f_i^1/\partial v_{k1}) + p^\beta(\partial f_i/\partial v_{k2}) > 0$ for all $i \neq k$. For any sufficiently small and positive Δt, therefore, not only will $\dot{v}_{i1}(t + \Delta t, \boldsymbol{v}^1(p^\alpha, \boldsymbol{v}))$ remain negative for $i = m_0 + 1, \ldots, m$ but it will be negative again for $i = 1, \ldots, m_0$. Thus the solution point either remains in the boundary of S, as when $I(\boldsymbol{v}^1(t_1, \boldsymbol{v}^1(p^\alpha, \boldsymbol{v})))$ contains all m indices, or it rebounds into S. In either case, it remains in the set $S \cup \partial S$. □

Theorem 5 tells us that factor allocations are monotone functions of time, but it tells us nothing about the time profiles of real factor rewards. In fact, without additional assumptions it does not seem possible to prove that real rentals respond monotonically to changes in commodity prices, even when $m = 2$ and even when the price changes are small. We proceed to develop a sufficient condition for the local monotonicity of real rentals when $m = 2$.

Suppose again that the relative price of the two commodities increases from p^α to p^β, and let the stationary value of $\boldsymbol{k}^j \equiv k^j$ be written $k^j(p, \boldsymbol{v})$, $j = 1, 2$. In view of the first degree homogeneity of $f^j(\boldsymbol{v}^j)$, (26) can be re-written as

$$\begin{aligned} \dot{v}_{11} &= \phi_1((g^1 - k^1 Dg^1) - p(g^2 - k^2 Dg^2)), \\ \dot{v}_{21} &= \phi_2(Dg^1 - pDg^2), \end{aligned} \tag{26'}$$

where $Dg^j \equiv dg^j/dk^j$. Linearizing (26′) at the stationary values $k^j(p^\beta, \boldsymbol{v})$, we obtain

$$\begin{aligned} \dot{\boldsymbol{x}} &\equiv \begin{pmatrix} \dot{x}_1 \\ \dot{x}_2 \end{pmatrix} \\ &= \begin{pmatrix} 1/v_{11} & 0 \\ 0 & 1/v_{12} \end{pmatrix} \begin{pmatrix} D^2g^1 \cdot (\phi_2' + (k^1)^2\phi_1') & -pD^2g^2 \cdot (\phi_2' + k^1k^2\phi_1') \\ -D^2g^1 \cdot (\phi_2' + k^1k^2\phi_1') & pD^2g^2 \cdot (\phi_2 + (k^2)^2\phi_1') \end{pmatrix} \begin{pmatrix} x_1 \\ x_2 \end{pmatrix} \\ &\equiv \begin{pmatrix} 1/v_{11} & 0 \\ 0 & 1/v_{12} \end{pmatrix} \boldsymbol{Bx} \\ &\equiv \boldsymbol{Ax}, \end{aligned} \tag{29}$$

where $x_i \equiv k^i - k^i(p^\beta, \boldsymbol{v})$, $\phi_i' \equiv \phi_i'(0) > 0$, and $D^2g^j \equiv d^2g^j/d(k^j)^2$.

We begin our study of the system (29) by establishing some relevant properties of the coefficient matrix $\boldsymbol{A}$. Consider the matrix $\boldsymbol{B}$, in terms of which $\boldsymbol{A}$ is defined. The restrictions placed on ϕ_i and g^j imply that $\boldsymbol{B}$ is Metzlerian. Moreover, a direct calculation shows that

$$\det \boldsymbol{B} = p \cdot D^2g^1 \cdot D^2g^2 \cdot \phi_1'\phi_2' \cdot (k^1 - k^2)^2 > 0.$$

Thus $-\boldsymbol{B}$ is a P-matrix, with negative off-diagonal elements. It follows from the Hawkins–Simon Theorem (Theorem 7 of Chapter 1) that $\boldsymbol{B}^{-1} < \boldsymbol{0}$

and $\boldsymbol{B}$ has a ndd. From Corollary 4 of Chapter 4, therefore, $\boldsymbol{A}$ has the following properties:

i. every off-diagonal element is positive;
ii. $\boldsymbol{A}^{-1} < \boldsymbol{0}$; and
iii. $\boldsymbol{A}$ has a ndd and hence is D-stable.

Let $\lambda_i, i = 1, 2$, be the eigenvalues of $\boldsymbol{A}$ and let $\boldsymbol{\xi}_i, i = 1, 2$, be the associated eigenvector. Property (i) of $\boldsymbol{A}$, together with Lemma 5 of Chapter 2 and the hypothesis that $m = 2$, implies that λ_1 and λ_2 are real and distinct. Without loss, therefore, it can be assumed that λ_1 is the Frobenius root of $\boldsymbol{A}$. Then, by virtue of the above Lemma again, ξ_1 can be taken to be positive. Applying Lemma 4* of Chapter 4, $\boldsymbol{X} \equiv (\boldsymbol{\xi}_1, \boldsymbol{\xi}_2)$ is nonsingular. It follows that $\boldsymbol{\xi}_i e^{\lambda_i t}$, $i = 1, 2$, are linearly independent solutions to (16); that is, for any t, $\sum_{j=1}^{2} c_j \boldsymbol{\xi}_j e^{\lambda_j t} = \boldsymbol{0}$ implies that $c_1 = c_2 = 0$. In view of Lemma 1 of Chapter 4, therefore, the general solution to (29) can be written as

$$\boldsymbol{x}(t) = \boldsymbol{X} \boldsymbol{e}^{\lambda t} \boldsymbol{y}, \tag{30}$$

where $\boldsymbol{e}^{\lambda t}$ is a diagonal matrix with the ith diagonal element $e^{\lambda_j t}$ and $\boldsymbol{y} = \boldsymbol{X}^{-1} \cdot \boldsymbol{x}(0)$.

Consider now the sign of $\dot{\boldsymbol{x}}(t)$. From (30),

$$x_i(t) = \sum_{j=1}^{2} \xi_{ij} y_j e^{\lambda_j t},$$

where ξ_{ij} is the (ij)th element of $\boldsymbol{X}$. Hence

$$\dot{x}_i(t) = \sum_j \lambda_j \xi_{ij} y_j e^{\lambda_j t} \tag{31}$$

and, in particular,

$$\dot{x}_i(0) = \sum_j \lambda_j \xi_{ij} y_j. \tag{32}$$

Substituting from (32) in (31) and arranging terms,

$$\dot{x}_i(t) = e^{\lambda_1 t}[\dot{x}_i(0) e^{(\lambda_2 - \lambda_1)t} + \lambda_1 \xi_{i1} y_1 (1 - e^{(\lambda_2 - \lambda_1)t})].$$

Now $1 > e^{(\lambda_2 - \lambda_1)t} > 0$ for all t; moreover, $\lambda_1 \xi_{i1} < 0$. Hence $\boldsymbol{x}(t)$ is monotone for all t if $\dot{\boldsymbol{x}}(0) y_1 < \boldsymbol{0}$. It will be shown that this inequality necessarily holds if ϕ_i and g^j are subject to the additional restrictions

$$\begin{aligned} \frac{k^2(g^2 - k^2 Dg^2)}{Dg^2} &> \frac{\phi_2'}{\phi_1'} > \frac{k^1(g^2 - k^2 Dg^2)}{Dg^2} \\ \frac{k^2(g^2 - k^2 Dg^2)}{Dg^2} &< \frac{\phi_2'}{\phi_1'} < \frac{k^1(g^2 - k^2 Dg^2)}{Dg^2}. \end{aligned} \tag{33}$$

If initially the economy is in the equilibrium state defined by p^α, $p^\alpha \in (\underline{p}, \bar{p})$, and if the price change $\Delta p \equiv p^\beta - p^\alpha$ is sufficiently small then the initial state of the economy can be approximated by

$$\boldsymbol{x}(0) = -\boldsymbol{A}^{-1} \cdot \boldsymbol{b} \cdot \Delta p \tag{34}$$

where

$$\boldsymbol{b}' \equiv (k^1(g^2 - k^2 Dg^2)\phi_1' - Dg^2\phi_2', Dg^2\phi_2' - k^2(g^2 - k^2 Dg^2)\phi_1').$$

Now from (29) and (34)

$$\dot{\boldsymbol{x}}(0) = \boldsymbol{A}\boldsymbol{x}(0) = -\boldsymbol{b} \cdot \Delta p. \tag{35}$$

Let $\boldsymbol{\eta}_i$ denote the row eigenvector of $\boldsymbol{A}$ associated with λ_i. Lemma 4* of Chapter 4 again ensures the nonsingularity of the matrix

$$\boldsymbol{Y} = \begin{pmatrix} \boldsymbol{\eta}_1 \\ \boldsymbol{\eta}_2 \end{pmatrix}.$$

Hence

$$\boldsymbol{Y}\boldsymbol{X} = \begin{pmatrix} \boldsymbol{\eta}_1 \xi_1 & 0 \\ 0 & \boldsymbol{\eta}_2 \xi_2 \end{pmatrix}$$

is nonsingular. A direct calculation then confirms that

$$\boldsymbol{X}^{-1} = \begin{pmatrix} \boldsymbol{\eta}_1 \xi_1 & 0 \\ 0 & \boldsymbol{\eta}_2 \xi_2 \end{pmatrix}^{-1} \cdot \boldsymbol{Y}$$

Moreover, by virtue of Lemma 5 of Chapter 2, $\boldsymbol{\eta}_1 > \boldsymbol{0}$ and $\boldsymbol{\eta}_1 \cdot \xi_1 > 0$. It follows that the first row of $\boldsymbol{X}^{-1}$ is positive. This, in conjunction with (34), (35) and property (ii) of $\boldsymbol{A}$, implies that if $\boldsymbol{b}$ is either positive or negative then $\dot{\boldsymbol{x}}(0) y_1 < \boldsymbol{0}$. And, as is easily seen, $\boldsymbol{b}$ is either positive or negative under (33).

Thus we have established

Theorem 6 (Local Monotonicity of Real Factor Rewards). *Suppose that a 2×2 economy is initially in equilibrium at the price ratio p^α, $p^\alpha \in (\underline{p}, \bar{p})$ and that (33) is satisfied. Then any sufficiently small change in price produces a monotone response of real factor rewards.*

From Theorems 4 and 6 we have the following

Corollary. *Suppose that a 2×2 economy is in equilibrium at the price ratio p^α, $p^\alpha \in (p, \bar{p})$, and that the price changes to p^β, $p^\beta \in (p, \bar{p})$. Suppose further that (33) is satisfied at a time point t_0 such that $k^j(t_0)$ is sufficiently close to $k^j(p^\beta, \boldsymbol{v})$. Then the convergence of $k^j(t)$ to $k^j(p^\beta, \boldsymbol{v})$ is monotone for all $t \geqq t_0$.*

Notes to Chapter 6

1. An alternative *tâtonnement* compatible with the nonnegativity of every revised price vector is

$$\dot{q}_i = \begin{cases} h_i(E_i(\boldsymbol{q})) & \text{if } q_i > 0 \text{ or } E_i(\boldsymbol{q}) \geqq 0 \\ 0 & \text{if } q_i = 0 \text{ and } E_i(\boldsymbol{q}) < 0 \end{cases} \quad (i \neq k) \tag{3'}$$
$$q_k = 1,$$

where $h_i(\cdot)$ is of course as in the text. Thus in the system of (3′) and (4), the nonnegativity of prices is secured by the *tâtonnement* rule to which the auctioneer is subject, while in the *tâtonnment* of the text all market participants are assumed to so behave that prices are always nonnegative (in fact, positive). However, the discontinuity involved in (3′) makes it cumbersome to verify the existence of a solution of that system.
2. ρ, j, $\boldsymbol{S}$, and λ^* are dependent on $\boldsymbol{q}$. If necessary, this dependence will be indicated, as, for example, in $\lambda^*(\boldsymbol{q})$. If either j is odd or $\boldsymbol{W}$ contains at least one row or column that is nonnegative then $\lambda^* > 0$. Thus we implicitly assume either one of these conditions. Moreover, the vector $\boldsymbol{x}$ appearing in (5) is of course $(N - 1)$-dimensional.
3. Since we assume that assumption (viii) holds for any $\boldsymbol{q} > \boldsymbol{0}$, $\boldsymbol{J}(\boldsymbol{q}^*)$ is quasi-negative definite. Benavie's proof (Benavie [1972; p. 239]) contains a slip in this respect.
4. In view of Lemma 2 of Appendix A, $\lim_{\|\boldsymbol{q}\|_x \to \infty} V(\boldsymbol{q}) = \infty$ where $\|\boldsymbol{q}\|_x$ denotes any vector norm of $\boldsymbol{q}$.
5. Assertion (1) of the lemma is a revised version of Arrow, Block and Hurwicz [1959; Lemma 1]. Our revision is due to Hotaka [1971; Lemma 1].
6. Noticing that $\{a_n\}$ and $\{b_n\}$ are respectively monotone decreasing and monotone increasing and that $a_n \geqq b_n$ for all n, a sufficient condition for the sequence $\{a_n\}$ ($\{b_n\}$) to be bounded below (above) is that b_1 (a_1) exist, where $b_n = \inf_{\boldsymbol{x} \in B_n} f(\boldsymbol{x})$.
7. Almost needless to say, j is not necessarily unique. Fastidiously, we should state that

$$p_i(t_1) < bp_i^* \quad \text{for } i \notin J_{t_1} = \{j \in \{1, \ldots, N\} : p_j(t_1) = bp_j^*\}.$$

8. Note that Theorem 18 of Chapter 4 is again applicable, since $\boldsymbol{p}(v)$ lies in $Q \subseteq \operatorname{int} \mathbb{R}^n_+$.
9. What is really argued above is the continuability of the solution of ordinary differential equations. Speaking in heuristic fashion, a local solution, such as $\boldsymbol{p}(t)$, defined only on a time interval $[0, v]$ can be continued unless its terminal state is on or very close to the boundary of the domain, for then the existence theorem by which the existence of the first local solution is guaranteed is again applicable. A general and rigorous discussion of this topic can be found in Coddington and Levinson [1955; Theorem 4.1 of Chapter 1].

   In stability analysis, local or global, our attention is focused on the behavior of a solution when t tends to infinity. Therefore, the existence of a solution defined for all nonnegative t is an indispensable prerequisite for the successful analysis of stability. Nevertheless, in Theorems 1 and 2 of this chapter and in Theorems 19, 20, and 21 of Chapter 4, nothing has been said about the continuability of the solution. Fortunately, it is known that the solution is continuable under some suitable conditions such as (a) and (b) of Theorem 19 of Chapter 4.

   Interested readers may consult Yoshizawa [1966; Theorem 3.4].
10. From the proof of (a), it immediately follows that $v_M(t) \geqq v_M(t')$ for $t \leqq t'$ and that $v_m(t) \leqq v_m(t')$ if $t \leqq t'$. However, what we really want to prove are (19) and (20) below.

11. Recall the well-known fact that any subsequence of a convergent sequence approaches the limit of the sequence.
12. More specifically, this can be shown as follows. With the continuity of $v_M(t)$, (14.1), (17), and Footnote 11 in mind, we at once see that

$$v_M(t) = \lim_{h_n \to 0} v_M(t + h_n) = \lim_{h_{n_i} \to 0} v_M(t + h_{n_i}).$$

Hence, in view of (15), (16), and the continuity of $p_k(t)$, we further obtain

$$v_M(t) = \lim_{h_{n_i} \to 0} v_M(t + h_{n_i}) = q_k(t).$$

13. Because of assertion (a), the condition stated in Footnote 6 is surely satisfied.

Appendix A

In this appendix, we state and prove several fundamental theorems concerning compact sets in a metric space. We begin with the properties of the real number system, presented without proof. Readers wishing to see rigorous proofs of these properties may consult Landau [1951] and Rudin [1964; especially Chapter 1].

Properties of the Real Number System

By the symbol $\mathbb{R}$ we mean the set of all real numbers.

I. $\mathbb{R}$ is a field.

More specifically,

I.A. To every pair, a and b, of $\mathbb{R}$ there corresponds a real number $a + b$, called the sum of a and b, such that

1. $a + b = b + a$,
2. $a + (b + c) = (a + b) + c$, where c is any real number,
3. there exists a unique element 0 of $\mathbb{R}$ such that $a + 0 = a$ for every a of $\mathbb{R}$, and
4. to every real number there corresponds a unique real number $-a$ such that $a + (-a) = 0$.

I.B. With every pair, a and b, of $\mathbb{R}$ there is associated a real number $a \cdot b$, called the product of a and b, such that

5. $a \cdot b = b \cdot a$,
6. $a \cdot (b \cdot c) = (a \cdot b) \cdot c$, where c is any real number,

7. there exists a unique nonzero real number, denoted by 1, such that $a \cdot 1 = a$ for every $a \in \mathbb{R}$, and
8. if a is a nonzero real number, then there exists a unique real number a^{-1} (or $1/a$) such that $a \cdot a^{-1} = 1$.

I.C. For any a, b and c of $\mathbb{R}$

$$a(b + c) = a \cdot b + a \cdot c.$$

II. Let a and b be any real numbers. Then exactly one of the following alternatives holds:

$$a > b, \qquad a = b, \qquad a < b.$$

III. *The denseness of real numbers.* If a and b are real numbers such that $a < b$, then there exists a real number c such that $a < c < b$.

IV. *The continuity of real numbers.* Let R_1 and R_2 be two subsets of $\mathbb{R}$ such that

i. $R_1 \cap R_2 = \emptyset$,
ii. $R_1 \cup R_2 = R$,
iii. $R_1 \neq \emptyset$, $R_2 \neq \emptyset$,
iv. If $a \in R_1$, and $b \in R_2$, then $a < b$.

Then exactly one of the following alternatives holds:

R_1 contains a largest number; or

R_2 contains a least number.

The final property, (IV), which is also called the *completeness theorem* for the real numbers, plays an especially important role in our discussion below.

Definition 1. A subset E of $\mathbb{R}$ is said to be *bounded above* (*respectively, below*), if there exists $M \in \mathbb{R}$ such that

$$M \geqq a \quad (\text{respectively, } M \leqq a) \quad \text{for all } a \in E.$$

Such a real number M, if it exists, is called an *upper* (*respectively, lower*) *bound of* E. Furthermore, E is said to be *bounded* if it is bounded above and below.

Definition 2. The minimal (respectively, maximal) upper (respectively, lower) bound of E is called *the supremum* (*respectively, infimum*) of E and denoted by sup E (respectively, inf E).

Let α denote sup E. Then it is clear that

$$\alpha \geqq a \quad \text{for all } a \in E \tag{1}$$

and that for any positive ε there exists an element a_ε of E such that

$$\alpha - \varepsilon < a_\varepsilon. \tag{2}$$

Conversely, a real number α satisfying (1) and (2) is equal to sup E. The same, of course, holds for inf E, *mutatis mutandis.*

Definition 3. Let X and Y be any given nonempty subsets. A function (mapping) of X into Y is a rule that associates an element $\boldsymbol{x}$ of X with an element $\boldsymbol{y}$ of Y, which is called the image of $\boldsymbol{x}$ under f and denoted by $\boldsymbol{y} = f(\boldsymbol{x})$. We often denote the function by $f: X \mapsto Y$. The sets X and Y are called the domain and the range of f, respectively. The set consisting of the images of all $\boldsymbol{x} \in X$ is denoted by $f(X)$. If with a given $\boldsymbol{y}$ of Y there is associated an element $\boldsymbol{x}$ of X such that $\boldsymbol{y} = f(\boldsymbol{x})$, then $\boldsymbol{x}$ is called an inverse image of $\boldsymbol{y}$ under f and denoted by $f^{-1}(\boldsymbol{y})$. Let V be any subset of Y. Then, the set $\{f^{-1}(\boldsymbol{y}): \boldsymbol{y} \in V\}$ is denoted by $f^{-1}(V)$.

Theorem 1. *Let E be a nonempty subset of* $\mathbb{R}$. *Then there exists* sup E *(respectively,* inf E*) if E is bounded above (respectively, below).*

This theorem follows from the postulated property (IV) of real numbers, and plays an essential role in the proof of Theorem 20 of Chapter 1.

Careful consideration of the familiar notion of the distance between two points leads us to the investigation of metric spaces.

Definition 4. A nonempty set X, the elements of which we shall call points, is said to be *a metric space* if with any ordered pair of points $(\boldsymbol{x}, \boldsymbol{y})$ there is associated a real number $d(\boldsymbol{x}, \boldsymbol{y})$ such that

1. $d(\boldsymbol{x}, \boldsymbol{y}) \geqq 0$ and $d(\boldsymbol{x}, \boldsymbol{y}) = 0$ if and only if $\boldsymbol{x} = \boldsymbol{y}$,
2. $d(\boldsymbol{x}, \boldsymbol{y}) = d(\boldsymbol{y}, \boldsymbol{x})$, and
3. $d(\boldsymbol{x}, \boldsymbol{y}) \leqq d(\boldsymbol{x}, \boldsymbol{z}) + d(\boldsymbol{z}, \boldsymbol{y})$ for any $\boldsymbol{z} \in X$ (triangle inequality)

$d(\boldsymbol{x}, \boldsymbol{y})$ is called a *metric* defined on X.

Let X and E be a metric space and a subset of X, respectively. All points mentioned below are understood to be elements of X, unless otherwise specified.

Definition 5. Let ε be any positive number.

1. An *ε-neighborhood* of a point $\boldsymbol{x}$ is the set $B(\boldsymbol{x}, \varepsilon)$ consisting of all points $\boldsymbol{y}$ such that $d(\boldsymbol{x}, \boldsymbol{y}) < \varepsilon$. Symbolically,

$$B(\boldsymbol{x}, \varepsilon) = \{\boldsymbol{y} \in X : d(\boldsymbol{x}, \boldsymbol{y}) < \varepsilon\}.$$

2. A point $\boldsymbol{p}$ is a *limit point* of E if every ε-neighborhood of $\boldsymbol{p}$ contains a point $\boldsymbol{q}$ of E such that $\boldsymbol{p} \neq \boldsymbol{q}$. Alternatively, $\boldsymbol{p}$ is a limit point of E if every ε-neighborhood contains infinitely many points of E.
3. Denote the set of all limit points of E by L_E. If $L_E \subseteq E$, then E is said to be *closed* (or a closed set).

4. E is said to be *open* (or an open set) if with every point $\boldsymbol{p}$ of E there is associated a δ-neighborhood $B(\boldsymbol{p}, \delta)$ such that $B(\boldsymbol{p}, \delta) \subseteq E$. Such a point $\boldsymbol{p}$ of E, if it exists, is called an *interior point* of E.

   Moreover, let E^c denote the complementary set of E. Then, a point $\boldsymbol{q} \in E^c$ is said to be an *exterior point* of E, if there exists a $\delta > 0$ such that $B(\boldsymbol{q}, \delta) \subseteq E^c$. A point $\boldsymbol{x}$ which is neither an interior nor an exterior point of E is called a *boundary point*. The set of all boundary points is called the *boundary set* of E and is often denoted by ∂E.
5. By an *open cover* of E we mean a collection $\{G_i\}$ of open subsets of X such that $E \subseteq \bigcup_i G_i$.
6. If every open cover $\{G_i\}$ of E contains a *finite subcover*, that is, if there are finitely many indices $i_1, \ldots, i_m$ for which

$$\bigcup_{i_j = i_1}^{i_m} G_{i_j} \supseteq E,$$

   then E is said to be *compact* (or a compact set).
7. An ε-*net* of E is the set of finitely many points $\boldsymbol{p}_1, \ldots, \boldsymbol{p}_m$ of E such that

$$\bigcup_{i=1}^{m} B(\boldsymbol{p}_i, \varepsilon) \supseteq E,$$

   where ε is an arbitrary positive number.
8. E is said to be *bounded* if $\sup \{d(\boldsymbol{x}, \boldsymbol{y}) : \boldsymbol{x} \text{ and } \boldsymbol{y} \in E\}$ is finite. The supremum of the distances between pairs of points of E is called the *diameter* of E. Hence E is bounded if it has a finite diameter.

Remark Concerning Definition 5(8). E is bounded if and only if there exists a real number r and a point $\boldsymbol{p}$ of E such that

$$d(\boldsymbol{x}, \boldsymbol{p}) < r \quad \text{for all } \boldsymbol{x} \in E.$$

Definition 6. Let X and Y be two metric spaces. A function (mapping) f of X into Y is said to be *continuous* at a point $\boldsymbol{x}_0$ of X if for any given $\varepsilon > 0$ there exists a δ-neighborhood of $\boldsymbol{x}_0$ such that

$$f(B(\boldsymbol{x}_0, \delta)) \subseteq B(f(\boldsymbol{x}_0), \varepsilon).$$

Moreover, if f is continuous at every point of X then we call f *continuous on* X.

Having completed the definitions of various notations necessary to our analysis to follow, we can now establish several theorems concerning compact sets in metric spaces. In the statements of Theorems 2, 3, and 4, X denotes the metric space to be considered.

Theorem 2. *A compact subset E of X is bounded and closed.*

Theorem 3. *A closed subset of a compact set E of X is also compact.*

Theorem 4. *Let $\{K_i\}$ be a collection of compact subsets of X. If the intersection of every finite subcollection of $\{K_i\}$ is nonempty then $\bigcap_i K_i \neq \varnothing$.*

Corollary 4.1. *If $\{K_i\}$ is a collection of closed subsets of a compact subset E of X and if the intersection of every finite subcollection of $\{K_i\}$ is nonvoid, then $\bigcap_i K_i \neq \varnothing$.*

Theorem 5. *A subset E of a metric space X is closed if and only if E^c is open.*

Theorem 6. *If f is a real-valued continuous function defined over a metric space X and α is any given real number then the set $A_\alpha^f = \{\boldsymbol{x} \in X : f(\boldsymbol{x}) \geqq \alpha\}$ is closed.*

Theorem 7. *A mapping f of a metric space X into a metric space Y is continuous on X if and only if $f^{-1}(V)$ is open in X for every open set V of Y.*

Theorem 8. *Let f be a continuous mapping of a compact metric space X into a metric space Y. Then $f(X)$ is compact.*

Theorem 9. *A real-valued continuous function f defined over a compact subset K of a metric space attains its maximum and minimum on the set K.*

We have so far investigated compact sets of a metric space in general. However, it is of interest to characterize compact sets of an n-dimensional real vector space $\mathbb{R}^n$, which appears most frequently in economic analysis. We begin by defining a vector space precisely.

Definition 7. Let $\mathscr{F}$ be a given field, the elements of which we shall call *scalars*. A *vector (linear) space over* $\mathscr{F}$ is a set $\mathscr{V}$ of elements called *vectors* satisfying the following conditions:

A. With every pair, $\boldsymbol{x}$ and $\boldsymbol{y}$ of $\mathscr{V}$ there is associated a vector $\boldsymbol{x} + \boldsymbol{y}$ of $\mathscr{V}$, called the *sum* of $\boldsymbol{x}$ and $\boldsymbol{y}$, such that

1. $\boldsymbol{x} + \boldsymbol{y} = \boldsymbol{y} + \boldsymbol{x}$,
2. $\boldsymbol{x} + (\boldsymbol{y} + \boldsymbol{z}) = (\boldsymbol{x} + \boldsymbol{y}) + \boldsymbol{z}$, where $\boldsymbol{z}$ is any vector of $\mathscr{V}$,
3. there exists an element $\mathbf{0}$ (called the origin or nullvector) of $\mathscr{V}$ such that $\boldsymbol{x} + \mathbf{0} = \boldsymbol{x}$ for any $\boldsymbol{x}$ of $\mathscr{V}$, and
4. to every $\boldsymbol{x}$ of $\mathscr{V}$ there corresponds a vector $-\boldsymbol{x}$ such that $\boldsymbol{x} + (-\boldsymbol{x}) = \mathbf{0}$.

B. To every pair, $\alpha \in \mathscr{F}$ and $\boldsymbol{x} \in \mathscr{V}$, there corresponds a vector $\alpha \cdot \boldsymbol{x}$ of $\mathscr{V}$ called the *product* of α and $\boldsymbol{x}$ in such a way that

1. $\alpha \cdot (\beta \cdot \boldsymbol{x}) = (\alpha \cdot \beta) \cdot \boldsymbol{x}$ for any scalar β,
2. $1 \cdot \boldsymbol{x} = \boldsymbol{x}$ for every $\boldsymbol{x} \in \mathscr{V}$,
3. $(\alpha + \beta) \cdot \boldsymbol{x} = \alpha \cdot \boldsymbol{x} + \beta \cdot \boldsymbol{x}$ for all $\alpha, \beta \in \mathscr{F}$ and every $\boldsymbol{x} \in \mathscr{V}$, and
4. $\alpha \cdot (\boldsymbol{x} + \boldsymbol{y}) = \alpha \cdot \boldsymbol{x} + \alpha \cdot \boldsymbol{y}$ for every scalar α and vectors $\boldsymbol{x}$ and $\boldsymbol{y}$ of $\mathscr{V}$.

Definition 8. Let $\mathscr{V}$ be a vector space over a given field $\mathscr{F}$, let k be an integer, and let $\boldsymbol{x}_i$ and α_i $(i = 1, \ldots, k)$ be k vectors and k scalars of $\mathscr{V}$ and $\mathscr{F}$ respectively. Then

1. the vector $\sum_{i=1}^{k} \alpha_i \cdot \boldsymbol{x}_i$ of $\mathscr{V}$ is called a *linear combination* of vectors $\boldsymbol{x}_1, \ldots, \boldsymbol{x}_k$,
2. if $\sum_{i=1}^{k} \alpha_i \cdot \boldsymbol{x}_i = \mathbf{0}$ implies that $\alpha_1 = \alpha_2 = \cdots = \alpha_k = 0$ then $\boldsymbol{x}_1, \ldots, \boldsymbol{x}_k$ are said to be *linearly independent*,
3. if $\boldsymbol{x}_1, \ldots, \boldsymbol{x}_k$ are linearly independent and every vector $\boldsymbol{y}$ of $\mathscr{V}$ can be expressed as a linear combination of them then the set of vectors $\boldsymbol{x}_1, \ldots, \boldsymbol{x}_k$ is called a *basis* for $\mathscr{V}$, and
4. a vector space which contains a finite basis is said to be *finite-dimensional* (or a *finite-dimensional vector space*).

Definition 9. Let f be a function of a finite-dimensional vector space $\mathscr{V}$ over $\mathscr{F}$ into $\mathbb{R}$ such that for any $\boldsymbol{x}$ and $\boldsymbol{y}$ of $\mathscr{V}$

1. $f(\boldsymbol{x}) \geqq 0$, and $f(\boldsymbol{x}) = 0$ if and only if $\boldsymbol{x} = \mathbf{0}$,
2. $f(\boldsymbol{x} + \boldsymbol{y}) \leqq f(\boldsymbol{x}) + f(\boldsymbol{y})$, and
3. $f(\alpha \cdot \boldsymbol{x}) = |\alpha| \cdot f(\boldsymbol{x})$ for any $\alpha \in \mathscr{F}$, where $|\alpha|$ denotes the absolute value of α supposed to be suitably defined.

Then $f(\boldsymbol{x})$ is called a norm of $\boldsymbol{x}$ and is often denoted by $\|\boldsymbol{x}\|$.

From Definition 9, $\|\boldsymbol{x} - \boldsymbol{y}\|$, a norm of the difference between vectors $\boldsymbol{x}$ and $\boldsymbol{y}$, is easily seen to serve as a metric $d(\boldsymbol{x}, \boldsymbol{y})$. Thus $\|\boldsymbol{x} - \boldsymbol{y}\|$ is said to be the distance between $\boldsymbol{x}$ and $\boldsymbol{y}$ induced by the underlying norm $\|\boldsymbol{\xi}\|$ on $\mathscr{V}$.

Definition 10. Let $\mathscr{V}_n(\mathscr{F})$ be the set of all n-tuples $(x_1, \ldots, x_n) = \boldsymbol{x}$, where the x_i $(i = 1, \ldots, n)$ are elements of $\mathscr{F}$. If we specify that

1. $\boldsymbol{x} = \boldsymbol{y}$ if and only if $x_i = y_i$ for $i = 1, \ldots, n$,
2. $\boldsymbol{x} + \boldsymbol{y} = (x_1 + y_1, \ldots, x_n + y_n)'$,
3. $\mathbf{0} = (0, \ldots, 0)'$ and
4. $\alpha \cdot \boldsymbol{x} = (\alpha x_1, \ldots, \alpha x_n)'$,

then $\mathscr{V}_n(\mathscr{F})$ clearly satisfies conditions (A) and (B) of Definition 7 and is called an *n-dimensional coordinate vector space over* $\mathscr{F}$, where $\boldsymbol{x}$ and $\boldsymbol{y}$ are arbitrary vectors of $\mathscr{V}_n(\mathscr{F})$.

As is shown in Halmos [1958; p. 15], every n-dimensional vector space over $\mathscr{F}$ is *isomorphic* to $\mathscr{V}_n(\mathscr{F})$, that is, there exists a mapping (or transformation) T of $\mathscr{V}$ into $\mathscr{V}_n(\mathscr{F})$ such that

1. $T(\mathscr{V}) = \mathscr{V}_n(\mathscr{F})$,
2. if $\boldsymbol{\xi} \neq \boldsymbol{\eta}$, then $T(\boldsymbol{\xi}) \neq T(\boldsymbol{\eta})$ for any $\boldsymbol{\xi}$ and $\boldsymbol{\eta}$ of $\mathscr{V}$, and

3. $T(\alpha \cdot \xi + \beta \cdot \boldsymbol{\eta}) = \alpha \cdot T(\xi) + \beta \cdot T(\boldsymbol{\eta})$ for any α and β of $\mathscr{F}$ and any ξ and $\boldsymbol{\eta}$ of $\mathscr{V}$, so that, without loss of generality, we can identify an n-dimensional vector space over $\mathscr{F}$ with $\mathscr{V}_n(\mathscr{F})$.

In particular, when $\mathscr{F} = \mathbb{R}$ we denote $\mathscr{V}_n(\mathbb{R})$ by $\mathbb{R}^n$, and when $\mathscr{F} = \mathbb{C}$ (the set of all complex numbers) we denote $\mathscr{V}_n(\mathbb{C})$ by $\mathbb{C}^n$.

Definition 11. By an *n-cell* we mean a set of points $\boldsymbol{x} = (x_1, \ldots, x_n)'$ of $\mathbb{R}^n$ such that

$$a_j \leq x_j \leq b_j \qquad (j = 1, \ldots, n).$$

Theorem 10. *Let $\{I_m\}$ be a sequence of n-cells such that $I_m \supseteq I_{m+1}$ $(m = 1, 2, \ldots)$. Then $\bigcap_{j=1}^{\infty} I_j \neq \emptyset$.*

Lemma 1. *Define a norm on $\mathbb{R}^n$ by $\|x\| = \max_{1 \leq i \leq n} |x_i|$. Then an n-cell in $\mathbb{R}^n$ is compact with respect to this norm.*

Lemma 2. *In $\mathbb{R}^n$, any two norms $\|x\|$ and $\|x\|'$ are equivalent to each other in the sense that there exist positive numbers α and β such that $\alpha \cdot \|x\|' \leqq \|x\| \leqq \beta \cdot \|x\|'$.*

Theorem 11. *An n-cell in $\mathbb{R}^n$ is compact.*

Theorem 12. *A subset E of $\mathbb{R}^n$ is compact if and only if it is bounded and closed.*

Appendix B

This appendix is designed to supplement Chapter 4 and a part of Chapter 1, and consists of three sections, the first two of which are further divided into several subsections.

1

The purpose of this section is twofold. In the first subsection we establish some properties of the function e^z, where z is a complex number, and then turn to the classical results of Routh and Hurwitz concerning necessary and sufficient conditions for the stability of a system of linear differential equations and/or of a higher order differential equation in a single variable.

1.1

For simplicity, let x and y be respectively $\text{Re}(z)$ and $\text{Im}(z)$ of a complex number z. Then we define the function e^z by

$$e^z = e^x(\cos y + i \cdot \sin y) \tag{1}$$

where $i = \sqrt{-1}$. By definition, $|\cos y + i \cdot \sin y| = \sqrt{\cos^2 y + \sin^2 y} = 1$. Hence,

$$|e^z| = |e^x| = e^x \quad \text{for all } z \in \mathbb{C} \tag{2}$$

where $\mathbb{C}$ is the set of all complex numbers. In particular, if z is purely imaginary, from (1)

$$e^z = e^{iy} = (\cos y + i \cdot \sin y) \tag{3}$$

Hence,

$$|e^z| = 1 \quad \text{for any } z \text{ which is pure imaginary,} \tag{4}$$

$$e^z = e^x \cdot e^{iy} \quad \text{for any } z \in \mathbb{C}. \tag{5}$$

Moreover, if t is a real number, (1) and (5) imply that

$$e^{zt} = e^{xt}(\cos yt + i \cdot \sin yt) = e^{xt} \cdot e^{iyt}. \tag{6}$$

Since, by definition, $i^2 = -1$, direct differentiation yields

$$\frac{d(e^{zt})}{dt} = ze^{zt}. \tag{7}$$

Thus we have established a result analogous to that for z real.

1.2

In this subsection, we briefly state the so-called Routh–Hurwitz Theorem. Bearing in mind Theorem 2 of Chapter 4 as well as (iii) of the remark concerning the theorem, the equilibrium states of those differential equations with which we are mainly concerned are stable if and only if every root of a polynomial equation possesses a negative real part. The Routh–Hurwitz Theorem provides a necessary and sufficient condition for a polynomial equation to have no root with nonnegative real part.

Let $f(x) \equiv \sum_{j=0}^{n} a_j x^{n-j}$, where a_j is a given real constant and $a_0 \neq 0$. In order to facilitate the statement of Hurwitz's version of the theorem, we introduce an $n \times n$ matrix $\boldsymbol{B}$ consisting of elements b_{ij} such that

$$b_{ij} = \begin{cases} a_{2i-j} & \text{for } j = 1, \ldots, 2i \\ 0 & \text{for } j = 2i+1, \ldots, n \end{cases} \qquad (i = 1, 2, \ldots, n).$$

Theorem 1 (Hurwitz [1895]). *A necessary and sufficient condition for the polynomial equation $f(x) = 0$ to possess no root with nonnegative real part is that every leading principal minor of* det $\boldsymbol{B}$ *be positive. More concretely,*

$$\det \boldsymbol{B}_r = \begin{vmatrix} a_1 & a_0 & 0 & 0 & 0 & \cdots & 0 \\ a_3 & a_2 & a_1 & a_0 & 0 & \cdots & 0 \\ a_5 & a_4 & a_3 & a_2 & a_1 & \cdots & 0 \\ \vdots & \vdots & \vdots & \vdots & \vdots & & \vdots \\ a_{2r-1} & a_{2r-2} & a_{2r-3} & a_{2r-4} & a_{2r-5} & \cdots & a_r \end{vmatrix} \tag{8}$$

$$> 0 \qquad (r = 1, \ldots, n).$$

Routh's version of the theorem is complicated. To facilitate the statement of the theorem we begin by defining the coefficients c_{ij}. For any real x let $\langle x \rangle$ be the least integer not less than x and let $m = \langle n/2 \rangle$. We then define

$$c_{11} = a_0,$$

$$c_{ij} = \begin{cases} a_{2(j-1)} & \text{for } j = 2, \ldots, m+1, \\ 0 & \text{for } j > m+1, \end{cases}$$

$$c_{2j} = \begin{cases} a_{2j-1} & \text{for } j = 1, \ldots, m. \\ 0 & \text{for } j > m, \end{cases}$$

with the subsequent coefficients defined by the recurrence relation

$$c_{ij} = c_{i-2,j+1} - \left(\frac{(c_{i-2,1} - c_{i-1,j+1})}{c_{i-1,1}} \right) \quad (i = 3, 4, \ldots, n+1; j = 1, 2, 3, \ldots).$$

Theorem 2 (Routh [1877; 1908]). *Every root of* $f(x) = 0$ *has a negative real part if and only if*

$$c_{i1} > 0 \quad \text{for } i = 1, \ldots, n+1. \tag{9}$$

The reader should notice that the matrix consisting of the coefficients c_{ij} has $n + 1$ rows and an infinity of columns. By successive subtraction of the *columns of* det $\boldsymbol{B}_r$ *there can be found a close and well-known relationship between Theorems* 1 *and* 2.

Theorem 3 (Gantmacher [1959; Volume II, p. 203 *et seq.*]). *The conditions stated in Theorem* 1 *and* 2 *are connected by*

$$c_{11} = a_0, \qquad c_{21} = \det \boldsymbol{B}_1, \qquad c_{21} = \det \boldsymbol{B}_1, \tag{10}$$

and

$$c_{i1} = \frac{\det \boldsymbol{B}_{i-1}}{\det \boldsymbol{B}_{i-2}} \qquad (i = 3, 4, \ldots, n+1).$$

For this reason, the two theorems together are usually referred to as the Routh–Hurwitz Theorem.

Remark Concerning the Routh–Hurwitz Theorem. (i) Classical as the Routh–Hurwitz Theorem is, it has recently attracted renewed attention. See, in particular, Parks [1962] and Fuller [1968]. Our exposition is mainly based on the former, while the latter connects the Routh–Hurwitz condition (expressions (8) and (9)) directly to the elements of the coefficient matrix of differential equations under investigation (Fuller [1968; Theorem 9, p. 90]). Murata [1977; Subsection 3.4, pp. 87–93] provides an exposition (with a slight extension) of Fuller's Theorem.

Switching our attention to systems of linear difference equations and/or higher order difference equations in a single variable, we find an analogue of the Routh–Hurwitz Theorem, which is usually called the Schur Theorem.

Theorem 4 (Schur [1918]). *Let $\boldsymbol{S}_n$ be a $2n \times 2n$ matrix with elements s_{ij} such that*

$$s_{ij} = \left\{\begin{matrix} a_{j-1} & \text{for } j = 1, \ldots, n+1 \\ 0 & \textit{otherwise} \end{matrix}\right\} \qquad (i = 1, \ldots, n)$$

$$s_{ij} = \left\{\begin{matrix} a_{i-j} & \text{for } j = n-i, \ldots, i \\ 0 & \textit{otherwise} \end{matrix}\right\} \qquad (i = n+1, \ldots, 2n)$$

and, corresponding to any k of $\{1, \ldots, n\}$, let $J_k = \{1, \ldots, k, n+1, \ldots, n+k\}$. Further define $\boldsymbol{S}_k$ to be the submatrix with elements s_{ij} ($i \in J_k, j \in J_k$). Then all roots of the polynomial equation $f(x) = 0$ have moduli less than 1 if and only if

$$\det \boldsymbol{S}_k > 0 \quad \textit{for } k = 1, \ldots, n. \tag{11}$$

Remark Concerning Theorem 4. Schur [1918] approached the stability problem from the function theoretic viewpoint. Cohn [1922] approached the problem from the algebraic point of view.

2

The second part of Appendix B consists of three subsections. The first is devoted to the Minkowski norm and the matrix norm (to be defined). In the second, we introduce the concept of relative closedness (respectively, openness) and then establish some fundamental results concerning connected sets. Finally, the third subsection is devoted to the Mean Value Theorem and Taylor's Theorem of second order.

2.1

As an immediate consequence of (M.P.1) of Chapter 1, for any $m \times n$ matrix $\boldsymbol{A}$ with complex entries there exists an $\boldsymbol{x}_0$ such that $h(\boldsymbol{x}_0) = 1$ and $\|\boldsymbol{A}\| = h(\boldsymbol{A}\boldsymbol{x}_0)$, where $h(\boldsymbol{x})$ is the norm of vector $\boldsymbol{x}$ defined in Appendix A (see Definition 9 of Appendix A). Therefore

$$\|\boldsymbol{I}\| = 1, \tag{12}$$

$$\|\boldsymbol{A} + \boldsymbol{B}\| \leqq \|\boldsymbol{A}\| + \|\boldsymbol{B}\|, \tag{13}$$

where $\boldsymbol{I}$ is an identity matrix of arbitrary but finite order and $\boldsymbol{B}$ is an $m \times n$ matrix with complex elements.

Moreover, a glance at property (3) of the vector norm (see (3) of Definition 9 of Appendix A) reveals that

$$\|c \cdot \boldsymbol{A}\| = |c| \cdot \|\boldsymbol{A}\| \quad \text{for any scalar } c. \tag{14}$$

If $A = \mathbf{0}$ then clearly $\|A\| = 0$. Since a nonzero matrix A contains a nonzero row, say a^i, (M.P.1), together with property (1) of the vector norm, guarantees that

$$\|A\| \geq h(a^i/h(a^i)) > 0.$$

Hence

$$\|A\| = 0 \quad \text{if and only if} \quad A = \mathbf{0}. \tag{15}$$

Thus the Minkowski norm has been shown to fulfill, besides (M.P.3) of Chapter 1, all requirements of the vector norm.

So far we have considered matrices that are not necessarily square. If attention is confined to the family of square matrices, we can define the matrix norm.

Definition 1. Let $\|\cdot\|$ be a real-valued function defined on the family Ω of all square matrices with complex entries and consider the four conditions:

i. $\|A\| \geqq 0$ and $\|A\| = 0$ if and only if $A = \mathbf{0}$;
ii. $\|cA\| = |c| \cdot \|A\|$ for all $c \in \mathbb{C}$ and for all $A \in \Omega$;
iii. $\|A + B\| \leqq \|A\| + \|B\|$ for all square matrices A and B of the same order;
iv. $\|A \cdot B\| \leqq \|A\| \cdot \|B\|$ for all square matrices A and B of the same order.

Then

a. $\|\cdot\|$ is said to be *the generalized matrix norm* if it satisfies conditions (i) through (iii), and
b. $\|\cdot\|$ is said to be *the matrix norm* if it fulfills all conditions enumerated above.

Remarks Concerning the Norm of Matrices. (i) Condition (iv) of the above definition stands in fairly clear contrast to the corresponding fact, $|ab| = |a| \cdot |b|$, for complex numbers.

(ii) In view of the continuity of the underlying vector norm, the matrix norm induced by the vector norm is obviously continuous.

Contriving the matrix $E_{ij} = (e^i)'e^j$ and proceeding as in the case of the vector norm, we can easily deduce the continuity of the generalized matrix norm and the fact that for any distinct generalized matrix norms $\|\cdot\|$ and $\|\cdot\|'$ there exist positive numbers M and m such that

$$m \cdot \|A\|' \leqq \|A\| \leqq M \cdot \|A\| \quad \text{for all } A.$$

It is this inequality that guarantees the invariance of some fundamental concepts, such as the convergence of a sequence of matrices and the closedness (respectively, openness) of the set of matrices, with respect to the norm employed. See Lancaster [1969; Theorems 6.1.1 and 6.1.2, pp. 199–200].

(iii) The generalized matrix norms of frequent appearance are:

iii–1. $\|A\| = \max_{i,j}|a_{ij}|$,
iii–2. $\|A\|_p = \{\sum_{i,j}|a_{ij}|^p\}^{1/p}$ for $p \geqq 1$ (Hölder norm);
iii–3. $\|A\| = n \cdot \max_{i,j}|a_{ij}|$.
iii–4. $\|A\| =$ (the largest eigenvalue of $A^*A)^{1/2}$.

Needless to say, the generalized matrix norms (iii–3) and (iii–4) are also matrix norms. Moreover, it is well known that $\|A\|_p$ is a matrix norm if and only if $p \in [1, 2]$ (see Lancaster [1969; Example 12, p. 202]), and that (iii–4) equals the matrix norm induced by the Euclidean vector norm $h(x) = (x' \cdot x)^{1/2}$ (see also Lancaster [1969; Example 4, p. 210–211]).

2.2

Throughout this subsection X denotes exclusively an underlying metric space, while E and F signify any subsets of X such that $E \subseteq F$. For the proof of lemmas and theorems stated below, the reader may consult Arrow [1974], Nikaido [1968; pp. 13–15], and Rudin [1964; pp. 31–38].

Definition 2. E is said to be *open relative to F* if with any $x \in E$ there is associated a neighborhood $B(x, \delta)$ such that $(B(x, \delta) \cap F) \subseteq E$.

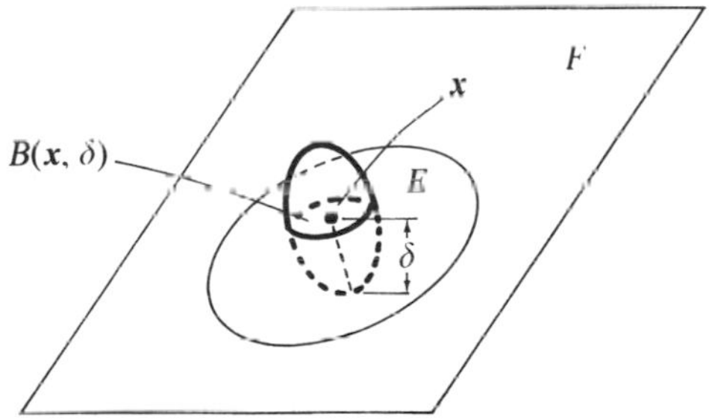

For ease of intuitive understanding of Definition 2, let X, F, and E be respectively $\mathbb{R}^3$, a plane of $\mathbb{R}^3$, and an open circle in F. (See the accompanying figure). Then E is by no means an open set in the sense of Definition 5(4) of Appendix A. However, if our attention is confined to the plane F then E is an open subset of F. Moreover, it is evident that an open set as defined in Appendix A is simply an open set relative to X, the whole space under consideration.

Definition 3. A sequence $\{x_n\}$ of E is said to be *convergent to* $x_0 \in F$ *relative to F* if for any $\varepsilon > 0$ there exists an index n_ε such that $n \geqq n_\varepsilon$ implies that $x_n \in (B(x, \varepsilon) \cap F)$, and a point $p \in F$ is said to be a *limit point of E relative to F* if to any $\varepsilon > 0$ there corresponds a point $q \in E$ with the properties that $q \neq p$ and $q \in (B(p, \varepsilon) \cap F)$.

Parallel to Definition 5(3) of Appendix A, we have the concept of relative closedness.

Definition 4. E is said to be *closed relative to* F if E contains all of its limit points relative to F.

We can now state the equivalence of seemingly different notions of a closed set relative to a given set.

Lemma 1. *The following three statements are mutually equivalent.*

i. *E is closed relative to F;*
ii. *if a sequence $\{\mathbf{x}_n\}$ of E converges to an $\mathbf{x}_0 \in F$ relative to F then $\mathbf{x}_0 \in E$;*
iii. *the set $F - E = \{\mathbf{x} \in F : \mathbf{x} \notin E\}$ is open relative to F.*

The next lemma clarifies the relationship between the relative closedness (respectively, openness) of a set and its *super set* (the set in which the former set is imbedded).

Lemma 2. (i) *If E is open relative to F then E is open relative to any F' of X such that $E \subseteq F' \subseteq F$.* (ii) *If E is closed relative to F then it is closed relative to any F' such that $F \subseteq F'$.*

The final concept necessary for our present purposes is that of connected set.

Definition 5. E is said to be *connected relative to* F if there exist no nonempty disjoint subsets G and H of F such that (a) $E = G \cup H$ and (b) G and H are open relative to F.

Remark Concerning Definition 5. Since condition (a) implies that both G and H are contained in E, a subset of F, Lemma 2(i) guarantees that G and H are open relative to E. This, in conjunction with (a), Lemma 2(ii), and Lemma 1 (the equivalence of (i) and (iii) of the lemma), further implies that G as well as H is closed relative to F. Therefore condition (b) of Definition 5 can be replaced by

(b′) G and H are closed relative to F.

Having completed the definition of concepts, we can proceed to the essential theorems concerning connected sets.

Theorem 5. *Let F' be any subset of X containing the set F. Then if E is connected relative to F it is also connected relative to F'.*

Remarks Concerning Theorem 5. (i) The notion of connectedness depends on the super set (for example, F). However, if condition (a) is relaxed, so that

(a′) $E \subseteq G \cup H$,

then the dependence is seen to disappear. For then the existence of disjoint nonempty subsets G and H with the properties that they are open relative to E, the smallest subset of X including E, and that $E = G \cup H$ enables us to construct disjoint nonempty open sets A and B such that $G = A \cap E$ and $H = B \cap E$.

(ii) As a straightforward consequence of the theorem, if E is connected relative to itself then it is connected relative to all subsets of X containing E. Therefore a set connected relative to itself is simply called a *connected set*.

Theorem 6. *A subset E of $\mathbb{R}$ is connected if and only if, for any x, y of E and for any real z, $x < z < y$ implies that z lies in E.*

As a direct implication of this theorem we have

Corollary 1. *A closed interval $[a, b]$ of $\mathbb{R}$ is connected.*

The following three theorems are essential to the understanding of Section 2 of Chapter 4.

Theorem 7. *If f is a continuous mapping of a connected set K into a metric space then $f(K)$ is connected.*

(For a proof of this theorem, see Rudin [1964; Theorem 4.22, p. 80] or Nikaido [1968; Theorem 1.7, p. 14].)

Theorem 8. *A closure $\bar{A}$ of a connected set A is also connected, where $\bar{A}$ is the union of A and the set L_A of all limit points of A.*

Theorem 9 (Nikaido [1968; Theorem 1.9, p. 15]). *Let J be an arbitrary but given set of indices and let $\{G_j\}$ be a family of connected sets G_j. If $\bigcap_{j \in J} G_j \neq \varnothing$ then $\bigcup_{j \in J} G_j$ is connected.*

With the aid of Corollary 1 and Theorems 7 and 9 we obtain

Corollary 2. *Every convex set is connected.*

From Theorems 6 and 7 and Corollary 2 an important result known as the Intermediate Value Theorem follows.

Theorem 10. *Let f be a real-valued continuous function defined on a convex set K, let $\boldsymbol{x}_0$ and $\boldsymbol{x}_1$ be any distinct points of K, and suppose that $f(\boldsymbol{x}_0) < f(\boldsymbol{x}_1)$. Then for any $c \in (f(\boldsymbol{x}_0), f(\boldsymbol{x}_1))$ there exists an*

$$\boldsymbol{x} \in (\boldsymbol{x}_0, \boldsymbol{x}_1) = \{\boldsymbol{x} = (1 - \lambda)\boldsymbol{x}_0 + \lambda \boldsymbol{x}_1, 0 < \lambda < 1\}$$

such that $f(\boldsymbol{x}) = c$.

2.3

First we state the Mean Value Theorem for real-valued functions of a single variable, from which the multivariate version follows.

Theorem 11. *Let f be real-valued and continuous on a closed interval $[a, b]$ and let f be differentiable on (a, b). Then there exists a point $x \in (a, b)$ such that*

$$f(b) - f(a) = (b - a) \cdot f'(x).$$

Remark Concerning Theorem 11. Note that differentiability is not required at the end points and that the point x whose existence is asserted in the theorem is not necessarily unique. However, the nonuniqueness of x does not reduce the usefulness of the theorem, for the Mean Value Theorem is used to obtain estimates which are valid no matter where x is in the interval (a, b). Elegant proofs of Theorem 11 and of its multivariate version (Theorem 12 below) can be found in Fleming [1965; especially p. 41 and p. 311].

We now turn to the multivariate version of the Mean Value Theorem.

Theorem 12. *Let f be a real-valued function and let $\mathbf{x}_0$ and $\mathbf{x}_1$ be any distinct points of the domain on which f is defined. If f is differentiable on the line segment joining $\mathbf{x}_0$ and $\mathbf{x}_1$ then there exists an $s \in (0, 1)$ such that*

$$f(\mathbf{x}_1) - f(\mathbf{x}_0) = \nabla f((1 - s)\mathbf{x}_0 + s \cdot \mathbf{x}_1) \cdot (\mathbf{x}_1 - \mathbf{x}_0).$$

Assuming differentiability of higher order, we obtain

Theorem 13 (Taylor's Theorem of Second Order). *Let f, $\mathbf{x}_0$, and $\mathbf{x}_1$ be as in Theorem 12. If, in addition, f is twice differentiable on the line segment connecting $\mathbf{x}_0$ and $\mathbf{x}_1$ then there exists an $s \in (0, 1)$ such that*

$$f(\mathbf{x}_1) - f(\mathbf{x}_0) - \nabla f(\mathbf{x}_0) \cdot (\mathbf{x}_1 - \mathbf{x}_0) = (1/2!) \cdot (\mathbf{x}_1 - \mathbf{x}_0)' \cdot \mathbf{H}(\mathbf{x}_s) \cdot (\mathbf{x}_1 - \mathbf{x}_0)$$

where $\mathbf{H}(\mathbf{x}_s)$ of course denotes the Hessian of f evaluated at $\mathbf{x}_s = (1 - s)\mathbf{x}_0 + s \cdot \mathbf{x}_1$.

3

We bring this appendix to a close by providing the Cauchy criterion for the convergence of a series and for the uniform convergence of a sequence of functions. We begin by defining sequences and series.

Definition 6. (i) The function f of the set J of all positive integers into any given range is said to be a *sequence*. Let a_n be the image $f(n)$ of $n \in J$. Then it is customary to denote the sequence f by $\{a_n\}$. Moreover, a_n is often called the nth term (or simply a term) of the sequence.

(ii) With a given sequence $\{a_n\}$ we associate a sequence with the nth term $s_n = \sum_{i=i}^{n} a_i$. The sequence $\{s_n\}$ thus constructed is said to be an *infinite series*, or just a *series* and each term of the series is called *the partial sum*. To emphasize the dependence of $\{s_n\}$ on $\{a_n\}$, we usually employ the symbol $\sum a_n$ for $\{s_n\}$. If the sequence $\{s_n\}$ is convergent (respectively, divergent) the series is said to be convergent (respectively, divergent).

Next, we introduce the notions of Cauchy sequence and of complete metric space.

Definition 7. Let $\{\mathbf{x}_n\}$ be any sequence of a metric space X. If for any $\varepsilon > 0$ there exists an integer n_ε such that $\min\{m, n\} \geqq n_\varepsilon$ implies $d(\mathbf{x}_m, \mathbf{x}_n) < \varepsilon$, then the sequence is said to be a *Cauchy sequence*. Moreover a metric space in which every Cauchy sequence is convergent is said to be a *complete metric space*.

A convergent sequence is a Cauchy sequence but the converse is not necessarily true. However, $\mathbb{R}^k$ is a complete metric space, a result which we state as a lemma.

Lemma 3 (Rudin [1964; Theorem 3.11(b), p. 46]). *Every Cauchy sequence in $\mathbb{R}^k$ converges.*

The Cauchy criterion for the convergence of a series is provided by

Theorem 14 (Rudin [1964; Theorem 3.22, p. 54]). *Let $\{a_n\}$ be a complex sequence. Then $\sum a_n$ converges if and only if for any $\varepsilon < 0$ there exists an integer n_ε such that if $n \geqq m \geqq n_\varepsilon$ then*

$$\left| \sum_{j=m}^{n} a_j \right| \leqq \varepsilon. \tag{16}$$

In particular, by taking $m = n$, (16) *reduces to*

$$|a_n| \leqq \varepsilon \quad \textit{for all } n \geqq n_\varepsilon. \tag{17}$$

Hence, the convergence of the series implies that property of the underlying sequence.

Finally, we turn to the Cauchy criterion for uniform convergence.

Theorem 15. *Let Γ be the family of all functions of a given set E into $\mathbb{R}^k$ and let $\{f_n\}$ be a sequence of Γ. Then $\{f_n\}$ converges uniformly on E if and only if for every $\varepsilon > 0$ there exists an integer n_ε such that* $\min\{m, n\} \geqq n_\varepsilon$ *implies*

$$\|f_m(\mathbf{x}) - f_n(\mathbf{x})\| \leqq \varepsilon \quad \textit{for all } \mathbf{x} \in E.$$

References

Arrow, K. J. [1974]. Stability independent of adjustment speed, *in* G. Horwich and P. A. Samuelson, eds., *Trade, Stability, and Macroeconomics, Essays in Honor of Lloyd A. Metzler*, pp. 181–202. New York: Academic Press.

Arrow, K. J., and A. C. Enthoven [1961]. Quasi-concave programming, *Econometrica* **29**, 779–800.

Arrow, K. J., H. D. Block, and L. Hurwicz [1959]. On the stability of the competitive equilibrium, *Econometrica* **27**, 82–109.

Arrow, K. J., and M. Kurz [1970]. *Public Investment, the Rate of Return, and Optimal Fiscal Policy*. Baltimore: The Johns Hopkins Press.

Arrow, K. J., and M. McManus [1958]. A note on dynamic stability, *Econometrica* **26**, 448–454.

Arrow, K. J., and M. Nerlove [1958]. A note on expectations and stability, *Econometrica* **26**, 297–305.

Bellman, R. [1953]. *Stability Theory of Differential Equations*, New York: Dover Publications Inc.

Benavie, A. [1972]. *Mathematical Techniques for Economic Analysis*. Englewood Cliffs: Prentice-Hall.

Berge, C. [1963]. *Topological Spaces*. New York: Macmillan.

Brauer, A. [1946]. Limits for the characteristic roots of a matrix, *Duke Mathematical Journal* **13**, 387–395.

Brauer, A. [1961]. On the characteristic roots of power-positive matrices, *Duke Mathematical Journal* **28**, 439–445.

Brock, W. A., and J. A. Scheinkman [1975]. Some results on global asymptotic stability of difference equations, *Journal of Economic Theory*, **10**, 265–268.

Bruno, M., E. Burmeister, and E. Sheshinski [1966]. Nature and implications of the reswitching of techniques, *Quarterly Journal of Economics* **80**, 526–554.

Buck, R. C. [1965]. *Advanced Calculus* (2nd ed.). New York: McGraw–Hill.

Burmeister, E., and K. Kuga [1970]. The factor-price frontier in a neo-classical multi-sector model, *International Economic Review* **11**, 162–174.

Burmeister, E. [1976]. The factor-price frontier and duality with many primary factors, *Journal of Economic Theory* **12**, 496–503.

Burmeister, E., and A. R. Dobell [1970]. *Mathematical Theories of Economic Growth*. London: Macmillan.

Chipman, J. S. [1965–1966]. A survey of the theory of international trade, *Econometrica* **33**, 477–519 and 685–760; **34**, 18–76.
Chipman, J. S. [1969]. Factor price equalization and the Stolper–Samuelson theorem, *International Economic Review* **10**, 399–406.
Coddington, E. A., and N. Levinson [1955]. *Theory of Differential Equations*. New York: McGraw–Hill.
Cohn, A. [1922]. Über die Anzahl der Wurzeln einer algebraischen Gleichung in einem Kreise, *Mathematische Zeitschrift* **14**, 110–148.
Dorfman, R., P. A. Samuelson, and R. M. Solow [1958]. *Linear Programming and Economic Analysis*. New York: McGraw–Hill.
Enthoven, A. C., and K. J. Arrow [1956]. A theorem on expectations and the stability of equilibrium, *Econometrica* **24**, 288–293.
Ethier, W. [1974]. Some of the theorems of international trade with many goods and factors, *Journal of International Economics* **4**, 199–206.
Farkas, J. [1902]. Über die Theorie der einfachen Ungleichungen, *Journal für reine und angewandte Mathematik* **124**, 1–27.
Fleming, W. H. [1965]. *Functions of Several Variables*. Reading, Massachusetts: Addison–Wesley.
Frobenius, G. [1908]. Über Matrizen aus positiven Elementen I, *Sitzungsberichte der königlichpreussischen Akademie der Wissenschaften*, 471–476.
Frobenius, G. [1909]. Über Matrizen aus positiven Elementen II, *Sitzungsberichte der königlichpreussischen Akademie der Wissenschaften*, 514–518.
Frobenius, G. [1912]. Über Matrizen aus nicht negativen Elementen, *Sitzungsberichte der königlichpreussischen Akademie der Wissenschaften*, 456–477.
Fuller, A. T. [1968]. Conditions for a matrix to have only characteristic roots with negative real parts, *Journal of Mathematical Analysis and Applications* **23**, 71–98.
Gale, D. [1960]. *The Theory of Linear Economic Models*. New York: McGraw–Hill.
Gale, D., and H. Nikaido [1965]. The Jacobian matrix and global univalence of mappings, *Mathematische Annalen* **159**, 81–93.
Gantmacher, F. R. [1959]. *Theory of Matrices*, Vols. I, II. New York: Chelsea.
Hadamard, J. [1903]. *Leçons sur la propagation des ondes et les équations de l'hydrodynamique*. Paris: Hermann.
Hadley, G., and M. C. Kemp [1971]. *Variational Methods in Economics*. Amsterdam: North-Holland Publishing Company.
Halmos, P. R. [1958]. *Finite-Dimensional Vector Spaces*. Princeton, N.J.: Van Nostrand. (Reprint, 1974, by Springer–Verlag.)
Halmos, P. R. [1960]. *Naive Set Theory*. Princeton, N.J.: Van Nostrand.
Hartman, P. [1964]. *Ordinary Differential Equations*. New York: Wiley.
Hawkins, D., and H. A. Simon [1949]. Note: some conditions of macroeconomic stability, *Econometrica* **17**, 245–248.
Hotaka, R. [1971]. Some basic problems on excess demand functions, *Econometrica*, **39**, 305–307.
Hurwitz, A. [1895]. Über die Bedingungen unter welchen eine Gleichung nur Wurzeln mit negativen realen Theilen besitzt, *Mathematische Annalen* **46**, 273–284.
Inada, K. [1966]. Factor intensity and the Stolper–Samuelson conditions. University of New South Wales, mimeographed.
Inada, K. [1971]. The production coefficient matrix and the Stolper–Samuelson condition, *Econometrica* **39**, 219–239.
Kalman, R. E., and J. E. Bertram [1960]. Control system analysis and design via the "second method" of Lyapunov. I. Continuous-time systems, *Journal of Basic Engineering* **82**, 371–393.
Katzner, D. W. [1970]. *Static Demand Theory*. New York: Macmillan.
Kemp, M. C. [1964]. *The Pure Theory of International Trade*. Englewood Cliffs, N.J.: Prentice–Hall.

Kemp, M. C. [1969]. *The Pure Theory of International Trade and Investment*. Englewood Cliffs, N.J.: Prentice–Hall.

Kemp, M. C. [1976]. *Three Topics in the Theory of International Trade. Distribution, Welfare and Uncertainty*. Amsterdam: North–Holland.

Kemp, M. C., and L. L. F. Wegge [1969]. On the relation between commodity prices and factor rewards, *International Economic Review* **10**, 407–413.

Kemp, M. C., and C. Khang [1974]. A note on steady-state price: output relationships, *Journal of International Economics* **4**, 187–197.

Kemp, M. C., C. Khang, and Y. Uekawa [1978]. On the flatness of the transformation surface. *Journal of International Economics*, forthcoming.

Kemp, M. C., and H. Y. Wan, Jr. [1976]. Relatively simple generalizations of the Stolper–Samuelson and Samuelson–Rybczynski theorems, Chapter 4 of Kemp [1976].

Kemp, M. C., Y. Kimura, and K. Okuguchi [1977]. Monotonicity properties of a dynamical version of the Heckscher–Ohlin model of production, *Economic Studies Quarterly*, **28**, 249–253.

Kestelman, H. [1973]. *Modern Theories of Integration*. Oxford: Oxford University Press.

Khang, C. [1971]. On the strict convexity of the transformation surface in case of linear homogeneous production functions: a general case, *Econometrica* **39**, 587–589.

Khang, C., and Y. Uekawa [1973]. The production possibility set in a model allowing inter-industry flows, *Journal of International Economics* **3**, 283–290.

Kimura, Y. [1975]. A note on matrices with quasi-dominant diagonals, *The Economic Studies Quarterly* **26**, 57–58.

Kimura, Y., and Y. Murata [1975], Amendments of "An alternative proof of the Frobenius theorem", *Journal of Economic Theory* **19**, 110–112.

Koopmans, T. C. [1957]. *Three Essays on the State of Economic Science*. New York: McGraw–Hill.

Kuhn, H. W. [1968]. Lectures in mathematical economics, *in* G. B. Dantzig and A. F. Veinott, eds., *Mathematics of the Decision Sciences, Part 2*, pp. 49–84. Providence, R.I.: American Mathematical Society.

Kuhn, H. W., and A. W. Tucker [1951]. Non-linear programming, *in* J. Neyman, ed., *Proceedings of the Second Berkeley Symposium on Mathematical Statistics and Probability*, pp. 481–492. Berkeley, California: University of California Press.

Lancaster, P. [1969]. *Theory of Matrices*. New York: Academic Press.

Landau, E. G. H. [1951]. *Foundations of Analysis*. New York: Chelsea.

Luenberger, D. G. [1969]. *Optimization by Vector Space Methods*. New York: Wiley.

Mangasarian, O. L., and J. Ponstein [1965]. Minmax and duality in nonlinear programming, *Journal of Mathematical Analysis and Applications* **11**, 504–518.

Mangasarian, O. L. [1969]. *Nonlinear Programming* (McGraw–Hill Series in System Science). New York: McGraw–Hill.

McKenzie, L. W. [1960A]. Matrices with dominant diagonals and economic theory, *in* K. J. Arrow, S. Karlin, and P. Suppes, eds., *Mathematical Methods in the Social Sciences, 1959*, pp. 17–61. Stanford: Stanford University Press.

McKenzie, L. W. [1960B]. Stability of equilibrium and the value of positive excess demand, *Econometrica* **28**, 606–617.

Melvin, J. R. [1968]. Production and trade with two factors and three goods, *American Economic Review* **58**, 1249–1268.

Metcalfe, J. S., and I. Steedman [1972]. Reswitching and primary input use, *Economic Journal* **82**, 140–157.

Metzler, L. A. [1945]. Stability of multiple markets: the Hicks conditions, *Econometrica* **13**, 277–292.

Minkowski, H. [1896]. *Geometrie der Zahlen*. Leipzig and Berlin: B. Teubner.

Morishima, M. [1952]. On the laws of change of the price-system in an economy which contains complementary commodities, *Osaka Economic Papers* **1**, 101–113.

Morishima, M. [1964]. *Equilibrium, Stability and Growth. A Multi-sectoral Analysis.* London: Oxford University Press.

Morishima, M. [1970]. A generalization of the gross substitute system, *Review of Economic Studies* **37**, 117–186.

Mukherji, A. [1972]. On complementarity and stability, *Journal of Economic Theory* **4**, 442–457.

Mukherji, A. [1973]. On the sensitivity of stability results to the choice of the numeraire, *Review of Economic Studies* **40**, 427–433.

Mukerji, A. [1974]. Stability in an economy with production, *in* G. Horwich and P. A. Samuelson, eds., *Trade, Stability, and Macro-Economics. Essays in Honor of Lloyd A. Metzler*, pp. 243–258. New York: Academic Press.

Murata, Y. [1972]. An alternative proof of the Frobenius theorem, *Journal of Economic Theory* **5**, 285–291.

Murata, Y. [1977]. *Mathematics for Stability and Optimization of Economic Systems.* New York: Academic Press.

Negishi, T. [1974]. Stability of markets with public goods: a case of gross substitutability, *in* G. Horwich and P. A. Samuelson, eds., *Trade, Stability, and Macro-Economics. Essays in Honor of Lloyd A. Metzler*, pp. 259–268. New York: Academic Press.

Newman, P. K. [1959]. Some notes on stability conditions, *Review of Economic Studies* **27**, 1–9.

Nikaido, H. [1968]. *Convex Structures and Economic Theory.* New York: Academic Press.

Ohyama, M. [1972]. On the stability of generalized Metzlerian systems, *Review of Economic Studies* **39**, 193–203.

Okuguchi, K. [1975]. Power-positive matrices and global stability of competitive equilibrium, *Keio Economic Studies* **12**, 37–40.

Okuguchi, K. [1976]. Further note on matrices with quasi-dominant diagonals, *Economic Studies Quarterly* **27**, 151–154.

Ostrowski, A. M. [1955]. Note on bounds for some determinants, *Duke Mathematical Journal* **20**, 95–102.

Parks, P. C. [1962]. A new proof of the Routh-Hurwitz stability criterion using the second method of Liapunov, *Proceedings of the Cambridge Philosophical Society* **58**, 694–702.

Pearce, I. F. [1974]. Matrices with dominating diagonal blocks, *Journal of Economic Theory* **9**, 159–170.

Perron, O. [1907]. Zur theorie der matrices, *Mathematische Annalen* **64**, 248–263.

Quirk, J. and R. Ruppert [1965]. Qualitative economics and the stability of equilibrium, *Review of Economic Studies* **32**, 311–326.

Routh, E. J. [1877]. *Stability of a Given State of Motion* (together with: Routh, E. J. [1908]. *Advanced Rigid Dynamics*) (6th ed.). London: Macmillan.

Rudin, W. [1964]. *Principles of Mathematical Analysis* (2nd ed.). New York: McGraw–Hill.

Rybczynski, T. N. [1955]. Factor endowments and relative commodity prices, *Economica* **22**, 336–341.

Samuelson, P. A. [1942]. The stability of equilibrium: linear and nonlinear systems, *Econometrica* **10**, 1–25.

Samuelson, P. A. [1947]. *Foundations of Economic Analysis.* Cambridge, Mass.: Harvard University Press.

Samuelson, P. A. [1953]. Prices of factors and goods in general equilibrium, *Review of Economic Studies* **21**, 1–20.

Samuelson, P. A. [1957]. Wages and interest: a modern dissection of Marxian economic models, *American Economic Review* **47**, 884–912.

Samuelson, P. A. [1962]. Parable and realism in capital theory: the surrogate production function, *Review of Economic Studies* **29**, 193–206.

Samuelson, P. A. [1975]. Steady-state and transient relations: a reply on reswitching, *Quarterly Journal of Economics* **89**, 40–47.

Samuelson, P. A. [1975]. Trade pattern reversals in time-phased Ricardian systems and intertemporal efficiency, *Journal of International Economics* **5**, 309–363.

Sato, R. [1972]. The stability of the competitive system which contains gross complementary goods, *Review of Economic Studies* **39**, 495–499.

Sato, R. [1973]. On the stability properties of dynamic economic systems, *International Economic Review* **14**, 753–764.

Schur, J. [1918]. Über Potenzreihen, die im Innern des Einheitskreises beschränkt sind, *Journal für die reine und angewandte Mathematik*, **148**, 122–145.

Simmons, G. F. [1963]. *Introduction to Topology and Modern Analysis.* New York: McGraw–Hill.

Solow, R. [1952]. On the structure of linear models, *Econometrica* **20**, 29–45.

Solow, R. M. [1962]. Comment, *Review of Economic Studies* **29**, 255–257.

Stiemke, E. [1915]. Über positive Lösungen homogener linearer Gleichungen, *Mathematische Annalen* **76**, 340–342.

Stolper, W., and P. A. Samuelson [1941]. Protection and real wages, *Review of Economic Studies* **9**, 58–73.

Takeuchi, K. [1966]. *Linear Mathematics* (*Senkei Sugaku*, in Japanese). Tokyo: Baifukan.

Taussky, O. [1949]. A recurring theorem on determinants, *American Mathematical Monthly* **56**, 672–676.

Tucker, A. W. [1956]. Dual systems of homogeneous linear relations, *in* H. W. Kuhn and A. W. Tucker, eds., *Linear Inequalities and Related Systems.* Princeton, N.J.: Princeton University Press.

Uekawa, Y. [1971]. Generalization of the Stolper–Samuelson theorem, *Econometrica* **39**, 197–217.

Uekawa, Y., M. C. Kemp, and L. W. Wegge [1973]. P- and PN-matrices, Minkowski- and Metzler-matrices, and generalizations of the Stolper–Samuelson and Samuelson–Rybczynski theorems, *Journal of International Economics* **3**, 53–76.

Wegge, L. L. F., and M. C. Kemp, Generalizations of the Stolper–Samuelson and Samuelson–Rybczynski theorems in terms of conditional input-output coefficients, *International Economic Review* **10**, 414–425.

Wolf, P. [1961]. A duality theorem for nonlinear programming, *Quarterly of Applied Mathematics* **19**, 239–244.

Yoshizawa, T. [1966]. *Stability Theory by Liapunov's Second Method* (Number 9 of Publications of the Mathematical Society of Japan). Tokyo: The Mathematical Society of Japan.

Name Index

I

K

L

M

N

O

P

Q

R

S

T

U

W

Y

Subject Index

A

B

C

D

E

F

G

H

I

J

K

L

M

N

O

P

Q

R

S

T

U

V

W

Z

List of Symbols

$\mathbb{R}$	The set of all real numbers
$\mathbb{C}$	The set of all complex numbers
$A \subseteq B$	Set inclusion (every element of A belongs to B)
$A \subset B$	Proper set inclusion, that is, $A \subseteq B$ but $A \neq B$
$\sup(\inf) E$	The supremum (infimum) of a subset E of R (Definition 2 of Appendix A)
$f: X \to Y$	A function (mapping) of X into Y (Definition 3 of Appendix A)
$f(x)$	The image of $x \in X$ under f
$f(A)$	The image set of $A \subseteq X$ under f, that is, $f(A) = \{f(x) : x \in A\}$
$f^{-1}(V)$	The inverse image set of $V \subseteq Y$ under the mapping f; symbolically $f^{-1}(V) = \{x : f(x) \in V\}$
$d(x,y)$	The distance between two points x and y (Definition 4 of Appendix A)
$B(x,r)$	An r-neighborhood of a point x (Definition 5 of Appendix A)
L_E	The set of all limit points of a given point set E (Definition 5 of Appendix A)
∂E	The set of all boundary points of E (Definition 5 of Appendix A)
$\mathcal{V}_n(\mathcal{F})$	An n-dimensional coordinate vector (linear) space over a field $\mathcal{F}$ (Definitions 7 and 10 of Appendix A)
$\mathbb{R}^n(\mathbb{C}^n)$	$\mathbb{R}^n(\mathbb{C}^n) = \mathcal{V}_n(\mathbb{R})(\mathcal{V}_n(\mathbb{C}))$
$[x,y]$	A closed line segment connecting x and y, that is, $[x,y] = \{z = (1-u)x + uy : 0 \leqslant u \leqslant 1\}$; replacement of the left (right) bracket with the corresponding

	parenthesis means that the inequality $0 \leqslant u$ $(u \leqslant 1)$ is replaced by the strict inequality $0 < u$ $(u < 1)$
$[a,b]$	A closed interval of real numbers a and b, that is, $[a,b]=\{x : a \leqslant x \leqslant b\}$; the replacement rule stated above is also applicable to the interval between two real numbers
$J^c, \bar{J}, U-J$	The complementary set of J relative to the underlying universal set U; symbolically, $J^c = \bar{J} = U - J = \{j \in U : j \notin J\}$
$J(i)$	A subset obtained from J by deleting an element i of J
$\#J$	The number of all elements of J
$\varnothing$	The empty set
$\mathrm{Re}(x)$	The real part of a complex number x
$\mathrm{Im}(x)$	The imaginary part of a complex number x
$\bar{x}$	The conjugate of x, that is, $\bar{x} = \mathrm{Re}(x) - i\,\mathrm{Im}(x)$, where $i = \sqrt{-1}$
$\boldsymbol{e}_j$	The jth unit column vector of any finite order; $\boldsymbol{e}_j = (0, \ldots, 0, 1, \ldots, 0)'$, where the element one is in the jth coordinate, and the prime applied to a vector (matrix) indicates the transpose of the vector (matrix)
$\boldsymbol{e}^j = \boldsymbol{e}_j'$	The jth unit row vector of any finite order; in general, a numbered vector with subscript (superscript) is a column (row) vector
$\mathbf{1}$	A vector consisting of suitably many ones
$\boldsymbol{I}$	The identity matrix of any finite order; if necessary, the order is indicated by a subscript with parentheses—for example, $\boldsymbol{I}_{(k)}$
$\mathbb{R}^n_+ = \{\boldsymbol{x} \in \mathbb{R}^n : \boldsymbol{x} \geqslant \mathbf{0}\}$	The nonnegative orthant of the n-dimensional real vector space
$S_k = \{\boldsymbol{x} \in \mathbb{R}^n_+ : \mathbf{1}'\boldsymbol{x} = 1\}$	A k-dimensional simplex
$\rho(\boldsymbol{A})$	The rank of matrix $\boldsymbol{A}$
$\mathrm{adj}\,\boldsymbol{A} = (A_{ij})$	The adjugate matrix of a square matrix $\boldsymbol{A}$, where A_{ij} is the cofactor of a_{ji}, the jith element of $\boldsymbol{A}$
$\boldsymbol{A}_{IJ}$	The submatrix of $\boldsymbol{A}$ with elements a_{ij} $(i \in I, j \in J)$, where $I = \{i_1, \ldots, i_k\}$ $(J = \{j_1, \ldots, j_r\})$ is a subset of the set $\{1, \ldots, m\}$ $(\{1, \ldots, n\})$ of all row (column) indices; if necessary, $\boldsymbol{A}_{IJ}$ is denoted by $\boldsymbol{A}\begin{pmatrix} i_1, \ldots, i_k \\ j_1, \ldots, j_r \end{pmatrix}$; furthermore, $\boldsymbol{A}(J)$ is an abbreviated symbol of $\boldsymbol{A}_{IJ}$ with $I = J$

$A\binom{i}{j}$	The submatrix obtained from A by deleting a_{ij}, the ijth element of A
N_A	The annihilator space of A; symbolically, $N_A=\{x:Ax=0\}$
$(Ax)_j$	The jth element of Ax, where A is an $m\times n$ matrix and x is an $n\times 1$ vector
$[p]$	The diagonal matrix with p_j $(j=1,\ldots,m)$ in the jth diagonal place
$d[S]$	The dimension of a linear space S, that is, the number of linearly independent vectors of S which any vector of S can be expressed as a linear combination of
A^*	The transposed conjugate matrix of an $m\times n$ complex matrix A, that is $A^*=(\bar{a}_{ij})'$
$\grave{A}$	$\grave{A}$ denotes $(A+A^*)/2$
A^{**}	A matrix obtained from a square complex matrix A by replacing the diagonal element a_{jj} by $\mathrm{Re}(a_{jj})$
$\det A$	The determinant of a square matrix A
$\Vert\cdot\Vert$, $N(\cdot)$	The vector and/or matrix norm (Definition 9 of Appendix A, Minkowski norm of a matrix (p. 14), generalized matrix norm (Definition 1 of subsection 2.1 of Appendix B))
$\partial f/\partial x_i\vert_{x_0}$	The partial derivative of a function $f(x)$ with respect to x_j, evaluated at $x=x_0$, or simply the jth partial derivative of f evaluated at the point x_0
$\nabla f(x_0)$	The row vector consisting of $\partial f/\partial x_1\vert_{x_0},\ldots,\partial f/\partial x_n\vert_{x_0}$
$H_f(x_0)=(f_{ij}(x_0))$	The Hessian of f evaluated at $x=x_0$, where $f_{ij}(x_0)$ denotes the jth partial derivative of the ith partial derivative of f, evaluated at x_0; if no confusion arises, $H_f(x_0)$ can be replaced with $H(x_0)$ or H
$J_f(x_0)$, $J(x_0)$	The Jacobian of a vector-valued function $f(x)=(f_1(x),\ldots,f_n(x))'$, evaluated at $x=x_0$, that is, $$J_f(x_0)=\begin{bmatrix}\nabla f_1(x_0)\\ \vdots\\ \nabla f_n(x_0)\end{bmatrix}$$
$\limsup\limits_{x\to a} f(x)$	Upper limit of $f(x)$ (Definition 1 of Chapter 6)
$\liminf\limits_{x\to a} f(x)$	Lower limit of $f(x)$ (Definition 1 of Chapter 6)